Springer Earth System Sciences

More information about this series at http://www.springer.com/series/10178

Germán Mariano Gasparini
Jorge Rabassa · Cecilia Deschamps
Eduardo Pedro Tonni
Editors

Marine Isotope Stage 3 in Southern South America, 60 ka B.P.–30 ka B.P.

Springer

Editors
Germán Mariano Gasparini
División Paleontología Vertebrados, Museo de La Plata
Facultad de Ciencias Naturales y Museo, Universidad Nacional de La Plata
La Plata, Buenos Aires
Argentina

Jorge Rabassa
Laboratorio de Geomorfología y Cuaternario
CADIC-CONICET, Universidad Nacional de Tierra del Fuego
Ushuaia, Tierra del Fuego
Argentina

Cecilia Deschamps
División Paleontología Vertebrados, Museo de La Plata
Facultad de Ciencias Naturales y Museo, Universidad Nacional de La Plata
La Plata, Buenos Aires
Argentina

Eduardo Pedro Tonni
División Paleontología Vertebrados, Museo de La Plata
Facultad de Ciencias Naturales y Museo, Universidad Nacional de La Plata
La Plata, Buenos Aires
Argentina

ISSN 2197-9596 ISSN 2197-960X (electronic)
Springer Earth System Sciences
ISBN 978-3-319-39998-0 ISBN 978-3-319-40000-6 (eBook)
DOI 10.1007/978-3-319-40000-6

Library of Congress Control Number: 2016943439

Printed on acid-free paper

This Springer imprint is published by Springer Nature
The registered company is Springer International Publishing AG Switzerland

Dr. Enrique J. Schnack (1941–2016)

Contents

Introduction

Germán Mariano Gasparini, Jorge Rabassa, Cecilia Deschamps and Eduardo Pedro Tonni

Abstract This volume was conceived during the Symposium "El Estadio Isotópico 3 en la Argentina y el sur de América del Sur: 60.000 a 25.000 años atrás" (The Marine Isotope Stage 3 (MIS 3) in Argentina and southern South America: 60,000 to 25,000 years ago) held in June 2013, in La Plata, Argentina. The main purpose of this meeting was to promote the interaction of the leading scientists in various disciplines of the Geological and Paleontological Sciences of the Late Cenozoic of South America in order to update the existing knowledge on the core issues cited in the title of the symposium (e.g., geology, geomorphology, vertebrate and invertebrate paleontology, palynology, paleomagnetism, paleoenvironmental and paleoclimatic studies, etc). This was the first time ever that these topics related to MIS 3 were publicly discussed in Argentina. The Quaternary geological history is characterized by cyclical climatic changes (glacial and intervals). These cyclical changes generated periodic reorganizations of the landscape and the environmental system. MIS 3 was an interstadial stage, a relatively warmer climatic period which

G.M. Gasparini · C. Deschamps
División Paleontologia Vertebrados, Museo de La Plata,
Facultad de Ciencias Naturales y Museo, Universidad Nacional de La Plata,
La Plata 1900, Buenos Aires, Argentina
e-mail: ceci@fcnym.unlp.edu.ar

J. Rabassa
Laboratorio de Geomorfología y Cuaternario, CADIC-CONICET,
Universidad Nacional de Tierra del Fuego, Ushuaia 9410, Tierra del Fuego, Argentina
e-mail: jrabassa@gmail.com

E.P. Tonni
Facultad de Ciencias Naturales y Museo, Universidad Nacional de La Plata,
La Plata 1900, Buenos Aires, Argentina
e-mail: eptonni@fcnym.unlp.edu.ar

G.M. Gasparini (✉) · J. Rabassa
Consejo Nacional de Investigaciones Científicas y Técnicas (CONICET),
Buenos Aires, Argentina
e-mail: germanmgasparini@gmail.com

C. Deschamps
Comisión de Investigaciones Científicas (CIC), La Plata, Buenos Aires, Argentina

G.M. Gasparini et al. (eds.), *Marine Isotope Stage 3 in Southern South America, 60 ka B.P.–30 ka B.P.*, Springer Earth System Sciences,
DOI 10.1007/978-3-319-40000-6_1

developed roughly between 60 ka B.P. and 25 cal. ka B.P. Several very cold periods (Heinrich events) developed during MIS 3, and several paleoclimatic moments with relatively warmer conditions (Dansgaard-Oeschger events) took place in between. We wish that this first attempt to compile data about MIS 3 in the southern part of South America will be appreciated by our colleagues around the world who are interested in this fascinating period of our climatic history, that the available evidence will be thoroughly analyzed and discussed, and that our data and interpretations will be compared with the existing information emerging from the rest of the planet, and particularly, from other regions of the Southern Hemisphere.

Keyword Southern South America • Quaternary • Marine Isotope Stage 3 • Dansgaard-Oeschger events • Heinrich events

The idea about compiling this volume was conceived during the Symposium "*El Estadio Isotópico 3 en la Argentina y el sur de América del Sur: 60.000 a 25.000 años atrás*" (The Marine Isotope Stage 3 (MIS 3) in Argentina and southern South America: 60,000–25,000 years ago) held in June 2013, in La Plata, Argentina. This symposium gathered prestigious Argentine, Brazilian, and Uruguayan specialists who gave several conferences that were attended by many colleagues, graduate and undergraduate students, coming from many different disciplinary fields. The main purpose of this meeting was to promote the interaction of the leading scientists in various disciplines of the Geological and Paleontological Sciences of the Late Cenozoic of South America in order to update the existing knowledge on the core issues cited in the title of the symposium. In addition, the idea was to provide an adequate scope for an open and exhaustive debate, in which the future research and academic interaction were among the main issues. It should be noted that this was the first time ever that these topics related to MIS 3 were publicly discussed in Argentina.

The organizing committee provided the possibility of publishing the presented papers in a special volume in which different approaches to the study of MIS 3 were thus updated, *e.g.*, geology, geomorphology, vertebrate and invertebrate paleontology, palynology, paleomagnetism, paleoenvironmental and paleoclimatic studies, etc.

The Quaternary geological history is characterized by cyclical climatic changes of different frequency and intensity comprising the alternation of colder periods (glacial intervals) and more temperate-warmer spans (interglacial intervals). These cyclical changes generated periodic reorganizations of the landscape and the environmental system.

Particularly, MIS 3 is a very interesting period for different reasons. MIS 3 was an interstadial stage, a relatively warmer climatic period which developed roughly between 60 and 25 cal. ka B.P. Several very cold periods, known as the Heinrich (H) events, developed during MIS 3. As well, several paleoclimatic moments with relatively warmer conditions, known as the Dansgaard–Oeschger (D–O) events, took place in between the Heinrich (H) events. These H and D–O cycles would have been very short in general terms, high intensity, and frequent extreme events.

In this respect, the Greenland ice cores contain a record of abrupt climatic changes during the Late Pleistocene, involving rapid warming events (with mean annual temperature increases of up to 8–15° C). They correspond to around 25 centennial-scale climatic oscillations (interstadials and stadials), fifteen of which occurred during the 60–25 ka interval. In the Northern Hemisphere, these changes are clearly reflected in terrestrial settings by pollen assemblages that suggest the alternation of colder and warmer phases of variable duration. In the Southern Hemisphere, the Antarctic ice cores also register a millennial scale climatic variability. The comparison of the climatic conditions at both hemispheres, however, suggests a complex global pattern, when Antarctica appears to warm up, Greenland becomes colder; these opposite responses are thought to be consistent with a mechanism involving ocean heat transport. In most of South America, as well as in many areas of Asia and northern Africa, the environmental and climatic reconstruction of MIS 3 is greatly restricted by the scarcity of higher resolution (centennial-scale) terrestrial records and, generally speaking, the problem of dating techniques.

The general paleoclimatic conditions of MIS 3 are considered in an introductory chapter by *Rabassa and Ponce*, in which the limiting ages, the number of short climatic episodes and the magnitude and nature of the changes have been analyzed. Additionally, since MIS 3 climates could be very important to understand the environmental conditions related to the arrival of humans to Beringia, Alaska and northern Canada, several paleoclimatic records available for these regions have been briefly discussed.

The origin of the Dansgaard–Oeschger (D/O) events and their variability between different isotope stages is at present under debate. Evidence is herein presented by *Duhau and De Jagger*, suggesting that during the Holocene the "Bond cycle", which is a reduced version of the D/O one, is related to a signal on solar activity known as the "Long Trend." A mechanism by which solar storms, which effect on the Earth's atmosphere strongly depends on the geomagnetic field morphology, may explain this relationship is herein proposed. Likewise, the manner how the differences between the D/O events along different isotope stages and between the two hemispheres may help in disentangle, at least partially, the ultimate origin of climate change, is discussed.

The mechanisms by which the variations of the internal magnetic field could trigger climate change are treated in the chapter by *Orgeira et al.*; this would be produced by the influence of the internal magnetic field on Galactic Cosmic Rays (GCR), since the geomagnetic field (GF) provides shielding to such radiation.

The abrupt climate changes during MIS3 are treated in the chapter by *Agosta and Compagnucci* who analyze the various theories on the causes of these changes (e.g., dynamic of ice sheets, changes in the solar flux, changes in the Atlantic Meridional Overturning Circulation). They discuss the bipolar pattern with warming conditions in the Northern Hemisphere and cooling in the Southern Hemisphere.

A general review about the main processes that are associated with uplift, regional subsidence, and exhumation of vast sectors of the Southern Andes and their foreland area throughout the Quaternary are presented in the chapter by *Folguera et al.* These regional geological processes may have had strong influence on continental climate.

Blasi et al. study the middle fluvial valley of the Undulated Pampa region of the Buenos Aires Province (locally known as the "Pampa Ondulada Bonaerense"). They registered the consequences of climate variations occurred during MIS3 upon the depositional units of this Central Argentina region.

Isla and Schnack have analyzed the relationships between climatic change and sea-level fluctuation considering data collected from the continental shelf of Argentina.

Using digital elevation models and available curves of sea-level change during the cited period, *Ponce and Rabassa* present and discuss paleotopographic maps which provide a general idea of the extension of the present submarine shelf that was abandoned by the sea during MIS 3 and the resulting paleomorphology of the South American coast.

MIS 3 studies include the aeolian deposits which occur broadly distributed across central Argentina (between ~30–40° S); together with the Late Pleistocene fluvial records, they offer a general insight into the major responses of the environmental systems between 60–25 ka (see *Zárate et al.).*

This interval is also studied in the paleontological site of the Sopas Formation, Uruguay, by *Ubilla et al.*, in which the fossil assemblage indicates open habitats, savannahs, and woodlands including gallery forests and perennial rivers. This assemblage includes living representatives of taxa related to benign climatic conditions (mostly tropical to temperate climates), as well as arid to semiarid environments, migrants, and seasonality indicators. A replacement versus mixed faunal models is discussed in the light of available evidence.

Dantas and Cozzuol review information about the extinct fauna that lived in the Brazilian Intertropical Region (BIR) between 64 and 10 ka B.P. They present absolute dates (^{14}C, ESR, U-series) and paleodiet reconstruction for some taxa of this Region, as well as paleoenvironmental reconstructions of two climatic moments, one at 64 ka, and another between 27 and 10 ka B.P.

Palaeontological sites in Argentina with continental vertebrates corresponding to the MIS 3 interval are scarce or poorly known (see *Gasparini et al.*). This situation is mainly due to the lack of absolute dating for Pleistocene fossil remains or their bearing sediments that would allow the verification of the chronology established for this interval. However, isolated evidence shows that continental vertebrates undoubtedly responded to abrupt temperature changes that characterized MIS 3.

The limit of the radiocarbon dating method (presently about 50,000 years B.P.) is also debated. Some contributions (e.g., *Rojas and Martínez* chapter) refer that there is conflicting evidence when comparing ages obtained by different methods. In certain cases, while ^{14}C dating suggests younger ages (related to MIS 3), OSL, where available, indicate older times (mostly related to MIS 5).

Terrestrial pollen records of Patagonia have been studied at both sides of the Andean ranges by *Borromei and Mussotto*. The study is focused in some profiles of central Chile, the Southern Lake District and northern Isla Grande de Chiloé, and in the Argentine sector in NW Patagonia, southern Patagonia, and Tierra del Fuego. Comparisons of the resulting data show that this interval was different in both areas.

Diatoms, treated by *Espinosa*, are very useful proxy indicators to reconstruct past climate changes. Their study suggests abrupt climate changes between *ca.* 60 and 30 cal. ka B.P. in the Southern Hemisphere. The future integration of diatom datasets constructed from different environments will solve the analogy problems between fossil and modern assemblages and increase the potential for reliable quantitative reconstructions of Late Quaternary climate in southern South America.

A synthesis of silicophytolith studies on pedosedimentary sequences of MIS 3 age in South America (especially in Mesopotamia and the Pampean Plain) is herein presented in chapter by *Osterrieth et al.* Within MIS 3, frequent climatic environmental variations during the Late Pleistocene may have led to a fluctuation in biogeographical connections between the Mesopotamian region and other parts of South America, which was closely linked to the Chaco-Pampean plain and, at other times, to intertropical regions.

We wish to acknowledge the authorities of the Centro Científico y Tecnológico (Science and Technology Center, CCT- La Plata-CONICET) for permission to use the excellent facilities in which this symposium was held. We also thank the Facultad de Ciencias Naturales y Museo (College of Natural Sciences and Museum), Universidad Nacional de La Plata; Consejo Nacional de Investigaciones Científicas y Técnicas (National Research Council of Argentina; CONICET); Centro Austral de Investigaciones Científicas (CADIC-CONICET); Asociación Geológica Argentina; Centro de Investigaciones Geológicas (CIG-UNLP-CONICET); Asociación Paleontológica Argentina (APA); Asociación Argentina de Cuaternario y Geomorfología (AACYG); International Association of Geomorphologists (IAG/AIG); and International Union for Quaternary Research (INQUA) for sponsoring this symposium in various ways.

We wish that this first attempt to compile data about MIS 3 in the southern part of South America will be appreciated by our colleagues around the world who are interested in this fascinating period of our climatic history, that the available evidence will be thoroughly analyzed and discussed, and that our data and interpretations will be compared with the existing information emerging from the rest of the planet, and particularly, from other regions of the Southern Hemisphere. We fully understand the serious problems pertaining to the availability of precise dating techniques applicable to this period, but we hope that this first effort will contribute to focus the attention of our colleagues on this interstadial period. Hopefully, sooner than later, this volume will be followed by other contributions enlightening the complex sequence of climatic events during this tantalizing epoch of the recent Earth history.

During the edition of this book Dr. Enrique J. Schnack passed away on March 30, 2016. He had a long and fruitful career as founder and Director of the Centro de Geología de Costas, Universidad Nacional de Mar del Plata, Professor of Quaternary Geology in the Universidad Nacional de La Plata, and Senior Researcher of the Comisión de Investigaciones Científicas y Técnicas de la Provincia de Buenos Aires. Being one of the most important coastal geology experts of Latin America, Dr. Schnack was a Honorary Member of the Argentine Association of Geomorphology and Quaternary Studies (AACYG). Many of the contributors to this book were fortunate to have worked with him in different projects and regions of South America. A deep friend of most of us, his death caused deep grief and sorrowfulness to the scientific community of the entire continent. We would like to dedicate this book to his memory.

The Heinrich and Dansgaard–Oeschger Climatic Events During Marine Isotopic Stage 3

Jorge Rabassa and Juan Federico Ponce

Abstract The Marine Isotope Stage 3 (MIS 3) was an interstadial stage, a relatively warm climatic period which developed roughly between 60 and 50 and 30 cal. ka B. P. Several very cold periods, known as Heinrich (H) events, developed during MIS 3 as a result of partial collapse of the North American ice sheet margins, with formation of huge amounts of icebergs which, after melting in more temperate latitudes, would have inundated the North Atlantic Ocean with low salinity waters which would have impeded the penetration of the Gulf Stream into the North Atlantic Ocean. Several paleoclimatic moments with relatively warmer conditions, known as the Dansgaard–Oeschger (D-O) events, took place in-between the Heinrich (H) events, throughout MIS 3. These H and D-O cycles would have been very short in geological terms (perhaps even only around 1 kiloyears (kyr) each in some cases) and quite intense, with mean annual temperatures, for instance in the area of Beringia (the land bridge between Siberia and North America) ca. 5–8 °C higher than those active at the Last Glacial Maximum (LGM; ca. 24 cal. ka B.P.) and perhaps close to those occurring in past interglacial periods, respectively. Even though climate was warmer than during the LGM, total melting of the continental ice sheets did not take place; thus, global sea level was perhaps lower than today during the entire MIS 3. It was low enough to allow the persistence of Beringia, without any interruptions throughout the whole of MIS 3. The aim of this paper is to present basic paleoclimatic and paleogeographic information about MIS 3, which may be useful to understand the nature and evolution of the South American terrestrial and marine ecosystems later on during the LGM.

Keywords Late Quaternary paleoclimate · Marine Oxygen Isotope 3 · Dansgaard/Oeschger climatic events

J. Rabassa (✉) · J.F. Ponce
Laboratorio de Geomorfología y Cuaternario, CADIC-CONICET, Bernardo Houssay 200, 9410, Ushuaia, Tierra Del Fuego, Argentina
e-mail: jrabassa@gmail.com

J.F. Ponce
e-mail: jfedeponce@gmail.com; jfponce@cadic-conicet.gov.ar

J. Rabassa · J.F. Ponce
Universidad Nacional de Tierra del Fuego, Ushuaia, Argentina

G.M. Gasparini et al. (eds.), *Marine Isotope Stage 3 in Southern South America, 60 ka B.P.–30 ka B.P.*, Springer Earth System Sciences,
DOI 10.1007/978-3-319-40000-6_2

1 Introduction

Middle and Late Pleistocene paleoclimates are characterized by climatic cycles that include glacial (colder) and interglacial (warmer) stages, with a total duration of ca. 100 kiloyears (kyr) each. The colder glacial periods, around 80–90 kyr duration, are longer than the interglacial ones, which are shorter and warmer, averaging 10–20 kyr. Glacial periods show significant paleoclimate variations, with colder events which are called "stadials" and warmer periods named as "interstadials." Full glacial conditions concerning both very low global temperatures and sea level stands are achieved only during stadials. Interstadials are characterized by warmer temperatures than those during the stadials, recession of the continental ice sheets (but not total vanishing), and rising sea levels to intermediate positions in between full glacial and interglacial times.

These cycles are very well exposed by the relative content of ^{18}O isotopes ($\partial^{18}O$), or other proxy elements or substances contained in ice from polar ice cores, such as those in Greenland and Antarctica, as well as the variations of the same isotopes in foraminifera and/or ostracoda found in marine sedimentary cores [for explanation of the $\partial^{18}O$ method, see Andrews (2000), and Wright (2000)]. In Fig. 1, the $\partial^{18}O$ variations during the last glacial–interglacial cycle are depicted, starting with the final phases of the last interglacial. In this figure, isotope peaks pointing upwards correspond to warmer periods, whereas those pointing downwards are colder events.

The periods showing a specific trend of $\partial^{18}O$ content are called "marine isotope stages" (MIS), and they correspond to moments with distinct global temperatures and climates. MIS 5 is the last full interglacial period and MIS 1 is the present interglacial. MIS 4 and 2 are colder, stadial events, the earlier corresponding to the process of building up of the continental ice sheets, and the latter representing the maximum of the Last Glaciation (LGM, ca. 24 cal. ka B.P.) and extending until the end of the Pleistocene (10 ^{14}C ka B.P.). MIS 3 corresponds to a long interstadial epoch which was much warmer than the following stadial period, that is, the LGM. MIS 3 lasted at least 25 kyr, perhaps even 30–35 kyr, between approximately 60–50 and 30 cal. ka B.P. The contents of this review paper are partly a summary of the corresponding sections in Rabassa and Ponce (2013).

2 The Climate of North America, Beringia, and the North Atlantic Ocean During MIS 3: The Heinrich and Dansgaard–Oeschger Events

During MIS 3, the North American and European ice caps receded from their outer positions achieved in MIS 4, and extensive portions of the landscape were abandoned by the ice; sea level stood between −55 and −90 m below present sea level (Figs. 3b, c; see Lambeck and Chappell 2001), only half to three quarters of the

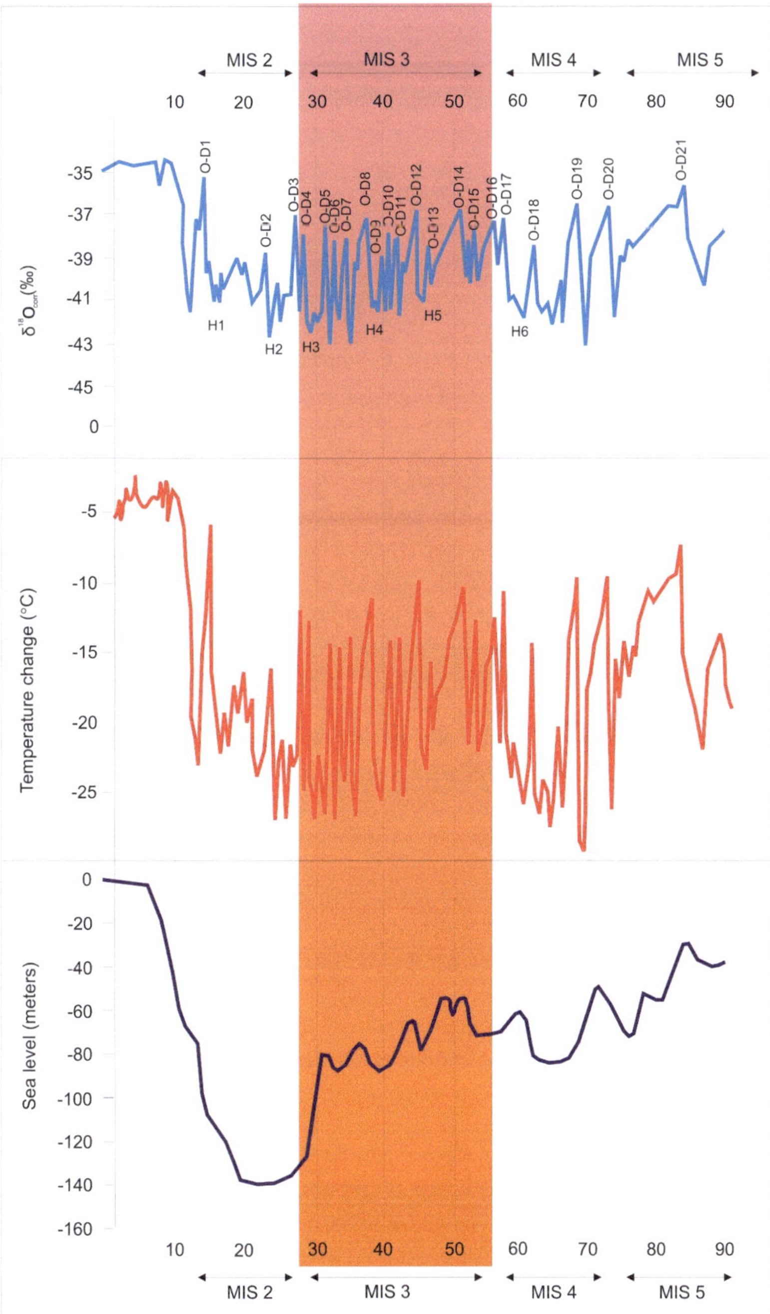
MIS 2
MIS 3
MIS 4
MIS 5
10 20 30 40 50 60 70 80 90
δ18Ocorr (‰)
-35 -37 -39 -41 -43 -45 0
O-D1 O-D2 O-D3 O-D4 O-D5 O-D6 O-D7 O-D8 O-D9 O-D10 O-D11 O-D12 O-D13 O-D14 O-D15 O-D16 O-D17 O-D18 O-D19 O-D20 O-D21
H1 H2 H3 H4 H5 H6
Temperature change (°C)
-5 -10 -15 -20 -25
Sea level (meters)
0 -20 -40 -60 -80 -100 -120 -140 -160

◀ **Fig. 1** *upper* ∂^{18}O contents in Greenland for MIS 5 to MIS 2 following Uriarte Cantolla (2003); *middle* strong climatic variations in Greenland of up to 16 °C for MIS 5 to MIS 2 (in mean annual temperature; IPCC 2007), lower: global sea level curve, in meters, according to Lambeck and Chappell (2001). The *shaded area* corresponds approximately to the actual extent of MIS 3 (see Rabassa and Ponce 2013)

maximum sea level depression during MIS 2. As the ice was then receding in the Northern Hemisphere, the Gulf Current was able to penetrate to higher latitudes, bringing warmer and moister air to the North Atlantic Ocean, favoring the temporary restoration of milder climates and more temperate environments (Uriarte Cantolla 2003).

However, climate was neither stable nor homogeneous during MIS 3. Very strong, intense, and fast climatic changes took place during this period, indicated by significant ∂^{18}O variations. MIS 3 was a period of moderate insolation that is in sharp contrast with the insolation troughs of MIS 4 and MIS 2 (Andrews and Dyke 2007). Very cold periods named as Heinrich (H) events (named after the famous paleoclimatologist Hartmut Heinrich) alternated with much warmer and moister periods called as Dansgaard–Oeschger (D-O) events (named after the prestigious geochemists and glaciologists Willi Dansgaard, from Danemark, and Hans Oeschger, from Germany) (Heinrich 1988; Uriarte Cantolla 2003; Labeyrie et al. 2007).

During the Last Glaciation, there were at least six paleoclimatic episodes in which large amounts of glacial debris were ice-rafted and deposited at the bottom of the ocean in an area between 40° N and 55° N. The thickness of the resulting bottom sediments diminishes from west to east, and the dominant lithology types are those coming from North America and, particularly, the Hudson Bay (Heinrich 1988). Some of the icebergs reached up to 3000 km from their place of origin. The most appropriate explanatory theory is that the North American ice sheets outgrew their stable boundaries during certain moments of the Last Glaciation, reaching the outer edges of the continental shelves where they became unstable and collapsed, throwing huge amounts of icebergs into the North Atlantic Ocean. The high iceberg discharge would have interrupted the thermohaline circulation in the North Atlantic (Denton 2000). Other opinions suggest that the ice collapse was forced by subglacial melting due to ground heat trapped under the huge ice sheets (between 2 and 3 km thick), or even that the enormous pressure of the ice sheet during maximum expansion triggered local earthquakes (Uriarte Cantolla 2003).

The exceptional abundance of fresh water due to iceberg melting would have forced changes in the North Atlantic deep water production and limited the northernmost reach of the Gulf Stream, allowing the southward displacement of polar waters and subsequent temperature lowering (Bard et al. 2000). Once the iceberg discharge was completed, the size of the glaciers releasing along the North American coasts dramatically diminished, lowering also the supply of fresh water to the Northern Atlantic Ocean; thus, the Gulf Stream was reestablished. Therefore, a sharp increase in middle-to-high latitude temperatures took place, leading into a warm interstadial stage. These are known as the Dansgaard–Oeschger (D-O) events, in which mean annual temperature would have risen between 5 and 8 °C, perhaps during only a century or even less. Usually, D-O events are characterized by a rapid warming up to ca. 3–5 °C per century (Labeyrie et al. 2007). Moreover, during the D-O 19 event, around 70 ka ago, in MIS 4, the rise of temperature would have been up to 16 °C (Lang et al. 1999; Uriarte Cantolla 2003).

During these warmer events, a much larger evaporation rate and atmospheric moisture export from the Atlantic Ocean to the Pacific Ocean, across Middle America, would have taken place. These events would have forced an increase in the Atlantic Ocean salinity, and therefore a reinforcement of the thermohaline circulation and the Gulf Stream, which would have warmed the whole of the northern Atlantic Ocean, including Greenland (Peterson et al. 2000), heating very rapidly the atmosphere of the Northern Hemisphere, but without pushing a sudden rise in sea level because of their short duration. For instance, MIS 3 warm events in Greenland had rapid (even less than 1 kyr) changes of up to 15 °C between peaking H and D-O subsequent episodes (Fig. 1; see Labeyrie et al. 2007, for a detailed discussion). It is of high interest to note that there are clear paleoclimatic signals about the H and D-O events both in higher and lower latitudes, and in both the North Atlantic and North Pacific oceans at almost identical times (Labeyrie et al. 2007).

Sea-surface temperature (SST) during the D-O events was at least 4–6 °C higher than those of the LGM (ca. 24 cal. ka B.P.), suggesting that terrestrial temperatures were higher as well, indicating that mean annual temperature during these events was perhaps slightly colder than present conditions, but much warmer than full glacial conditions.

Six major Heinrich (H) events have been identified in MIS 2, MIS 3 and MIS 4, at approximately 17, 22, 29, 39, 45, and 61 cal. ka B.P., H1 being the younger episode and H6 the older one. H2 corresponds to the LGM. It has been suggested that the "Younger Dryas" cold event could be considered as the youngest Heinrich event, thus becoming a sort of H0. Likewise, at least 14 Dansgaard–Oeschger (D-O) events have been detected in the $\partial^{18}O$ curves, roughly at ca. 14, 23, 27, 29, 32, 33.5, 34, 38, 40, 41, 43, 45, 47, and 52 cal. ka B.P.; not all of them were of identical magnitude, but they were clearly warmer than the LGM in all cases (Table 1). As with Heinrich events, D-O 1 is the youngest event. The total length of each H or D-O events during MIS 3 is variable, but each whole cycle was probably around 1–2 kyr long in average. The longer and more intense are the D-O 8, 12 and 14 events, at ca. 38, 45 and 52 cal. ka B.P., respectively. However, violent secular transitions between events, of not more than 1–2 centuries long, have been quantified in the $\partial^{18}O$ curves. A detailed record of these variations during the last part of MIS 3, between 30 and 46 cal. ka B.P. is presented in Fig. 4. Particularly, Vidal et al. (1999) have found two very warm D-O events at ca. 43 and 33 cal. ka B.P., which would be triggered following H5 and H4 episodes, respectively. This figure illustrates the paleoclimatic and paleoenvironmental conditions both in Greenland and Antarctica, proving the global and possibly synchronous impact of these climatic changes. There are diverging opinions about such synchronicity (see, for instance, Blunier and Brook 2003; White and Steig 1998; Vidal et al. 1999; Rabassa 2008),

Table 1 List of Dansgaard–Oeschger events during MIS 3, with the relative position of sea level corresponding to each event. Chronology and sea level data from the literature cited in the text

Dansgaard-Oeschger (D-O) events	Age (cal. ka B.P.)	Sea level position (m below present sea level)
2	23.5	−140
3	27	−135
4	29	−128
5	31	−80
6	33.5	−87
7	34	−85
8	38	−78
9	40	−88
10	41	−85
11	43	−80
12	45	−65
13	47	−72
14	52	−55

but this is not a matter of discussion here. Isotopic data from caves in China as well as Greenland paleotemperatures for MIS 3 (Alley 2004) confirmed the regional impact of the MIS 3 climatic changes, both in temperature and precipitation, in such areas which are geographically related to northern North America. The region of Beringia, a huge land bridge that developed between Siberia and Alaska when sea level was lower than today, has been specially discussed because this area is highly relevant to the problem of the peopling of the Americas, perhaps as early as during MIS 3 (see Rabassa and Ponce 2013).

The palynological record for MIS 3 in NE Siberia shows that in the northern lowlands tundra predominated, under moderately warm environmental conditions (Lozhkin and Anderson 2007), probably together with isolated larch-birch tree communities. Based on pollen records, MIS 3 is represented in Siberia by the Karginski interstadial, which is separated into five environmental phases, as follows: ca. 50–45 ka B.P., warmer; ca. 45–43 ka B.P., cooler; ca. 43–33 ka B.P., maximum warmth, the so-called Malokhetski sub-horizon; ca. 33–30 ka B.P., cooler, the Konotzelski sub-horizon; and ca. 30–22 ka B.P., warmer, the Lipovskoy–Novoselovski sub-horizon (Lozhkin and Anderson 2007). Likewise, in northwestern North America, pollen records dating to MIS 3, between ca. 60 and 30 ka B.P., indicate widespread tundra across Alaska, perhaps with minor amounts of spruce in interior Alaska and Yukon (Bigelow 2007). In western Alaska, MIS 3 was characterized by grass, sedge, and *Artemisia*, with minor amounts of willow and birch. In the Yukon, spruce pollen frequencies of less than 20 % suggest the existence of scattered trees within widely extended birch/graminoid tundra (Bigelow 2007).

Similarly, paleoentomological studies in NW North America and NE Siberia have shown that climate was much milder during MIS 3 than during the LGM. Elias (2007) identified a "long MIS 3 interstadial complex in eastern Beringia," with warming intervals at 46.4, 36, and 33.6 ka B.P. The latter one is known from the Totaluk River beetle fauna, with an association indicating that the maximum temperature during this event was between 0.5 and 2.0 °C warmer than present. Other sites have indicated maximum temperatures only between 0.5 and 2.0 °C cooler than today. However, there are also faunas suggesting much cooler conditions, up to 7.0–8.0 °C cooler than present, proving the existence of dramatic temperature oscillations in this period. The identified beetle faunas are characterized by species which correspond to open-ground habitats, not necessarily occupied by trees. The open-ground or steppe tundra environments were maintained throughout MIS 3. Elias (2007) stated that, for this area, MIS 3 "oscillations coincide with climatic patterns inferred from oxygen isotope records in Greenland ice cores."

Elias and Brigham-Grette (2007) have identified substantial differences between western and eastern Beringia during MIS 3, dated between 48 and 28 ka B.P. In western Beringia, temperatures reached near present-day levels and the forest migrated northwards near its present position. Contrarily, in eastern Beringia there is little evidence of coniferous forest expansion during MIS 3. The Arctic regions reached temperatures perhaps up to 1.5 °C above present times at ca. 35 ka B.P., whereas in sub-arctic environments, temperature was ca. 2 °C below current conditions at the same time (Elias and Brigham-Grette 2007). In any case, the environmental conditions were not too different from those found today, and the region would have been fully accessible for humans during MIS 3.

Finally, the transition between MIS 3 to MIS 2 in eastern Beringia, from interstadial to full glacial conditions, has been dated in ca. 32–31 ka B.P.

Likewise, Sher and Kuzmina (2007) found, for the 35–40 ka B.P. period in northeastern Asia, beetle associations with a high percentage of arboreal, mostly shrub, pollen, in contrast to the completely grass-herb dominated spectra of the LGM. Though they acknowledged the possibility of radiocarbon dating problems, they identified xerophilic beetle species for the 48–34 ^{14}C ka B.P. period; then, an increase of Arctic species by 34 ^{14}C ka B.P., almost equating the LGM levels; and again xerophilic beetles with mesic tundra insects between 34 and 24 ^{14}C ka B. P. After this age, the beetle species indicated a gradual decrease in temperature towards the LGM. They also stated that the regional climate was much more continental during MIS 3 times than it is today, a condition which they assigned to a lower sea level (Table 1).

The vertebrate record of the Late Pleistocene in northern Asia shows evidence of faunas of a warm interval called the Briansk or Dunaevo interstadial, dated between 33 and 24 ka B.P., the last of a series of warm episodes along MIS 3 (Markova and Puzachenko 2007). The faunas characteristic of Beringia for this period are included in the Arctic sub-assemblage of the Mammoth I assemblage, with woolly mammoth, woolly rhinoceros, reindeer, Pleistocene bison, horse, rare saiga (an Asian antelope), arctic fox, cave hyena, cave bear, steppe pika (a small mammal of Asia and North America), arctic hare, several lemming species, and voles (small rodents from the northern hemisphere). No forest animals are found in this assemblage (Markova and Puzachenko 2007). The distribution and composition of this assemblage shows differences with present faunas, indicating a climate cooler than today; but this faunal association became later severely restricted in surficial terms during the LGM epoch. The abundance of large Pleistocene herbivores and cave carnivores during MIS 3 also depicts the large variations from today's faunas (Markova and Puzachenko 2007).

Relevant to the topics developed in the present paper, careful analysis of recently published glacial geological evidence in northern and northeastern Asia and northwestern North America is also pointing towards very warm conditions during MIS 3. Andrews and Dyke (2007) established that very soon after the MIS 4/MIS 3 transition rising insolation forced the retreat of the Laurentide Ice Sheet from the western Canadian lowlands, a condition that probably lasted for more than ca. 30–40 years, until the end of MIS 3, allowing the availability of the Yukon corridor for human displacements. Velichko et al. (2011) have identified a warm period during the Late Pleistocene in NE Europe which they named as the Middle Valdai. During this period, the ice had receded as far north as at least 68° N, with milder climates and temperate environments such as mixed forest of conifers and broad-leaved trees and southern taiga along the Arctic coasts around 38–40 ka B. P. They have called this period as a "mega-interstadial" or the "Leningrad or Bryansk mega-interval," with alternating warm and cool phases, rhythmic climate changes, and a main warm phase between 40 and 34 ka B.P., with environmental conditions quite close to that of Mulikino, the last interglacial or MIS 5e. They consider this period as "a short interglacial," extending over D-O 10 and 12 events, with a new warm pulse in the Dunaevo warming episode, dated at 31–25 ka B.P., the terminal phase of MIS 3.

Likewise, Vorren et al. (2011) have shown that the ice sheet of Northern Siberia was restricted only to the Kara Sea and Novaja Zemlya islands between 55 and 45 ka B.P., and that the Barents Sea was mostly ice free between 48 and 26 ka B. P. Moller et al. (2011) determined that there was no ice on the Taymir Peninsula and the Severnaya Zemlya island between 50 and 25 ka B.P. In NE Asia, sedimentology studies have shown that no major changes in the environment had occurred between 60 and 12 ka B.P., with a mosaic of arctic tundra and tundra-steppe communities dominating during the Karginsky interstadial (MIS 3) and the Sartan ice age (MIS 2) (Glushkova 2011). According to this author, climate was continental, with summer not colder than today but with colder winters. Glaciation was restricted to the mountain cirques during MIS 3, and much more reduced than during previous glaciations as well as in MIS 2. In contrast to the viewpoint of previous researchers, glaciers during the LGM were located then only in a few regions of the highest mountains. Therefore, it may be deduced that there were no physical restrictions to human displacement toward Beringia and Alaska during both cited isotope stages.

In the Verkhoyansk Mountains, an important orographic barrier across easternmost Siberia, the youngest proven glaciations dated back to ca. 50 ka B.P., and no LGM glaciations have been identified. East of this mountain range, restricted MIS 2 glaciation has been found, this being explained as a result of atmospheric paleocirculation and differential moisture availability (Stauch and Lehmkuhl 2011). It should then be noted that there were no ice barriers here for human displacement during MIS 3 and 2. It is also interesting to note that, similarly, there were no glaciers in the Central Alaska lowlands throughout the entire Late Pleistocene (Kaufman et al. 2011). Glaciation in the high mountains of easternmost Siberia was characterized by expansion of the glaciers between late MIS 5 and MIS 4, and significant retreat during MIS 3, with a short ice advance ca. 45–40 ka B.P. and a major readvance of the ice in MIS 2.

In northwestern North America, Clague and Ward (2011) presented a model of glaciation of British Columbia with glaciers limited to the summits and uppermost valleys for the 35–30 ka B.P. period, and full glacial conditions and closing of the Yukon corridor only ca. 25 ka B.P. A "non-glacial Olympia interval" is described for the 50–25 ka B.P. period, correlated with MIS 3. Thus, this information confirms that the Yukon corridor would have been available for humans entering from Beringia during a very long period before 25 ka B.P. In western Alberta, Jackson et al. (2011) have described nonglacial, fossil bearing sediments dated between 39 and 24 ka B.P., and correlated them to MIS 3, of clearly interglacial nature. In eastern Alberta, Barendregt (2011) mentioned "mid-Wisconsin interglacial sediments," well dated between 65 and 23 ka B.P., also corresponding to MIS 3. Much farther south, the paleoclimate pattern is similar. Gillespie and Clark (2011) have identified very intense D-O events as far south as the Sierra Nevada of California, from D-O 8 (ca. 41–40 ka B.P.) to D-O 4 (31 ka B.P.), with intermediate warm episodes as D-O 7, 6 and 5, at ca. 36, 35 and 34 ka B.P., respectively.

The paleogeography of Beringia during MIS 3 has been analyzed and discussed by Rabassa and Ponce (2013). Sea level during the Late Pleistocene is one of the key questions concerning human population of the Americas, since the availability of a terrestrial path across Beringia allowed the eastwards displacement of Siberian humans into the new continent. The position of sea level during MIS 3 is crucial to understand that the Beringia land bridge was available for humans not only during the LGM but during many thousands of years before as well. The sea level curves by Lambeck and Chappell (2001) and Lambeck et al. (2002) are clearly indicating which would have been the position of the coastline during different times of MIS 3. Lambeck et al. (2002) estimated that, using data from Papua New Guinea and Australia, sea level was never below −50 m between 50 and 30 cal. ka B.P., thus fully supporting the paleogeographic reconstructions presented by Rabassa and Ponce (2013).

The last closure of the Bering Straits started at around 82 cal. ka B.P., when sea level lowered 45 m below its present position. The straits remained closed continuously until 11.5 cal. ka B.P. The figures presented here show that the closure period was perhaps as long as 70 kyr.

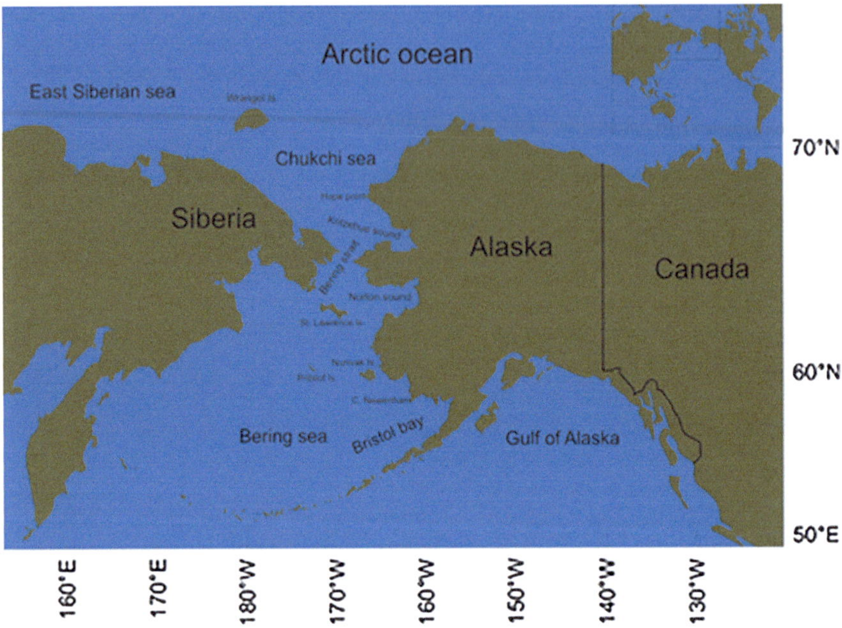

Fig. 2 Beringia, location map. See Rabassa and Ponce (2013)

During most of MIS 3 (between 55 and 30 cal. ka B.P.), sea level oscillated between a maximum elevation of −55 m (between 48,000 and 52,000 cal. year B.P., which would roughly correspond to the D-O 14 event) and a minimum stand at −90 m (at around 32,000 and 40,000 cal. year B.P., approximately, in coincidence with the H3 and H4 events).

During its lowermost position, an enormous plain connected both continents (Rabassa and Ponce 2013). Southwards, this plain would include the present Pribilof Islands and, towards the north, the present Wrangell Island (Fig. 2). Towards the NE, the plain would follow the northern coast of Alaska and parts of Canada, with a mean width of at least 80 km; it extended continuously until approximately longitude 128° W. Toward the NW, the plain formed a narrow wedge at the latitude of Wrangell Island (Fig. 2), to later on become in contact with another huge plain developed in northern Russia. This plain presented an enormous surface of approximately 1,200,000 km^2 between longitude W180° and W156°. It presented an extension as large as 1800 km in N–S direction at the longitude of the Bering Straits. Its relief was mostly flat, and the general slope was smaller than 0.2°, with a maximum local relief in the order of 60 m between its northern and southern extremes. According to the Lambeck and Chappell (2001) sea level curve, which

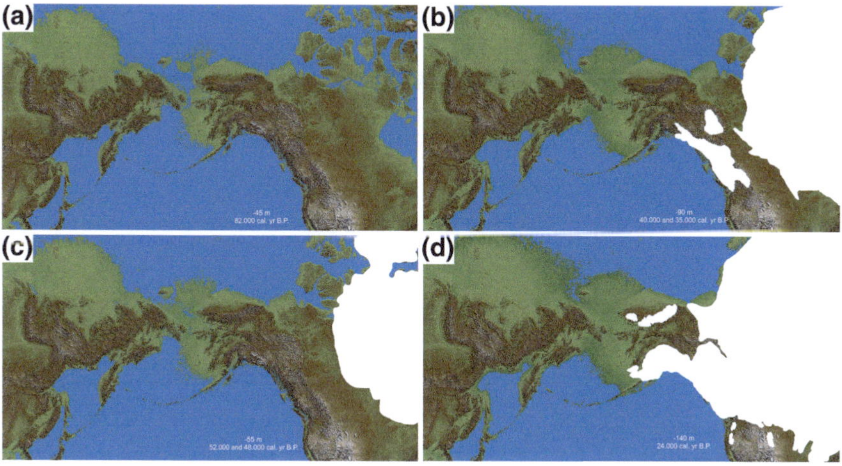

Fig. 3 **a–d** paleogeography of Beringia, with ancient positions of the coastline for lower sea levels during different moments of the Late Pleistocene (after Rabassa and Ponce 2013). Sea level information from Lambeck and Chappell (2001). *Whitish areas* correspond to the actual extent of the Laurentide and the Cordilleran ice sheets for the studied moments

relates sea level with ice volume on the continents, a sea level position at −90 m would be equivalent to the ice volume that existed around 13,000 cal. year B.P.

Dyke (2004) identified the existence of two main ice sheets in northern North America by 13 cal. ka B.P., the Cordilleran mountain ice sheet, developed on top of the Rocky Mountains and the Pacific Ranges, and the Laurentide Ice Sheet, of much larger size, extended over most of Canada and the whole of Greenland (Fig. 3b). This latter ice sheet had an approximate surface of 11 million km^2. In between them, there was an ice-free corridor which communicated Alaska with the rest of the continent (Fig. 3b). It had an approximate length of 1400 km in a NW–SE direction and a width which varied between 400 and 700 km (Fig. 3a). Immediately to the west of this corridor, another smaller ice sheet had developed over the Mackenzie Mountains. This situation would have represented the maximum ice extension during MIS 3, which would have been coincident with a sea level position of −90 m (Fig. 4).

The highest sea level position during MIS 3 was perhaps at −55 m. This condition took place in two periods, toward 52 and 48 ka B.P. and it is coincident with one of the warmest D-O oscillations (D-O event 14; Table 1) and with the smallest ice expansion during MIS 3. The geographical conditions of Beringia determined that the area lowlands were permanently devoid of ice during the Late Pleistocene; in spite of being cold enough, the land bridge was too dry to arid polar conditions to develop glaciers at low elevations (Elias and Brigham-Grette 2007). The absence of lowland glaciers allowed the availability of the land bridge for human displacement throughout MIS 3 and 2.

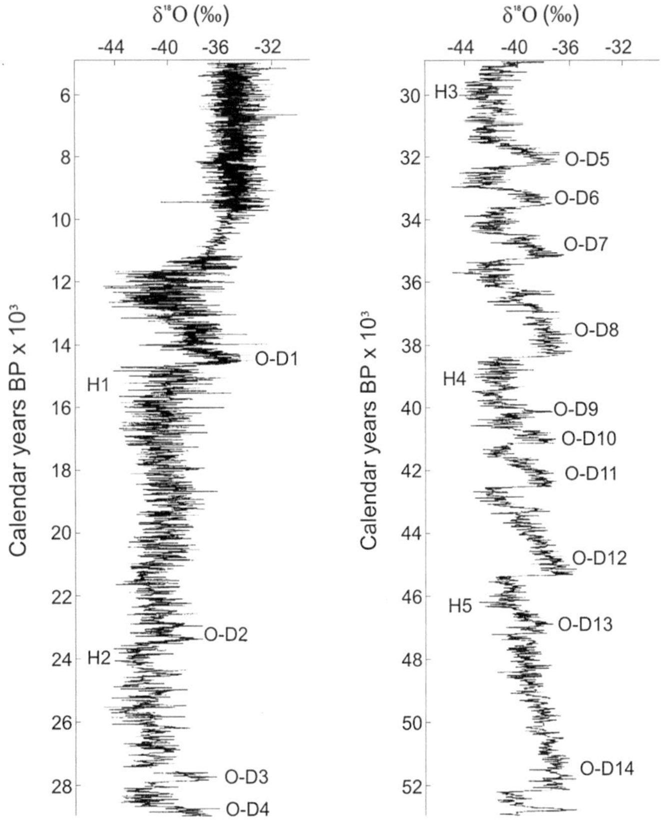

Fig. 4 Abrupt climatic oscillations during the last 90,000 years in GIPS II core from Greenland (after Denton 2000). *H1–H5* represents the cold Heinrich events. *O-D1–O-D14* corresponds to the Dansgaard–Oeschger warm events. See Rabassa and Ponce 2013

3 Final Remarks

Marine Isotope Stage 3 (MIS 3) was a period of very complex climatology, with previously unheard strong variability, with colder moments, named as Heinrich events, followed by warmer episodes, known as the Dansgaard/Oeschger events. Most authors agree that these paleoclimatic events are the result of higher discharge of icebergs and meltwater in the northern Atlantic Ocean, thus forcing a strong decrease in salinity, which would have impeded the penetration of the Gulf Stream into northern latitudes and therefore generating colder conditions. These are the Heinrich events. Later on, a gradual stabilization of the ocean/atmosphere system induced progressive entrance of the Gulf Stream, increasing the ocean temperature and, consequently, leading toward the Dansgaard/Oeschger events. These paleoclimatic changes were very fast and the duration of each event was quite short, between just a few centuries and a few millennia in each case. This highly complex paleoclimatology affected simultaneously the entire world, and their consequences are still very poorly known.

The period embraced by these changes is at the edge of the dating capability of the radiocarbon method, or even beyond it. This limitation has made very difficult to establish local and regional chronologies for this period, and a global record for these events is still uncertain. Future progress in dating techniques will certainly provide more reliable data, the impact of these paleoclimatic events will be better understood and their global correlation would be more accurate and meaningful.

Then and there, many paleoclimatic and paleobiogeographic circumstances will be properly interpreted and their influence upon MIS 2 and Holocene environments would be truthfully recognized.

Finally, the paleoclimatic history of Beringia during MIS 3 is particularly significant since it may have controlled the appropriate environmental conditions to allow the inbound displacement of the first waves of human peopling of the Americas, moving from Siberia to Alaska and northern Canada.

Acknowledgments The criticism of anonymous reviewers on earlier versions of this manuscript, which greatly improved the definitive text, is deeply acknowledged. Usual disclaimer applies.

References

Alley RB (2004) Abrupt climate changes: oceans, ice and us. Oceanography 17(4):194–206

Andrews JT (2000) Dating glacial events and correlation to global climate change. In: Noller JS, Sowers JM, Lettis WR (eds) (2000) Quaternary geochronology, methods and applications. American Geophysical Union, AGU Reference Shelf, vol 4, pp 447–455

Andrews JT, Dyke AS (2007) Late quaternary in North America. In: Elias SA (ed) Encyclopedia of quaternary science. Elsevier, Amsterdam

Bard E, Rostek F, Turon JL, Gendreau S (2000) Hydrological impact of Heinrich events in the subtropical Northeast Atlantic. Science 289:1321–1324

Barendregt RW (2011) Magnetostratigraphy of Quaternary sections in Eastern Alberta, Saskatchewan and Manitoba. In: Ehlers J, Gibbard P, Hughes P (eds) Quaternary glaciations—extent and chronology. A closer look, Elsevier, Developments in quaternary science, vol 15, pp 591–600

Bigelow N (2007) Pollen records. Late Pleistocene/Northern North America. In: Elias S (ed) Encyclopedia of quaternary sciences, 2633–2648, Elsevier, Amsterdam

Blunier T, Brook EJ (2003) Timing of millenial-scale climate change in Antarctica and Greenland during the last glacial period. Nature 291:110–112

Clague J, Ward B (2011) Pleistocene glaciation of British Columbia. In: Ehlers J, Gibbard P, Hughes P (eds) Quaternary glaciations—extent and chronology. A closer look, developments in quaternary science. Elsevier, Amsterdam, vol 15, pp 563–572

Denton G (2000) Does an asymmetric thermohaline ice-sheet oscillator drive 100,000 year glacial cycles? J Quat Sci 15:301–318

Dyke AS (2004) An outline of North American deglaciation with emphasis on central and northern Canada. In: Ehlers J, Gibbard P (eds) Quaternary glaciation-extent and chronology, Part II. Developments in quaternary science. Elsevier, Amsterdam, vol 2, pp 373–424

Elias SA (2007) Beetle records. Late Pleistocene of North America. In: Elias S (ed) Encyclopedia of quaternary sciences, pp 222–236. Elsevier, Amsterdam

Elias SA, Brigham-Grette J (2007) Glaciations. Late Pleistocene events in Beringia. In: Elias S (ed) Encyclopedia of quaternary sciences, pp 1057–1066. Elsevier, Amsterdam
Gillespie A, Clark D (2011) Glaciations of the Sierra Nevada, California, U.S.A. In: Ehlers J, Gibbard P, Hughes P (eds) Quaternary glaciations—extent and chronology. A closer look, developments in quaternary science. Elsevier, Amsterdam, vol 15, pp 447–462
Glushkova O (2011) Late Pleistocene glaciations in North-East Asia. In: Ehlers J, Gibbard P, Hughes P (eds) Quaternary glaciations—extent and chronology. A closer look, developments in quaternary science. Elsevier, Amsterdam, vol 15, pp 865–872
Heinrich H (1988) Origin and consequences of cyclic ice rafting in the Northeast Atlantic Ocean during the past 130,000 years. Quatern Res 29:142–152
Jackson LE, Andriashek LD, Phillips M (2011) Limits of successive middle and late Pleistocene continental ice-sheets, interior plains of southern and central Alberta and adjacent areas. In: Ehlers J, Gibbard P, Hughes P (eds) Quaternary glaciations—extent and chronology. A closer look, developments in quaternary science. Elsevier, Amsterdam, vol 15, pp 575–590
Kaufman DS, Young NE, Briner JP, Manley W (2011) Alaska paleo-glacier atlas (version 2). In: Ehlers J, Gibbard P, Hughes P (eds) Quaternary glaciations—extent and chronology. A closer look, developments in quaternary science. Elsevier, Amsterdam, vol 15, pp 427–446
Labeyrie L, Skinner L, Cortijo E (2007) Paleoclimate reconstructions. Sub-Milankovitch (DO/Heinrich) events. In: Elias S (ed) Encyclopedia of quaternary sciences, pp 1964–1974. Elsevier, Amsterdam
Lambeck K, Chappell J (2001) Sea level change through the Last Glacial cycle. Science 292 (27):679–686
Lambeck K, Yokoyama Y, Purcell T (2002) Into and out of the last glacial maximum: sea level change during oxygen isotope stages 3 and 2. Quatern Sci Res 21:343–360
Lang C, Leuenberger M, Schwander J, Johnsen S (1999) 16 °C rapid temperature variation in central Greenland 70,000 years ago. Science 286:934–937
Lozhkin AV, Anderson PM (2007) Pollen records. Late Pleistocene/Northern Asia. In: Elias S (ed) Encyclopedia of quaternary sciences, pp 2623–2633. Elsevier, Amsterdam
Markova A, Puzachenko A (2007) Vertebrate records. Late Pleistocene of Northern Asia. In: Elias S (ed) Encyclopedia of quaternary sciences, pp 3158–3175. Elsevier, Amsterdam
Moller P, Hjort C, Alexanderson H, Sallaba F (2011) Glacial history of the Taymyr Peninsula and the Severnaya Zemlya Archipelago, Arctic Russia. In: Ehlers J, Gibbard P, Hughes P (eds) Quaternary glaciations—extent and chronology. A closer look, developments in quaternary science, vol 15, pp 373–384. Elsevier, Amsterdam
Peterson L, Haug GH, Hughen KA, Röhl U (2000) Rapid changes in the hydrologic cycle of the Tropical Atlantic during the Last Glacial. Science 290:1947–1951
Rabassa J (ed) (2008) The Late Cenozoic of Patagonia and Tierra del Fuego, developments in quaternary science, 11, 513 pp. Elsevier, Amsterdam
Rabassa J, Ponce JF (2013) The Heinrich and Dansgaard-Oeschger climatic events during marine isotopic stage 3: searching for appropriate times for human colonization of the Americas. Quatern Int 299(2013):94–105
Sher A, Kuzmina S (2007) Beetle records. Late Pleistocene of Northern Asia. In: Elias S (ed) Encyclopedia of quaternary sciences, pp 246–267. Elsevier, Amsterdam
Stauch G, Lehmkuhl F (2011) Extent and timing of Quaternary glaciations in the Verkhoyansk mountains. In: Ehlers J, Gibbard P, Hughes P (eds) Quaternary glaciations extent and chronology. A closer look, developments in quaternary science, vol 15, pp 877–882. Elsevier, Amsterdam
Uriarte Cantolla A (2003) Historia del clima de la tierra. Servicio Central de Publicaciones del Gobierno Vasco, Victoria-Gasteiz, p 306
Velichko A, Faustova M, Pisareva V, Gribchenko Y, Sudakova N, Lavrentiev N (2011) Glaciations of the East European plain: distribution and chronology. In: Ehlers J, Gibbard P, Hughes P (eds) Quaternary glaciations—extent and chronology. A closer look, developments in quaternary science, vol 15, pp 337–360. Elsevier, Amsterdam

Vidal L, Schneider R, Marchal O, Bickert T, Stocker T, Wefer G (1999) Link between the North and South Atlantic during the Heinrich events of the last glacial period. Clim Dyn 15:909–919
Vorren T, Landvik J, Andreassen K, Laberg JS (2011) Glacial history of the Barents Sea region. In: Ehlers J, Gibbard P, Hughes P (eds) Quaternary glaciations—extent and chronology. A closer look, developments in quaternary science, vol 15, pp 361–372. Elsevier, Amsterdam
White JWC, Steig EJ (1998) Timing is everything in a game of two hemispheres. Nature 394: 717–718
Wright JD (2000) Global climate change in marine stable isotope records. In: Noller JS, Sowers JM, Lettis WR (eds) Quaternary geochronology, methods and applications. American Geophysical Union, AGU Reference Shelf 4, pp 427–433

On the Origin of the Dansgaard–Oeschger Events and Its Time Variability

Silvia Duhau and Cornelis de Jager

Abstract The origin of the Dansgaard–Oeschger (D/O) events and its variability between different isotope stages is at present under debate. Evidence is herein presented that during the Holocene the "Bond cycle," which is, a reduced version of the D/O one, is related to a signal on solar activity that it has been baptized as the "Long Trend." A mechanism by which solar storms, which effect on the Earth's atmosphere strongly depends on the geomagnetic field morphology, may explain this relationship is proposed, and how the differences between the D/O events along different isotope stages and between the two hemispheres may help in disentangle, at least partially, the ultimate origin of climate change is discussed.

Keywords Climate change · Dansgaard–Oescherger events · Solar activity · Geomagnetic excursions

1 Introduction

Bond et al. (1999) found that there exist, since Marine Isotope Stage 5 (MIS 5) through the present times, a 1–2 ky distinct cycle in the sub-Milankovich scale, which has a certain relationship with the colder Heinrich events (H). During the Holocene it appears to be a weakened version (called at present the "Bond cycle") of the Dansgaard/Oeschger (D/O) events that took place throughout the last glaciation. On that framework, the Little Ice Age (LIA) was the coldest phase of the most recent 1–2 ky cycle.

S. Duhau (✉)
Departamento de Física, Facultad de Ingeniería, Universidad de Buenos Aires, 1428, Buenos Aires, Argentina
e-mail: silvia.duhau@gmail.com

C. de Jager
SRON Laboratory for Space Research, Royal Netherlands Institute for Sea Research, Formerly Astronomical Institute, Sorbonnelaan 2, Utrecht, The Netherlands
e-mail: info@cdejager.com

G.M. Gasparini et al. (eds.), *Marine Isotope Stage 3 in Southern South America, 60 ka B.P.–30 ka B.P.*, Springer Earth System Sciences,
DOI 10.1007/978-3-319-40000-6_3

The origin of climate changes in the millennial time scale remains poorly understood not only in the case of the D/O events (see, for instance, Schulz 2002; Dokken et al. 2013) but also in the case of its Holocene version, the Bond cycles (Humlum et al. 2011). On the other hand, tenacious influence of solar activity was found by Bond et al. (2001) on climate during the course of the entire Holocene and he concluded that there must be a mechanism for amplifying the solar signals and transmitting them globally. In the present work, this mechanism is investigated and how it may lead not only to the Bond cycles but also to the D/O events themselves.

A relationship does exist between the Milankovitch cycles in Earth orbital parameters and the glacial and interglacial periods. As orbital parameters modulate the insolation, that is, the final distribution of solar irradiation over the Earth's surface, it is assumed that those cycles are the origin of these extreme climate variations. An assessment to this theory is given by the synchronicity of glaciations in the two hemispheres (Rutter et al. 2012 and references therein).

Solar irradiance is the only source of solar energy that is included in the models of the climate system (see, for example, the IPCC report 2013). Observations show that climate fluctuations are much stronger than the calculated variations; thus many mechanisms of amplification and damping (positive and negative feedback) of the impact of insolation variations in climate change have been proposed for explaining glaciations, but none of them has been proved to work by a consistent model (see Ave-Ouchi et al. 2013). Besides solar irradiance there are other sources of solar origin that may affect climate (for a review see Duhau 2005). As solar irradiation is the primordial source of atmospheric heating, the most relevant factors are those that may be contributing to mechanisms of feedback of insolation, as are cloud cover and ice-albedo feedback (see, for instance, Randall et al. 2007).

Cloud cover results from the balance between cloud nucleation and precipitation. Svensmark and Friis-Christensen (1997) found a high degree of correlation between total cloud cover and cosmic ray flux and suggested that it may due to the contribution of cosmic rays to cloud nucleation; in turn, cosmic ray impact on the Earth' environment is controlled by the solar magnetic open flux. As much as cloud nucleation is modulated by cosmic rays, cloud precipitation is controlled by the electric circuit between Earth and the ionosphere (Tinsley 2008), whose variability depends upon the frequency and intensity of solar flares and solar energetic particles produced by "coronal mass ejections," and their final impact on the Earth environment is called a "solar storm." Therefore, solar activity modulates cloud cover by intervening in cloud nucleation and in precipitation. However, when interpreting observations, only the first of these two processes is assumed to depend on solar activity. This may be the origin of the difficulty (Agee et al. 2012) in determining the final effect of solar activity on cloudiness.

Solar energetic particles are further accelerated by the compression of the Earth magnetosphere by solar coronal mass ejections in such a way that the flux of these particles on the atmosphere strongly depends not only on solar activity but also on geomagnetic field morphology. When the geomagnetic field is mainly dipolar and the geomagnetic poles are near the geographic ones, like is currently happening, its shielding effect against high energetic protons and relativistic electron events that are produced by flares and coronal mass ejections is the weakest at higher latitudes. As a result, it can be locally very strong in the polar cap regions (Velinov et al. 2013). Duhau and Martínez (2012) suggested that in these circumstances it contributes to heating the atmosphere at higher geomagnetic latitudes and thus, to the ice melting at the sub-auroral region and at the polar cap. Besides, Bucha (1976, 1983) found evidence that at the sub-auroral arc, relativistic electrons heat the atmospheric column from the Earth's surface to the thermosphere heights (about 300 km), thus controlling the opening of the polar vortex and, as a consequence, affecting the global circulation. In turn, changes in global circulation are known to be directly related to glacier response (Akçar et al. 2014) and to D/O events (Leuschner and Sirocco 2000; Dokken et al. 2013) and model computations indicate also that increases in polar cloud cover are significantly correlated to amplified Arctic warming (Holland and Bitz 2003). Accordingly, solar activity may regulate not only two of the feedback mechanisms that amplify the absorption of insolation by the atmosphere such as are ice-albedo and cloud cover, but also would drive the main mechanism that regulates their global transmission, by modulating polar vortex aperture. Thus, its variability may be driving the Bond cycle.

Summing up, the final impacts of all solar sources of energy on climate change, excepting irradiation, are modulated by the geomagnetic field strength and by the position of the geomagnetic pole with respect to the geographical one. Consequently, if these solar sources were indeed significantly contributing to climate change, a dependence of climate not only in solar activity, but also in geomagnetic field morphology would exist. In fact evidence does exist of a connection between the geomagnetic field and climate variability in a broad range of time scales (0.1–1000 ky) (Wollin 1971; Courtillot et al. 2007). The existence of this connection between climate changes and geomagnetic field variation gives strong support to the hypothesis that solar storms and cosmic rays flux significantly contribute to climate change not only in the sub-Milankovitch time scale but in the Milankovitch time scale too. As solar activity also undergoes strong variations in the secular to the millennial time scale, its variability must be taken into account when comparing geomagnetic field morphology and climate variability. This is the main objective of the present work.

In Sect. 2, the variables that measure the sources of solar origin that may contribute to climate change are discussed from data obtained from direct observations. A working summary of a signal processing technique (Duhau and de Jager Submitted 2016) that allows representing transient signals in the time domain is herein presented and applied to comparing their time variation. The same technique is applied in Sect. 3 to describe the nonlinear relationship between sunspot maxima and a temperature time series for the last two millennia. The Long Trend in solar activity, as defined by Duhau and De Jager (Submitted 2016), is followed by a similar trend in temperature, that been identified as the Bond cycle. In Sect. 4 the INCAL98 cosmogenic isotope time series provided by Stuiver et al. (1998) is applied to analyze the Long Trend in solar activity during the last 15.7 ky. Some remarkable signatures of this trend in the Gisp2 temperature time series from Mayewski and Bender (1995) for the last 50 ky are also described. In both Sects. 3

and 4, the relevance of variations in the geomagnetic field strength and the position of the geomagnetic pole with respect to the geographical one are investigated in terms of the modulation of the impact in temperature of the Long Trend in solar activity in the sub-Milankovitch as well as the Milankovitch time scales. Finally, in Sect. 5 the results and conclusions are summarized.

2 Solar Dynamo Transitions and the Gleissberg Cycle in Solar Activity as Seen in Data Obtained from Direct Observations

2.1 The Data

To obtain a precise description of the nature of solar activity, data from direct observations are needed. Solar activity is produced by variations in the solar magnetic field of the Sun. This field inverts its polarity approximately every ~ 11 years, which is called the Schwabe cycle; thus, it comes back to the same polarity approximately every ~ 22 years, known as the Hale cycle. The dynamo system that produces this cycle is nearly symmetrical with respect to the solar equator. As a consequence it has basically two components: the toroidal and the polar components. Different manifestations of solar activity are produced by time changes in the strength of these two components.

Sunspots are the manifestation of the strength of the toroidal component that has been directly observed in the solar surface for the longer time. Sunspot number does not depend upon the sign of the toroidal field and so it follows the Schwabe cycle (thin black lines in Fig. 1, upper panel), in such a way that sunspot number at maxima, SN_{max}, (starts in Fig. 1, upper panel) gives a measure of the strength of the toroidal field (Nagovitsyn 2005). The geomagnetic index introduced by Mayaud (1971, 1972) is a proxy for the polar field (Russell 1975; Duhau and Chen 2002). Maximum strength in toroidal field is delayed by half of a Schwabe cycle with respect to maxima strength in the polar one; thus, the value of geomagnetic index aa occurring at sunspot minima, aa_{min} (diamonds in Fig. 1, upper panel), gives a precise measure of the strength of the polar magnetic field (see Fig. 3 in De Jager and Duhau 2011) and therefore of the open magnetic flux from the Sun.

As it is readily apparent in the upper panel of Fig. 1, the amplitude of the two components of the solar magnetic field oscillates around a constant level, named as the 'transition level' (indicated by the green horizontal line in the top panel of Fig. 1). This level has the property that after passing the two components simultaneously by their respective transition levels, the oscillation undergoes sudden changes in amplitude and in periodicity, a phenomenon that has been called the 'transition' (Duhau and De Jager 2008). In that way, the solar dynamo oscillations evolve by a succession of three kind of episodes: the Grand Minimum (as it was the case of the Maunder Minimum, blue stars), Regular (green stars) and the Grand Maximum (as the one in the twentieth Century, red stars).

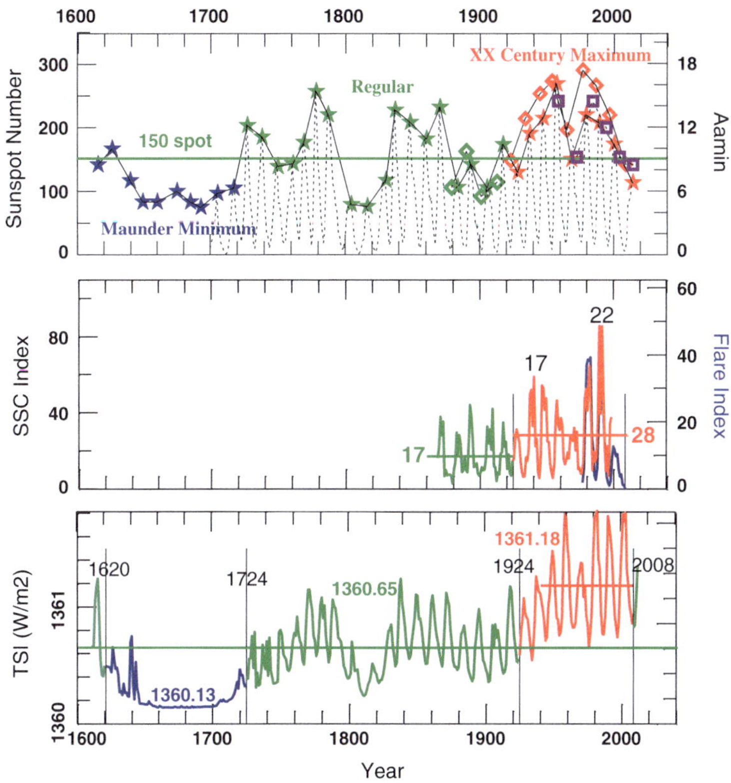

Fig. 1 *Upper panel* Sunspot number at maxima, SN_{max} (*stars*) and geomagnetic index at minima (Aa_{min}) (*diamonds*) in the intervals AD 1610–2014, 1860–2000, respectively. The *violet squares* are the prediction by Schove (1955) for the sunspot maxima from #19 to #24. The *vertical lines* are at the date of occurrence of each of the four solar dynamo transitions that occurred since AD 1600 and the sequence of *blue*, *green*, *red*, and again *green colors* differentiate the kind of episode that started after each of them. *Middle panel* Flare index (*blue line*) and the SSC Index, as defined by Duhau (2003) that quantifies the intensity and frequency of the solars storms. *Bottom panel* Total solar irradiance (*TSI*) inferred from sunspot number by Wang et al. (2005) and observed (Kopp 2015) for the time intervals AD 1610–1994 and 1995–2013, respectively. In the three panels the horizontal lines are at the average value along the Maunder Minimum (*blue*), Regular Oscillation Episode (*green*) and twentieth century Grand Maximum (*red*), respectively; with the exception of the *green line* in the top panel that correspond to the transition level (Adapted from Fig. 1 in Duhau and Martínez 2012, in which the upper panel SN_{max} has been updated, as it is quoted in Sect. 3)

The three kinds of solar dynamo episodes appear themselves in different ways in all variables shown in Fig. 1 (the sources of the data on the cited figure are given in Duhau and Martínez 2012). With the exception of the sunspot number, all of them are related to climate change. Aa_{min} (diamonds, upper panel) is proportional to the strength of the open magnetic flux of the sun that modulated cosmic ray flux. The Flare index (blue line, middle panel) quantifies the flux of energy in the X-ray wavelength produced by solar flares, that together with other components of the hard part of the solar spectrum has a central role in creating the ozone layer and the ionosphere. The SSC index (sudden storm commencement index; the succession of green and red lines in the middle panel) was defined by Duhau (2003) and quantifies the solar storms. It is strongly modulated by the geomagnetic field intensity and tilt and is based in the time of duration and intensity of the sudden impulse that precedes sudden commencement geomagnetic storms as defined by Mayaud (1975). Total solar irradiance (TSI; bottom panel) is the solar radioactive power per unit area normal to the rays, incident on the Earth's upper atmosphere. It has been obtained from direct observations for the last three Schwabe cycles (Kopp 2015). For previous cycles values were inferred from Sunspot Number according to Wang et al. (2005).

It has been observed that the behavior of aa_{min} is quite similar to that of SN_{max}, whereas SSC and TSI follow the three kinds of episodes that undergo the solar dynamo magnetic field in its own way. Its average value substantially changes from one episode to the next. Note that in the brief interval along which it is known the Flare index is qualitatively similar to the SSC one. It is remarkable that after the 1923 transition the average level of TSI has increased in less than 0.001 %, whereas that of aa_{min} and SSC indexes, in synchronicity with the global warming of the twentieth Century, increased in 43–64 %, respectively. Therefore, whereas TSI is the primordial source of energy that maintains temperature at its background level, it is plausible that the other sources of solar energy that vary several orders of magnitude as compared with TSI may play a central role not only in driving the fluctuations in temperature but also in suddenly changing, after each transition other variables related to climate change. Evidence of these changes does exist, as it has been summarized in the Introduction of this chapter.

The solar dynamo is a nonlinear system the general consensus is that it must have undergone a chaotic behavior (Weiss 1987; Usoskin 2008; Artl and Weiss 2014). As a result the usual practice is restraining the prediction of solar activity to a few years in advance (Pessnell 2007). However, Schove's successful prediction of sunspot maxima by six decades in advance is based in some regularity that he observed in sunspot maxima time series. This indicates that the fluctuations in sunspot maxima maybe the signature of well-defined oscillations. These oscillations may be quantified by the application of a suitable signal processing technique. With this purpose we have developed a signal processing technique (for a review see De Jager and Duhau 2011) based in wavelet analysis that has been applied for predicting sunspot maxima for the next 150 years (Duhau and De Jager Submitted 2016). In Sect. 2.2 an operative summary of this method is provided, which is applied in 2.3 to represent the Gleissberg cycle, that is, a cycle in the solar dynamo magnetic field that gives origin to the succession of the three kinds of episodes shown in Fig. 1.

2.2 A Summary of the Methodology Applied for Representing Solar Activity Variables

It has been noted in the data of Fig. 1 that time variations in solar activity appear to be the result of the superposition of oscillations which period and amplitude change on time, especially after the transitions, which indicate that this oscillation has a transient behavior. Fourier analysis, that presumes that the signal to be represented is stationary, is not suitable for analyzing this transient behavior in the time domain. Thus, most of the study of such kind of data searches for the signature of these oscillations in the presence of conspicuous peaks in the Fourier spectra of the involved time series (for a review see De Jager 2005). This is also the case for data related to climate change and in particular in the one usually applied to search the Bond cycles. The main problem of this procedure is that not necessarily the presence of two different peaks in the spectrum of a given time series represents the existence of two different natural modes of oscillation in the system, because in some cases each of them may be merely the signature of sudden changes in period that undergoes a unique mode of oscillation (Fig. 4 in Duhau and Martínez 2012). Therefore, the only way to detect these features is by the study of the time series in the time domain. It has been found that all the variables associated with solar activity may be represented by the superposition of several 'wave trains' (WTs), defined as a succession of wave packets, each of them having a nearly constant period which value alternatively changes from a wave packet to the next one (Duhau and de Jager Submitted 2016).

As a result, wavelets have been chosen for the analysis of the relevant data, which are functions of compact support and selected as the mother function of our signal processing technique, the Morlet wavelet (1981). It may be written as follows (Duhau and de Jager Submitted 2016):

$$\boldsymbol{M}(\boldsymbol{t}) = \frac{1}{\boldsymbol{\pi}^{\frac{1}{4}}} \mathrm{e}^{\boldsymbol{i}\left(\frac{2\pi}{T}\right)\boldsymbol{t}} \mathrm{e}^{-\frac{1}{2}\left(\frac{\boldsymbol{D}}{2\pi \boldsymbol{T}}\boldsymbol{t}\right)^2} \tag{1}$$

We conclude that (t) is a harmonic function of period T that has a Gaussian envelope with a variance $\boldsymbol{\sigma}$. The latter is related to T by:

$$\boldsymbol{\sigma} = \frac{\boldsymbol{D}}{2\boldsymbol{\pi}} \boldsymbol{T} \tag{2}$$

Once the mother function has been chosen, the parameters that define each of the functions of the basis must be selected. The value of D is chosen equal to 6 in order to fulfill the admissibility condition (Farge 1992). The periods are computed as powers of 2 (Torrence and Compo 1998):

$$T_n = T_o 2^{\frac{j}{n_o}}, \quad j = 0, 1, \ldots J, \tag{3}$$

where n_o, j and J are natural numbers, T_o the smallest period,

$$T_o = 2\delta t \tag{4}$$

being δt the time interval between two successive data points that must be constant and it is known once the time series to be analyzed has been chosen. Therefore, for the determination of the set of periods of the base functions from Eqs. 2 and 3 only t, n_o and J remain to be fixed. The maximum period must be in the order of the length of the time series, N, and so:

$$J = \mathrm{Int}\left(n_o \frac{\ln N}{\ln 2}\right) \tag{5}$$

From the above equation it can be seen that the total number of wavelets that must be applied for a time series containing N data points increases with increasing N and n_o. The minimum value of n_o that leads to a precise representation of the Sunspot Maxima time series has been studied and, at the same time, of the meaningful signals on which we may split it. This election reduces the numerical complexity and at the same time the value assigned to n_o may have a physical meaning. It has been found that this value is $n_o = 3$.

After the values of T_o, J and n_o are fixed the numerical values of the periods of the base functions are computed from Eq. 2. The signal is represented by summing up the transform of those functions, that for shortness they have been called ‘components’. These components are defined as the convolution of the time series with each of the basis function (see Farge 1998 and references therein). These computations have been performed with the program provided by Torrence and Compo (1998).

In the next subsection this formalism has been applied to compute the Gleissberg cycle, as defined by De Jager and Duhau (2012) from the data of Sect. 2.1.

2.3 The Gleissberg Cycle

As the data points of SN_{max} (diamonds in Fig. 1 top) are not equally spaced, and there are yearly values for the remainder data of Fig. 1, data point has been added for one year by a linear interpolation between two successive points. Thus, from 2.1, $T_o = 2$ years and, as the time series extend by 405 years, a data point per year is considered, $N = 405$, and from (4), $J = 24$. By introducing this values in Eq. 1 and setting $n_o = 3$ it may be computed the set of periods of the wavelets function (given by Eq. 2). Finally, the ‘Gleissberg cycle’ is defined as the addition of the linear trend to all the components whose periods $T_j > 80$ year (which are those computed from Eq. 2 for $j \geq 16$) (For details see Duhau and de Jager Submitted 2016) The result for the variables of Fig. 1, with the exception of aa_{min}, is shown in

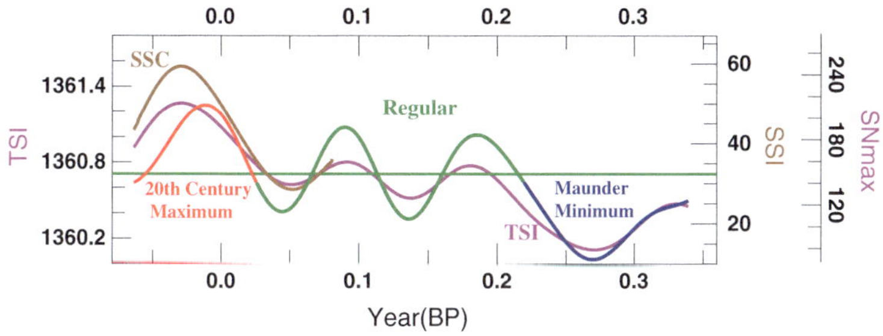

Fig. 2 The Gleissberg cycle from sunspot number time series (the sequence of *red*, *green* and *blue lines*, as in Fig. 1a), total solar Irradiation (TSI) (in *violet*) and the sudden storm commencement index (SSC) (in *brown*). The *horizontal green line* represents the transition level of 1365.65 W/m^2 (*left abscise*), and 17.0 and 150 spot (*the two right abscises*)

Fig. 2. In the case of aa_{min} its Gleissberg cycle is not plotted because it is similar to that cycle in SN_{max} along the time interval on which it has been observed.

It may be seen in Fig. 2 that during the last 0.4 ky the Gleissberg cycle is similar in the three variables related to solar activity. Thus, in spite that sunspot maxima time series does not give a direct measure of the flux of energy of solar origin that is impinging on the climate system, the sunspot number data for time scales at and above the secular time scale may be applied to describe the qualitative behavior of all the variables related to solar activity. The data applied with this purpose is presented in Sect. 3.1 and by applying to it the signal processing technique presented in Sect. 2.2, in Sect. 3.2 the signature on temperature of the Gleissberg cycle in solar activity for the last 1.7 ky is investigated. Also, in Sect. 3.3, the Bond cycle is represented and it is compared with the Long Trend in solar activity as defined by Duhau and De Jager (Submitted 2016).

3 Comparison of Solar Activity and Temperature for the Last 1.7 Ky and the Bond Cycle

3.1 *The Data*

The only successful prediction for more than a few years in advance is that from Schove (1955) that successfully predicted sunspot maxima #19 to #24 (violet squares in Fig. 1, top panel). This prediction was made based upon the regularity that he found in his time series of sunspot maxima. This time series was obtained from naked eye observations of sunspots and aurora occurrence provided by ancient archives and has a continuous record since 1.74 ka B.P. The Zurich yearly values of sunspot number that are obtained by direct observations started in AD 1705. The conventional 0.6 scale factor that is commonly applied to this time series prior to the nineteenth century was recently dropped, thus raising the scale of this time series to the level of modern sunspot counts (Clette et al. 2014). After dividing the (Schove 1955) (Sc55) time series by the 0.6 factor it blend with the sunspot maxima found from the new version of sunspot number including the six predicted maxima (see Fig. 1 in Duhau and de Jager Submitted 2016), which gives confidence in the qualitative features of the Sc55 time series. So we have extended backward, since year 1705 till year 290, the sunspot maxima time series as determined form the sunspot number yearly values given by Clette et al. (2014) by Schove (1955) sunspot maxima time series. The time series, SN_{max}, obtained by this procedure is plotted in Fig. 3, upper panel.

Thermometric observations of surface atmospheric temperature have started as recently as AD 1850. Accordingly, it is needed to resort on proxy data of this variable. The sunspot number is a measure of the effect of solar activity over the entire globe, so it would be optimal to compare it with global temperature data. Such a kind of average for a time interval long enough does not exist and thus research must rely on Moberg et al. (2005) time series (the black line in Fig. 3, bottom) that is the largest time series of atmospheric temperature averaged over the Northern Hemisphere. This time series has been obtained by combining tree-ring data with other much lower-resolution proxies as lake and ocean sediments.

The long-lasting episode of temperate climate appears to follows the Regular Episode in solar activity occurring prior to 1.2 ka. As much as the Medieval

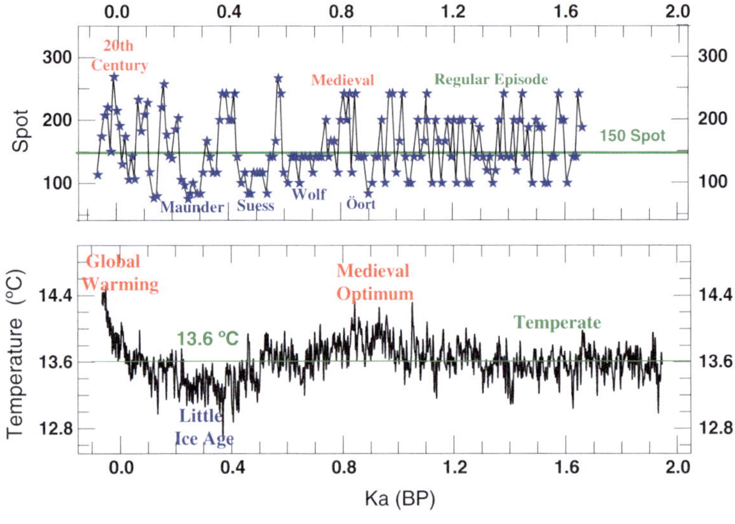

Fig. 3 *Upper panel* sunspot maxima time series (*blue stars*) obtained by Duhau and De Jager (Submitted 2016) by means of the procedure outlined in the text. The *green horizontal lines* indicate the 150 spot transition levels. The names of grand maxima and grand minima are indicated by *red* and *blue* characters, respectively. *Bottom panel* Northern hemisphere surface temperature yearly values derived from ^{14}C tree ring and land and ocean sediment dates (Moberg et al. 2005). The NOAA source is: http://www.ncdc.noaa.gov/paleo/Globalwarming/moberg.html. The *green line* is at the constant level of 13.6 °C, around which temperature fluctuates

Optimum and the Global Warming episodes followed its respective Grand Maximum in solar activity, the Little Ice Age (LIA) followed the succession of the three Grand Minima that occurred in the interval between these two Grand Maxima and the isolated Oort Minimum is barely noticed, a century later in terms of temperature.

To analyze the origin of the above behavior that indicates that a nonlinear relationship does exist between temperature and solar activity, in the next subsection are compared, by means of the signal processing technique summarized in Sect. 2.2, the oscillations that composes the Gleissberg cycle in sunspot maxima with the corresponding oscillations in temperature. The Bond cycle is also isolated from temperature and it is compared with the Long Trend in sunspot maxima as defined by Duhau and De Jager (Submitted 2016). Based upon these comparisons, the mechanism by which changes in solar activity and geomagnetic field morphology may explain those relationships is discussed.

3.2 The Signature of the Gleissberg Cycle in Temperature and the Origin of the Heinrich Events

The number of components that, when added, allow the representation of the Gleissberg cycle may be computed from the time series of Fig. 3. Likewise, it has been so presented in Sect. 2.2 for the time series of Fig. 1, with the difference that it is now $N = 1800$ instead of $N = 405$, and therefore, it is $J = 31$ instead of $J = 24$. (For details see Duhau and de Jager Submitted 2016) Note that the number of components is larger but the linear trend is quite smaller as compared with the time series of Fig. 1. The results are shown in Fig. 4. To interpret the similarities and differences between these two signals it must be taken into account that, if cosmic rays flux and solar storms were indeed affecting climate, temperature fluctuation must be following, in some way, variations in the geomagnetic field strength and in

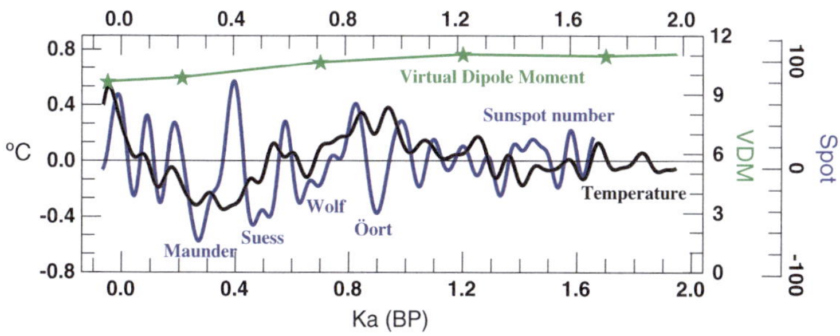

Fig. 4 The Gleissberg cycle (*blue line*) and the corresponding signal in temperature (*black line*) from the time series of Fig. 3 after subtracting their respective average values. The *green starts* are the average values of the VDM in the European region, taken from Table 1 in Mclhinny and Senayake (1982)

the position of the geomagnetic pole with respect to the geographical one. A measure of the geomagnetic pole is given by the virtual dipole moment (VDM) as for the geographical pole it is provided by the geomagnetic tilt (Constable 2005).The variation of VDM (green stars in Fig. 4) was quite similar in the two hemispheres since 50 ka ago [compare Fig. 5 in Mclhinny and Senanayake (1982), with Fig. 4 in Gallet et al. (2006)], and it has achieved its greater values during the interval 1.2–3 ka.

It may be observed in Fig. 4 that the twentieth Century maximum in sunspot maxima is amplified in temperature, leading to the Global Warming of the twentieth century when the highest temperatures of the last 2 ka have taken place (cf. Fig. 3). During the last 2 ka the geomagnetic field was dipolar and strong, solar storms strength has been enhanced as compared with epochs of weaker geomagnetic field. Regarding to the geomagnetic pole, while the South geomagnetic pole has been drifting to the equator, from 78° to 65° S, since AD 1750, contrarily the North geomagnetic pole has been moving towards the geographic pole from 70° N to 85° N at present times (see Fig. 11.2 in Korte and Mandea 2008; NOAA 2015). Since AD 1850 until today, the average temperature for higher latitudes (60° N–90° N and 60° S–90° S) has increased by 3 °C in the northern hemisphere and only 1.5 °C in the southern hemisphere (see bottom panel in Fig. 10 in Duhau and Martínez 2012). This seems to be consistent with the hypothesis that solar storms are able of contributing to ice melting at higher latitudes, because the drifting in the geomagnetic poles implies that there was more solar energy accumulating near the North geographic pole than near the South pole.

The steady approximation of the North geomagnetic pole to the geographic pole has strongly accelerated after AD 1900. The strong increase of TSI and SSC, associated to the AD 1923 solar dynamo transition (see Figs. 1 and 2) occurred not long after that. On the other hand, climate models, that consider solar irradiation as the only source of atmospheric heating, found that the Arctic sea ice is declining four times faster than forecasted (Stroeve et al. 2007; Rampal et al. 2011).

All the aforementioned circumstances suggest that not only LIA may be the Holocene version of a D/O event, as suggested by Bond et al. (1999), but also the fast melting of the polar ice that started about a decade ago may be the Holocene version of the Heinrich events that sporadically occurs along glaciations. According to our results this Holocene, reduced version, of a Heinrich events originates in a prolonged period of height solar activity (a strong solar magnetic field) that is synchronic with a strong nearly dipolar geomagnetic field. The fact that, while the North geomagnetic pole is drifting to the geographical pole, the South pole is instead wandering to the equator may explain why the increase of temperature in the northern hemisphere has been greater. Like is currently happening, Heinrich events may manifest in a different way in the two hemispheres (see e.g. Leuschner and Sirocco 2000). Moreover, the relative phase in the changes on the Solar magnetic and the geomagnetic fields is changing on time, which could explain also why the occurrence of Heinrich event is sporadic.

Other striking feature that may be observed in Fig. 4 is that the strong peak at ~0.4 ka on the Gleissberg cycle barely appears in temperature. This, again, indicates that the main effect of solar storms in climate is contributing to ice-albedo feedback, because in that case after the prolonged period of very low solar activity that started at about 0.7 ka, ice had enough time to thicken until this isolated peak on solar activity was only able to produce a tiny effect on ice melting (see the petite peak a century later, just when the Maunder Minimum was taking place). Giving support to this hypothesis, it is well known that the glacier retreat after LIA is occurring since around AD 1800–1850 and has not ended yet (Masiokas et al. 2010; Akasofu 2011), thus it is following, with nearly one century of delay, the emergence of solar activity from the Maunder Minimum.

To precisely determine the nonlinear dependence of temperature with solar activity, which is envisaged at a first glance on Fig. 4, in the next subsection the signal processing technique presented in Sect. 3.2 is applied to analyze its signature in different time scales.

3.3 *The Long Trend in Solar Activity and the Bond Cycle*

The Gleissberg cycle from the time series of Fig. 3 may be represented by the addition of a linear trend to five wave trains (WT). These are the following (Duhau and de Jager Submitted 2016): the Secular, Suess, Semi-Millennial, Eddy and Hallstatt WTs. With the exception of the Eddy WT that is found by adding four components, the other WTs are found by adding only three. The periods of these components are computed from Eq. 1 for $16 \leq j \leq 31$; for each of the five WTs, they are (expressed in ka): 0.08/0.102/0.128; 0.161/0.203/0.256; 0.322/0.406/0.512; 0.645/0.812/1.024/1.290; 1.625/2.048/2.580, respectively. These results are plotted in Fig. 5.

All the WTs in temperature whose oscillations have periods above the secular one follow those in solar activity with a variable time delay and relative amplitude. In case of the secular oscillations they appear as not bearing a regular relationship, which not necessarily means that solar variability on this time scale is not driving climate change. Instead, they indicate that on that time scale the climate system is reacting in a quite different manner at different moments, due to the changes in land and ice cover that accompanies the changes in temperature, global circulation and cloud cover in scales above the secular one. In other words, the involved mechanisms are of nonlinear nature. As an example of this behavior it has been noted that the strongest maximum in the Suess oscillation in solar activity was synchronic with the LIA, and followed the weakest one in temperature, consistently with the fact that ice-albedo feedback was the weakest of the entire period along the LIA. It is remarkable that, while the Suess WT is the strongest in solar activity, the stronger one in temperature is the Eddy WT. This is due to the fact that the effect of the latter on temperature is seven times larger than that of the Suess WT. It is concluded that the understanding of the origin of the differences between solar activity and temperature at different time scales may give valuable information about the feedback mechanism that is working on each of them.

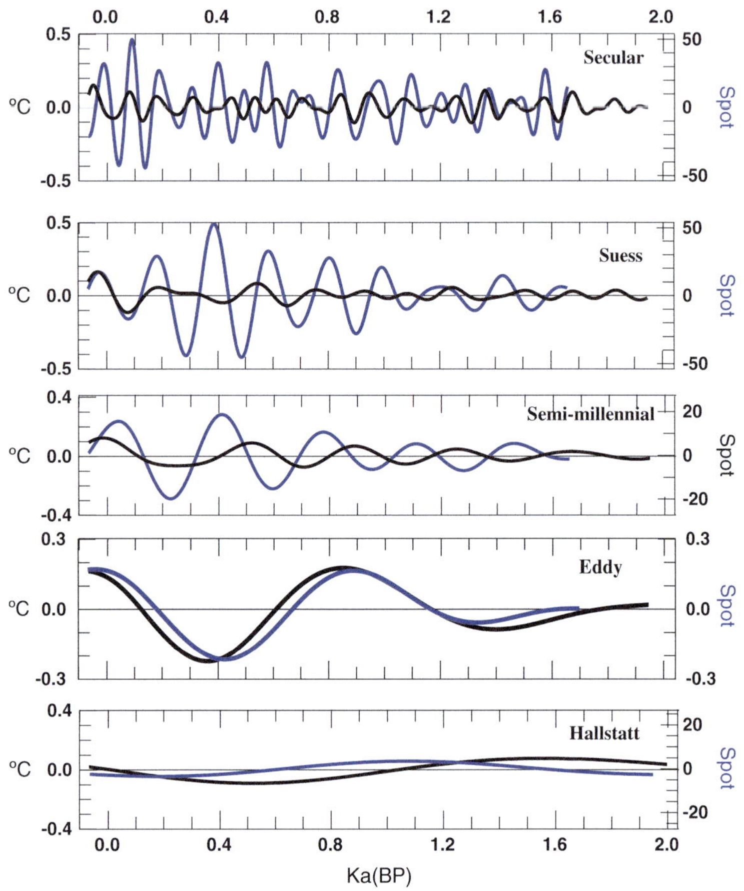

Fig. 5 The five wave trains (WT) whose names are at the upper right portion of the figure, which when added they comprise the Gleissberg cycle as computed from the data of Fig. 2 (*blue* and *black colors* refers to sunspot maxima and temperature time series, respectively)

In Fig. 6 the Gleissberg cycle is indicated, after subtracting its average level (blue line) and the Eddy (blue) and Hallstatt WTs (pink line). This figure illustrates how the relative signature of the Eddy and the Hallstatt oscillations, going on during the last 2 ka, determines the nature of the Gleisberg cycle. Explicitly, the Grand Maxima and Grand Minima only occurred throughout the negative phase of the Hallstatt oscillation; otherwise, a long-lasting episode of Regular oscillations takes place, like the one prior to ~1.2 ka. Moreover, when the Hallstatt and the Eddy oscillations are simultaneously negative, Grand Minima occur in clusters, whereas Grand Maxima take place when the Eddy oscillation is positive. Based upon these properties, the Long Trend has been defined as the addition to the linear trend to the Eddy and the Hallstatt WTs. Moreover, based on these properties, as the

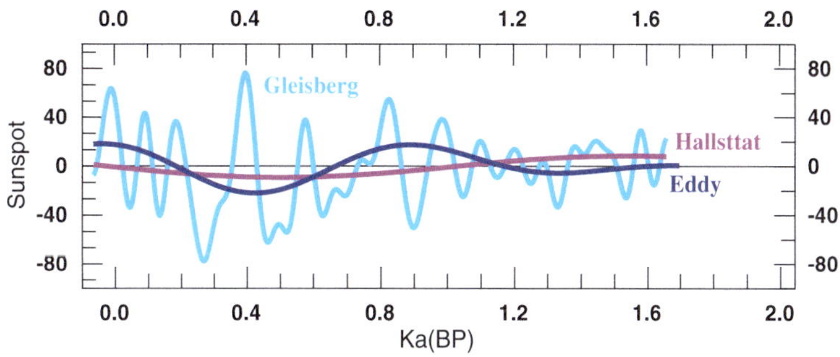

Fig. 6 The Gleissberg cycle and the Eddy and Hallstatt wave trains (the last two multiplied by 2 to make them clearly observable in the scale of the figure) computed from the sunspot maxima time series of Fig. 4 (*top*) by the procedure outlined in the text

last Hallstatt oscillation has recently passed from negative to positive, De Jager and Duhau (2012) have predicted that the episode on solar activity that started after the AD 2006 transition will be of the Regular type and will endure for the rest of the present millennium.

The WT with the largest period that may be determined from the data of Fig. 4 is the Hallstatt WT. The Long Trend found in this case by adding the Eddy and the Hallstatt WTs to the linear trend. As the Bond cycle has a length of 1–2 ka it has been identified with the signals that result from applying the same procedure to temperature. The results are shown in Fig. 7. The similitude between the two curves of Fig. 7 is beyond what is expected, especially if it is taken into account that sunspot maxima is only a proxy for the true variables and the relationship between temperature and solar activity is strongly non linear.

According to Bond et al. (2001), the LIA was the colder stage of the most recent 1–2 ka. It is noted in Fig. 7 that the global warming and the Medieval Optimum on one hand and the LIA on the other one follow the positive and the negative phase, respectively, of the strong oscillation in the Long Trend that started 1.24 ka with a

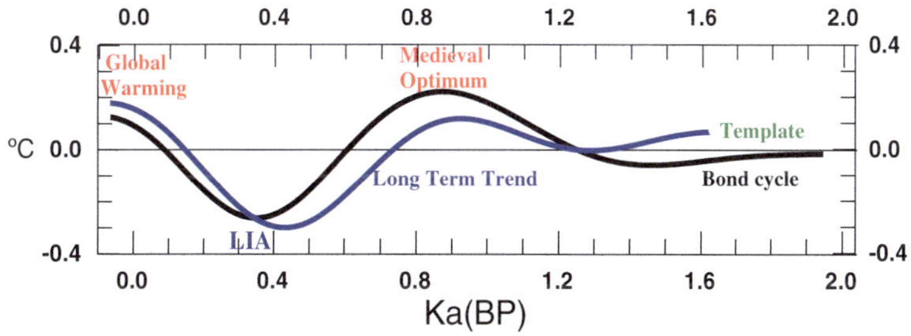

Fig. 7 The Bond cycle and the Long Trend in solar activity computed by adding the respective Eddy and Hallstatt WTs shown in Fig. 6 and linear trends, after subtracting 13.6° and 150 spot constants, respectively

time delay about 0.2 ka, while a temperate climate long-lasting episode occurred prior than that, when the oscillation in the Long Trend had quite smaller amplitude.

From the relevant data, in the next section the two components of the Long Oscillation for the last 15.7 years are defined. Temperature, TSI, geomagnetic virtual dipole moment (VDM), and geomagnetic excursions for the last 50 ka are compared. Based upon these results, the origin of the Younger Dryas event is discussed.

4 On the Younger Dryas, Mono Lake, and Laschamp Events

In the previous section, it has been shown that the Bond cycle is driven by the Long Oscillation in solar activity. Here, by studying the Eddy and the Hallstatt oscillations, which are the strongest WTs that comprise the Long Trend, the relationship of the Younger Dryas event with solar activity and geomagnetic field strength and tilt is investigated. Based upon the results, the origin of the strong decrease of temperature during the Mono Lake and Laschamp excursions is discussed.

The proxies of solar activity for which there is data for a longer period are the occurrence of cosmogenic radionuclides (such as ^{10}Be and ^{14}C) which are produced by cosmic rays in the Earth's atmosphere (for a review, see Usoskin 2008). The fact that the Gleissberg cycle in both sunspot number and geomagnetic index follows each other very closely, as discussed in Sect. 2, makes the cosmogenic radionuclides suitable for representing the solar activity WTs, whose periods are at or above the secular phases. The INTCAL98 tree-ring data set (see Fig. 8) has been used for this purpose.

By applying the same procedure as in Sect. 3.2 the Eddy and the Hallstatt WTs were found (blue and violet lines, respectively, in Fig. 9) from the INCAL98 time series of Fig. 8, and they were scaled to fit the corresponding WTs from SN_{max}

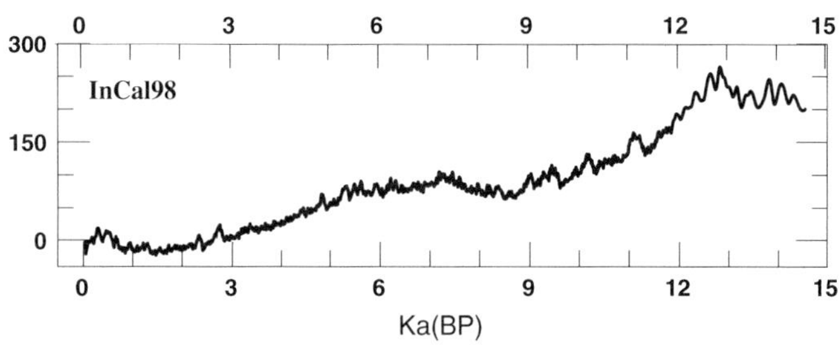

Fig. 8 The INCAL98 ^{14}C cosmogenic isotope data (Stuiver et al. 1998). *Source* University of Oxford, Radiocarbon Web-Info, at: https://c14.arch.ox.ac.uk/intcal98.14c. This calibration series is based upon a mix of mid-latitude northern hemispheres records (Germany, Ireland, and the states of Washington, Oregon and California in the U.S.A.)

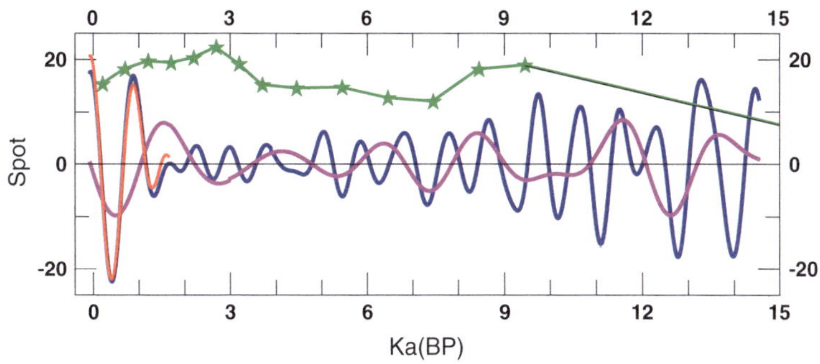

Fig. 9 The Virtual Dipole Moment, VDM (10^{22} A^2) in the European region (*green stars*) (from Table 1, in McElhinny and Senanayake 1982) and the Eddy (*blue*) and the Hallstatt (*pink*) wave trains from INCAL98 time series (cf. Fig. 8), scaled them to fit the corresponding WTs from SN_{max} (cf. Fig. 6) along contemporaneous times. The *red lines* are these two WTs computed from direct observations of solar activity (cf. Fig. 3, *upper panel*)

during contemporaneous times (the scaling factors are 1.55 and 0.76 for the Eddy and the Hallstatt WT, respectively). The period of the oscillations in the two WTs computed from INCAL98 (blue and pink lines for the Eddy and the Hallstatt WTs, respectively) fit quite well those computed from SN_{max} (red line). Regarding to the amplitude of the Eddy WT computed from SN_{max}, it is gradually becoming larger to the past when compared with that computed from INCAL98. This behaviour may be due to the fact that VDM decreased to the past in about 0.3 % along the time interval when SN_{max} is known, and thus the filtering of the cosmic rays by the geomagnetic field also decreased towards the past.

As VDM values around the LIA and the Younger Drays events are nearly the same and as small as the values sustained along the 3.8–7.5 ka period, and the Eddy and the Hallstatt oscillation were the strongest of the whole 15 ka interval around this two episodes, it is concluded that the strongest Long Trend occurring along the last 15 ka were the ones that leaded to the LIA and the Younger Dryas.

Past ice ages in the northern hemisphere correlate well with the summer insolation at the longitudinal band located at the 65° N latitude (Imbrie and Imbrie 1980). In Fig. 10 (upper panel) the variations in temperature, in geomagnetic field intensity as measured by VDM and summer insolation at a 5° latitudinal band around 65° N due to the Milankovitch cycles for the last 50 ka are shown. To facilitate the description of the phenomena illustrated in the upper panel, it has been plotted in the lower panel the average summer insolation over a 0° band around the equator. It has been observed that the lowest value of insolation at 65° N occurring around 29 ka is nearly synchronic with the highest one at the equator. According to the computations of Huybers and Eisenman (2006) the difference between the value of the maximum and minimum in the insolation along the last 50 ka, decreases toward 45° N changing of sing at this latitude (the same behavior is followed by average summer insolation at the southern hemisphere). Therefore, when the

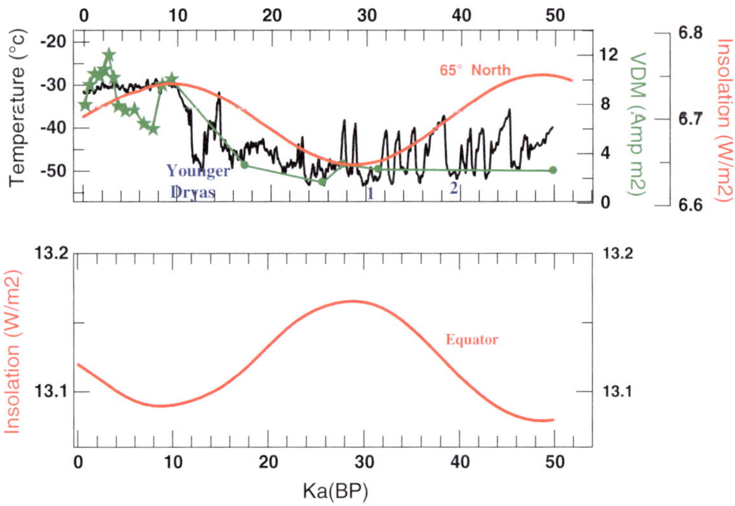

Fig. 10 *Upper panel* Gisp2 Temperature time series from ice cores, latitude 72.6° N and longitude 38.5° W (Mayewski and Bender 1995) (*black line*) and the Virtual Dipole Moment (VDM) in the European region (*green*) (until 15 ka they are as shown in Fig. 8 and after that, from Table 4 in McElhinny and Senanayake 1982). The *blue numbers* indicate the date of maxima approximation of the geomagnetic poles to the equator during the Mono Lake (1) and the Laschamp (2) geomagnetic excursions. The *red lines* represent the average summer insolation at 65° N longitudinal band (*upper panel*) and at the equator (*bottom panel*) (computed by Huybers and Eisenman 2006)

temperature was at its absolute minimum about 29 ka (observe the black line in the upper panel) the total amount of insolation over the entire globe was at its absolute maximum. Thus, the phenomenon that has strongly reduced temperature at Greenland must be strongly localized at higher latitudes. It has also been noted that, unless along geomagnetic excursions, the auroral oval cover a range of latitudes that goes from 60° to 75°. Thus, the fact that the best correlation between temperature and insolation occurs when insolation at the latitudinal band located at 65° N, as found by Imbrie and Imbrie (1980), indicates that one of the most relevant mechanisms able to amplify the effect of insolation on climate change is the ice-albedo feedback and the control of the global atmospheric circulation by the heating of the auroral oval and the polar cap by solar storms.

The deep valley in temperature (black line) around 11 ka that characterizes the Younger Dryas event cannot be attributed neither to a reduction of insolation, because it have had values that are above those sustained along the LIA, nor by a reduction of the geomagnetic field strength that was only slightly smaller than along the LIA. As a result, there is at present a consensus that the Younger Dyras event comes from a significant reduction of the North Atlantic Ocean thermohaline circulation due to a sudden flux of fresh water (Broecker 2006), but this interpretation remains controversial (Carlson 2013).

On the other hand, while along the LIA the north and south geomagnetic poles were at most 30° and 35° away from the geographic poles, as commented in Sect. 3, the Younger Dryas event was synchronic with a short geomagnetic excursion which has been observed in the northern hemisphere (see Fig. 1b in Bucha 1976) that started at 15 ka, reaching the maximum departure from the geographic North Pole around 12.5 ka, coming back to it at 11 ka. It has been noticed that the temperature (black line in Fig. 10) started a sudden decrease around one century after the starting date of the geomagnetic excursion, being its minimum value nearly synchronic with the maximum approaching of the geomagnetic pole to the equator. The Younger Dryas event occurred around one century after the coming back of the North geomagnetic pole closer to the geographical one. Thus, if heating by solar storms at higher latitudes were indeed the driver of the main feedback mechanism of insolation, the geomagnetic excursions that occurred during the Younger Dryas event would explain the drastic decrease of temperature along that event. In that framework the change in thermohaline circulation that occurred during the Younger Dryas event would not be the cause but instead one of the consequences of the loss of a principal driving of climate change during the geomagnetic excursion.

Besides the excursion reported by Bucha (1976) in the northern hemisphere, during the last 50 ka there were other two well-known geomagnetic excursions: the Mono Lake and Laschamp excursions. The moment of maximum departure of the geomagnetic poles from the geographical poles during these two excursions is indicated by the two blue numbers at the bottom of Fig. 10. As it happened during the Younger Dryas excursion, the Mono Lake and Laschamp excursions haves lead to episodes of prolonged valleys in the curve, but when the temperature reached even smaller values primarily following the lower values in both TSI and VDM along them. Many paleomagnetic determinations in the entire globe of the date of occurrence of the Mono Lake and Laschamp excursions demonstrate that they had been synchronous in the two hemispheres (Roberts 2008; Casatta et al. 2008). It may be expected then that the excursion found by Bucha (1976) in the northern hemisphere was also synchronous with a similar event in the southern hemisphere, which might explain why the Younger Dryas event was of worldwide nature too (Glasser et al. 2002; Carlson 2013). This possibility must be further investigated.

It may be seen in Fig. 10 that, during the last 49 ka, most of the time the minimum value attained at each of the oscillations in temperature follows VDM. Regarding to the amplitude of the oscillations most of them follow somehow the value of insolation, as much as the absolute minimum in temperature of the whole period occurred at around 25.7 ka when VDM was at its minimum of the entire period and insolation was nearly at its minimum as well. To give a full description of the relationship between temperature and the geomagnetic field morphology and insolation, it is needed to know first the evolution of the position of the geomagnetic poles with respect to the geographical one since, as suggested by Courtillot et al. (2007), the results presented in this paper indicate that the tilt of the dipole to lower latitudes plays a central role on climate change.

5 Summary and Conclusions

The only source of climate change external to the Climate System that is usually assumed to actually exist is the variation in insolation due both to solar activity and the Milankovitch cycles. However, when the geomagnetic field is mainly dipolar, and the geomagnetic poles are near the geographical poles, as it has been most of the time during the last 50 ka solar storms may contribute to heat the atmosphere at the auroral arc and the polar cap, leading to changes in global circulation (Bucha 1983). It also substantially contributed to melting the ice at higher latitudes, with increasing ice-albedo feedback of insolation (Duhau and Martínez 2012), leading to cloud cover variations (Svensmark and Friis-Christensen 1997; Tinsley 2008) that also may contribute to heat the higher latitudes by cloud cover feedback of solar irradiance, as model computations indicate (Holland and Bitz 2003).

The final impact of cosmic rays and solar storms on climate strongly depends upon geomagnetic field strength and the position of the geomagnetic poles with respect to the geographical ones. Along geomagnetic excursions, the deposition of the energy of the solar storms occurs at lower latitudes, and thus its contribution to ice-feedback of insolation is lost.

It is known (following Schove 1955) that solar activity has a recurrent behavior that may be represented in the time domain (Duhau and de Jager Submitted 2016) by the addition of eight wave trains (WTs), defined as a succession of wave packets, each of them having a nearly constant period which value alternatively changes from a wave packet to the next one. To represent these transient signals it has been introduced a signal processing technique based in Morlet (1981) wavelet; this procedure was applied for the analysis of the relevant data for the last 14.7 ky. Data for the last 49 ky were also analyzed by direct comparison. It was found that:

- There is further evidence, besides that reported by Bond et al. (2001), of the persistent 1–2 cycle in temperature found by Bond et al. (1999) which is driven by solar activity, since it follows with a time delay ($\sim$0.1–0.2 ky) and amplitude of the Long Trend in solar activity, defined as the addition of all WTs whose periods are at and above the Eddy time scale. The strongest of these periods are the Eddy and the Hallstatt WTs whose oscillations have an average periodicity of 1–2.4 ky, respectively.
- The LIA, which is the cold phase of the last Bond cycle, follows a strong negative oscillation in the Long Trend in solar activity with a time delay of about two centuries. This time delay is consistent with the hypothesis that the main mechanism of amplification of solar irradiation is ice-albedo feedback, since glacier retreat after the LIA started at around AD 1850 (Masiokas et al. 2010; Akasofu 2011), that is close to two centuries later than the moment at which solar activity started to emerge from the Maunder Minimum.
- Not only the LIA but also the Younger Drays events followed the negative phase of a strong oscillation in the Long Trend in solar activity; the main difference among these two cold events is that, during the LIA, the geomagnetic poles were not far away from its respective geographical poles (less than 30°).

While the Younger Dryas event was synchronous with a geomagnetic excursion along which (Bucha 1983) the North geomagnetic pole reached its maximum departure from the geographical pole, and passing the equator at around 12.5 ka. Episodes like the Younger Drays event occurred during the Mono Lake and Laschamp excursions with an amplitude that is also modulated by the strength of the geomagnetic dipolar field along these excursions and by variations in insolation due to the Milankovitch cycles.

- Besides colder events, and exceptionally warmer events, as the Medieval Climate Optimum and the twentieth century global warming, the positive phase of a strong oscillation in the Long Trend in solar activity follows, with variable time delay, while the long-lasting template episode occurring prior to the Medieval Climate Optimum was taking place with a very week oscillation in the Long Trend in solar activity.
- The sudden increase of temperature at latitudes above 60° N occurring after the year AD 1850, was twice the increase at latitudes above the 60° S and it was synchronous with a rapid drifting of the north geomagnetic pole towards the corresponding geographical pole, while the South geomagnetic pole was undergoing a rapid drifting to the equator. This may be explaining why at the North Pole the ice is melting four times faster (Stroeve et al. 2007; Rampal et al. 2011) than estimated by climate models, while the same does not seem to happen in the South pole.
- The Minimum in the background value and in the amplitude of the D/O events during a record of temperature at Greenland for the last 49 ky occurred when the minimum value in geomagnetic strength and in insolation were reached and while the background value of temperature followed the geomagnetic field strength, that of the maxima primordially follows the geomagnetic tilt and secondarily, the insolation.

All the above suggests that, as the Heinrich event may be due to the simultaneous occurrence of the positive phase of a strong oscillation in the Long Trend in solar activity, with a sudden drifting of the geomagnetic pole toward the geographical pole at the hemisphere where it took place, the Dansgaard–Oeschger events would be an amplified version of the Bond cycles as it was suggested by Bond et al. (2001). The way in which their amplitude is modulated by variations in insolation, solar storms intensity and frequency, geomagnetic field strength, and the wandering of the geomagnetic poles, must be further investigated.

It must be emphasized that the non-dipolar part of the geomagnetic field may contribute also to regional climate change by intervening in cloud cover (Svensmark and Friis-Christensen 1997) and by producing longitudinal variation of the average pressure in the troposphere leading to pressure depression at the geomagnetic anomalies (King 1974). As a result when interpreting worldwide climate events at different regions the possible contribution of the non-dipolar component of the geomagnetic field must be taken into account. In fact, a particular relationship between temperature and sunspot number time series was found, for instance, at the South Atlantic Ocean geomagnetic anomaly region (Frigo et al. 2013).

References

Abe-Ouchi A, Saito F, Kawamura K, Raymo ME, Okuno J, Takahashi K, Blatter H (2013) Insolation-driven 100,000-year glacial cycles and hysteresis of ice-sheet volume. Nature 500:190–193. doi:10.1038/nature12374

Agee E, Kiefer M, Cornett E (2012) Relationship of lower-troposphere cloud cover and cosmic rays: an updated perspective. J Clim 25:1057–1060

Akasofu S (2011) On the recovery from the Little Ice Age. Nat Sci 2:1211–1224

Akçar N, Yavuz V, Ivy-Ochs S, Reber R, Kubik PW, Zahno C, Schlüchter C (2014) Glacier response to the change in atmospheric circulation in the eastern Mediterranean during the last glacial maximum. Quat Geochr 19:27–41

Artl R, Weiss W (2014) Solar activity in the past and the chaotic behaviour of the dynamo. Sp Sc Rev, arXiv:1406.7628v1[astro-ph.SR]

Bond GC, Showers W, Elliot M, Evans M, Lotti R, Hajdas I, Bonani G, Johnson S (1999) The North Atlantic 1–2 ky climate rhythm: relation to Heinrich events, Dansgaard–Oeschger cycles and the Little Ice Age. Am Geophys Union Monogr 112:35–58

Bond GC, Kromer B, Beer J, Muscheler R, Evans MN, Showers W, Hoffmann S, Lotti-Bond R, Hajdas I, Bonani G (2001) Persistent solar influence on north Atlantic climate during the Holocene. Science 294:2130–2136. doi:10.1126/science.1065680

Broecker WS (2006) Was the younger dryas triggered by a flood? Science 312(5777):1146–1148. doi:10.1126/science.1123253

Bucha V (1976) The variations of the geomagnetic field, the climate and weather. Studia Geophysica and Geodaetica 20:149–167

Bucha V (1983) Conclusions. In: Bucha V (ed) Magnetic field and the processes in the earth's interior. Czechoslovak Academy of Science, Czech Republic

Carlson AE (2013) The younger Dryas Climate event. In: Elias SA (ed), Encyclopedia of quaternary science, vol 3, pp 126–134

Casatta W, Singer B, Cassidy J (2008) Laschamp and Mono Lake geomagnetic excursions recorded in New Zealand. Earth and Planet Sci Lett 268:66–78

Clette F, Svalgaard L, Vaquero JM, Cliver EW (2014) Revisiting the Suspot number. A 400-year Perspective on the Solar Cycle. Space Science Reviews 186:35–103. doi:10.1007/s11214-014-0074-2

Constable C (2005) Dipole moment variation. In: Gubbins D, Herrero-Bervera E (eds) Encyclopedia of geomagnetism and paleomagnetism, 159–161. Springer, Dordrecht

Courtillot V, Gallet Y, Le Mouël J-L, Fluteau F, Genevey A (2007) Are there connections between the Earth's magnetic field and climate? Earth Planet Sci Lett 253:328–339

De Jager C (2005) Solar Forcing on Climate 1: Solar Variability Space Science, Review 120:197–241. doi:10.1007/s11214.005-7046-5

De Jager C, Duhau S (2009) Forecasting the parameters of Sunspot cycle 14 and beyond. J Solar Terr Phys 71:239–435

De Jager C, Duhau S (2011) The variable solar dynamo and the forecast of solar activity, influence in terrestrial atmospheric temperature. In: Coscia JM (ed), Global warming of the twenty century. IBM 278-1-61728-980-4

De Jager C, Duhau S (2012) Sudden transitions and grand variations in solar dynamo, past and future. J Space Weather Space Clim 2:A07. doi:10.1051/2001

Dokken TM, Nisancioglu KH, Li C, Battisti DS, Kissel C (2013) Dansgaard-Oeschger cycles: interactions between ocean and sea ice intrinsic to the Nordic seas. Paleoceanography 28: 491–502

Duhau S (2003) Global Earth surface temperature changes induced by mean Sun dynamo magnetic field variations. In: Wilson A (ed), Solar variability as an input to the Earth's environment. International Solar Cycle Studies (ISCS) symposium. ESA SP-535, Noordwijk. ISBN 92-9092-845-X

Duhau S (2005) Long term variation in solar magnetic field, geomagnetic field and climate. In: Sutantyo W, Premadi PW, Mahasena P, Hidayat T, Mineshige S (eds) Proceedings of the 9th Asian-Pacific Regional IAU Meeting, pp 18–23

Duhau S, Chen ChY (2002) The sudden increase of solar and geomagnetic activity after 1923 as a manifestation of a non-linear solar dynamo. Geophys Res Lett, 29(13): 6-1–6-4

Duhau S, de Jager C (2008) The solar dynamo and its phase transitions during the last millennium. Solar Phys 50:1–15

Duhau S, Martínez E (2012) Solar dynamo transitions as drivers of sudden climate changes. In: Bharat RS (ed) Global warming impacts and future perspective, Chapter 7. ISBN 978-953-51-0755-2

Farge M (1992) Wavelet transforms and their applications to Turbulence. Annu Rev Fluid Mech 24:395–458

Frigo E, Pacca IG, Pereira-Filho AJ, Rampelloto PH, Rigozo NR (2013) Evidence for cosmic ray modulation in temperature records from the South Atlantic Magnetic Anomaly region Ann. Geophys 31:1833–1841

Gallet Y, Genevey A, Le Goff M, Fluteau F, Eshraghi A (2006) Possible impact of the earth's magnetic field on the history of ancient civilizations. Earth Planet Sci Lett 246:17–26

Glasser NF, Harrison S, Schnabel C, Fabel D, Jansson KN (2002) Younger Dryas and early Holocene age glacier advances in Patagonia. Quaternary Sc Rev 58:7–18

Geophysical Data Center (2012). ftp://ftp.ngdc.noaa.gov/STP/SOLAR_DATA/SOLAR_FLARES/FLARES_INDEX/Yearly/. Accessed 01 May 2012

Holland MM, Bitz ÆCM (2003) Polar amplification of climate change in coupled models. Clim Dyn 21:221–232

Humlum O, Solheim J, Stordahl K (2011) Identifying natural contributions to late Holocene climate change. Glob Planet Change 79:145–156

Huybers P, Eisenman I (2006) Integrated summer insolation calculations. IGBP PAGES/World Data Center for Paleoclimatology Data contribution series # 2006-079. NOAA/NCDC Paleoclimatology Program, Boulder CO, USA

Imbrie J, Imbrie JZ (1980) Modelling the climatic response to orbital variations. Science New Series, 207: 943–953 Published by: American Association for the Advancement of Science Stable. URL:http://www.jstor.org/stable/1683550

IPCC Climate Change (2013) Working group I contribution to ARI. http://www.climatechange2013.org/. Accessed 7 June 2015

ISGI (2014) International service of geomagnetic indices. http://isgi.unistra.fr/indices_aa.php. Accessed 15 June 2015

King JW (1974) Weather and the Earth's magnetic field. Nature 247(131):134. doi:10.1038/247131a0

Kopp G (2015) Kopps TSI page. http://spot.colorado.edu/~koppg/TSI/. Accessed 10 December 2014

Korte M, Mandea M (2008) Magnetic poles and dipole tilt variation over the past decades to millennia. Earth Planets Space 60:937–948

Leuschner DC, Sirocco F (2000) The low-latitude monsoon climate during Dansgaard–Oeschger cycles and Heinrich events. Quater Sci Rev 19:243–254

Masiokas MH, Luckman BH, Villalba R, Ripalta A, Rabassa J (2010) Little Ice Age fluctuations of Glaciar Río Manso in the north Patagonian Andes of Argentina. Quater Res 73:96–106

McElhinny MW, Senayake WE (1982) Variations in the geomagnetic dipole. 1: the past 50,000 years. J Geomag Geoelectr 34:39–51

Mayaud PN (1971) Une mesure planétaire d'activité magnétique basée sur deux observatoires antipodaux. Ann Géophys 27:67–70

Mayaud PN (1972) The aa geomagnetic index: a 100 years index characterizing the geomagnetic activity. J Geophs Res 77:6878–6970

Mayaud PN (1975) Analysis of storm sudden commencements for the years 1868–1967. J Geophys Res 80:111–122

Mayewski B, Bender M (1995) The GISP2 ice core record—paleoclimate highlights. Rev Geophisics 33:1287–1296. doi:10.1029/95RG00498

Moberg A, Sonechkin DM, Holmgren K, Datsenko NM, Kartén W (2005) Highly variable north hemispheric temperature from low and high resolution proxy data. Nature 433:613–617. doi:10.038/03265

Morlet J (1981) Sampling theory and wave propagation. Proceedings of the 51st Annual Meeting Society of Exploration Geophysics, Los Angeles

Nagovitsyn Y (2005) To the description of long-term variations in the solar magnetic flux: the sunspot area index. Astron Lett 31:557–562

National Oceanic and Atmospheric Administration (NOAA) (2015) Wandering of the geomagnetic poles. http://www.ngdc.noaa.gov/geomag/GeomagneticPoles.shtml. Accessed 10 Feb 2015

Pessnell WD (2007) Predictions of solar cycle 24. http://www.swpc.noaa.gov/sites/default/files/images/u2/May_24_2007_table.pdf. Accessed 9 June 2015

Randall DA, Wood RA, Bony S, Colman R, Fichefet T, Fyfe J, Kattsov V, Pitman A, Shukla J, Srinivasan J, Stouffer RJ, Sumi A, Taylor KE (2007) Climate models and their evaluation. In: Solomon S, Qin D, Manning M, Chen Z, Marquis M, Averyt KB, Tignor M, Miller HL (eds) Climate change 2007: the physical science basis. Contribution of working group I to the fourth assessment report of the intergovernmental panel on climate change. Cambridge University Press, Cambridge

Rampal P, Weiss J, Dubois C, Campin JM (2011) IPCC Climate models do no capture Arctic sea ice drift acceleration: Consequences in term of projected sea ice thinning and decline. J Geophys Res, 116 C00D07, 17 pp

Roberts A (2008) Geomagnetic excursions: knowns and unknowns. Geoph Res Lett 35:L17307. doi:10.1029/2008GL034719

Rosenfeld D, Rudich Y, Lahav R (2001) Desert dust suppressing precipitation: a possible desertification feedback loop. Proc Nat Acad Sci USA 98:5975–5980. doi:10.1073/pnas.101122798

Russell CT (1975) On the possibility of delivering interplanetary and solar parameters from geomagnetic records. Sol Phys 42:259–269

Rutter N, Coronato A, Helmens K, Rabassa J (2012) Glaciations in North and South America from the miocene to the last glacial maximum. comparison, linkages and uncertainties. Springer briefs in Earth system, Springer, Dordrecht, Geidelberg, New-York, London. Science. doi:10.1007/978-94-007-4399-1

Schove DJ (1955) The sunspot cycle, 649 B.C. to A.D. 2000. J Geoph Res 60:127–146

Schulz M (2002) On the 1470 year passing of Dansgaard-Oeschger warm events. Paleocenography, 17(2). 10.1029/2000PA000571

Steig EJ, Brook EJ, White JWC, Sucher CM, Bender ML, Lehman SJ, Morse DL, Waddington ED, Clow GD (1998) Synchronous climate changes in Antarctica and the North Atlantic. Science 282(5386):92–95

Stroeve J, Holland MM, Meier W, Scambos T, Serreze M (2007) Arctic sea ice decline: faster than forecast. Geophs Res Lett 34:L09501. doi:10.1029/2007GL029703

Stuiver M, Reimer PJ, Bard E, Beck JW, Burr GS, Hughen KA, Kromer B, McCormac G, van der Plicht J, Spurk M (1998) Intcal98 radiocarbon age calibration curves, 24,000–0 cal BP. Radiocarbon 3:1041–1083

Svensmark H, Friis-Christensen E (1997) Variation of cosmic ray flux and global cloud coverage —a missing link in solar-climate relationships. J At Solar Terr Phys 59:1225–1232

Tinsley BA (2008) The global atmospheric electric circuit and its effects on cloud microphysics. IOP Publishing Reports on Progress in Physics, vol 71, p 066801 (31 pp). doi:10.1088/0034-4885/71/6/066801

Torrence C, Compo GP (1998) A practical guide to wavelet analysis. Bull Amer Meteor Soc 79:61–78

The Geophysical Data Center (2012). ftp://ftp.ngdc.noaa.gov/STP/SOLAR_DATA/SOLAR_FLARES/FLARES_INDEX/Yearly/. Accessed 01 May 2012

Usoskin IG (2008) A history of solar activity over millennia. Living Rev Solar Phys, 5(3). http://www.livingreviews.org/lrsp-2008-3

Velinov P, Asenovski S, Kudela K, Lastovicka J, Mateev L, Mishev A, Tonev P (2013) Impact of cosmic rays and solar energetic particles on the Earth's ionosphere and atmosphere. J. Space Weather Space Clim 3:A14. doi:10.1051/swsc/2013036

Weiss NO (1987) Is the solar cycle an example of deterministic chaos? In: Stephenson FR, Wofendale AW (eds) Secular solar and geomagnetic variations in the last 10,000 years. Kluwer Academic press, USA, pp 69–78

Wang YM, Lean JL, Sheeley NR Jr (2005) Modelling the sun's magnetic field and irradiances since 1713. Astrophys J, 625:522–538. http://lasp.colorado.edu/sorce/tsi_data/TSI_TIM_Reconstruction.txt. Accessed 01 May 2012

Wollin G, Ericson DB, Ryan WBF (1971) Variations in Magnetic Intensity and Climate Changes Nature 232:549–550

The Influence of the Geomagnetic Field in Climate Changes

María Julia Orgeira, Ana María Sinito and Rosa Hilda Compagnucci

Abstract The present authors propose in this paper that a connection exists between the variations of the Earth's magnetic field during polarity reversal and climate change. The mechanism by which the variations of the internal magnetic field could trigger climate changes would be produced by the influence of the internal magnetic field on Galactic Cosmic Rays (GCR), since the geomagnetic field (GF) provides shielding to such radiation.

Keywords Geomagnetic field reversion · Climate changes · Galactic Cosmic Rays

1 Introduction

Global warming seems today to be an unquestioned and unmistakable phenomenon. It becomes clear from current direct observations of the average atmospheric and oceanic temperatures, melting of snow and ice, as well as the proposed rising of the average level of the oceans (Stocker et al. 2013). The vulnerability of certain areas of the planet to climate change in terms of ecological, economic, and social impacts is a contemporary cause of deep global concern.

M.J. Orgeira (✉)
IGEBA, Universidad de Buenos Aires/CONICET, Buenos Aires, Argentina
e-mail: orgeira@gl.fcen.uba.ar

A.M. Sinito
IFAS, Universidad del Centro de la Provincia de Buenos Aires/CONICET, Buenos Aires, Argentina

R.H. Compagnucci
Equipo de Clima, Ambiente y Sociedad [PEPACG], Facultad de Ciencias Físicomatemática e Ingeniería, Pontificia Universidad Católica Argentina [UCA], Buenos Aires, Argentina
e-mail: rosa_compagnucci@uca.edu.ar

G.M. Gasparini et al. (eds.), *Marine Isotope Stage 3 in Southern South America, 60 ka B.P.–30 ka B.P.*, Springer Earth System Sciences,
DOI 10.1007/978-3-319-40000-6_4

While the causes of global warming are multiple and include human activity as well, there is a natural variable that should be carefully assessed.

It is well known by Quaternary geologists that several natural warming and cooling events have been registered globally throughout geological time. The study of such past climate changes and their connection with alleged climate forcing is of great importance to assess the evolution, estimated time, and environmental impact of ongoing global warming.

Even though solar activity and the orbital cycles seem to be the most important and pertinent climate forcings, they have different frequency variations. As a consequence, depending on the analyzed time lapse only one of them will be relevant.

The action of other forcing variables is also relevant during the Quaternary. Among them, the Geomagnetic Field (GF) formed by an external field and an internal one will be the subject of review of this chapter. The external GF, as well as the ionosphere, are clearly affected by solar activity (Courtillot et al. 2007, among others). In this case, the action of the Sun would be modulating part of the GF; the external part is not linked to the main mechanisms of the internal field which are responsible for the polarity reversals of the GF.

Up to the present, there is no consensus as regards the mechanisms that generate the GF polarity reversals or the configuration that the GF adopts during those lapses.

The purpose of this paper is to briefly review the different hypotheses of the changes in solar variability and of the Total Solar Irradiance (TSI) received due to terrestrial orbital changes and mainly to GF changes as a climate forcing.

With this purpose, the climate changes taken place during the late Cenozoic period and their connection to variations of the previously mentioned forcings will be analyzed.

1.1 External Climate Forcings During the Quaternary

The external forcings of the Quaternary climatic variations are summarized in Fig. 1. They may produce changes of different intensities and duration.

As it was mentioned before, variations in global climate should be analyzed according to their corresponding timescale. There are small variations that produce differences of certain degree in global temperature (i.e., the Little Ice Age and the Medieval Climate Anomaly) as a product of solar variability among other forcings and relevant changes of greater magnitude such as glacial and interglacial events traditionally related to orbital changes. As a consequence, the analysis should be done taking into account only a specific period studied.

It is described below the external forcings that may have relevant influence according to the different periods.

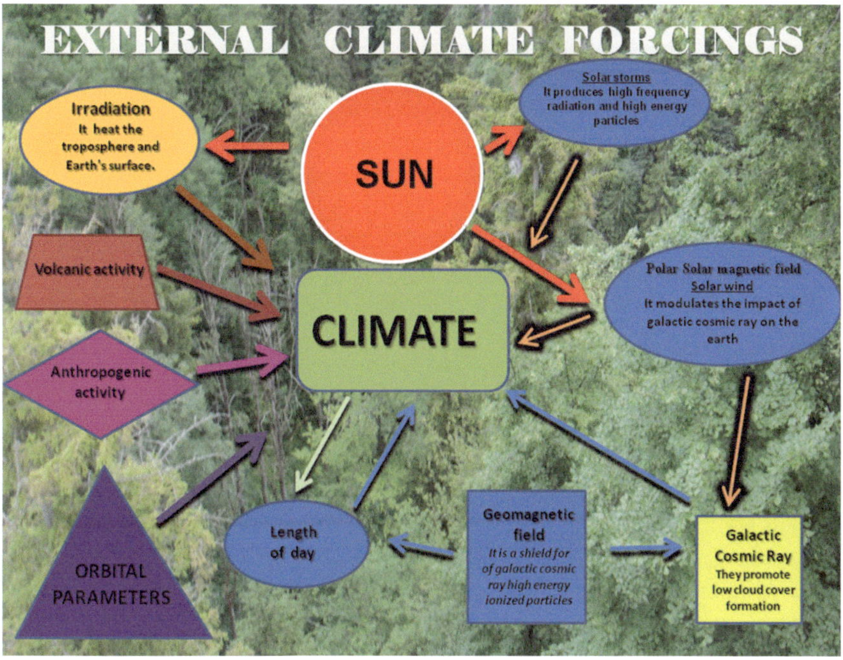

Fig. 1 Climate forcings

1.1.1 Solar Effects—Holocene

Duhau and Martinez (2012) suggested that the flow of energies due to solar storms have influence on high frequency climatic variations. They based their conclusions on the synchronicity between ice kinetic acceleration and the occurrence of the strongest solar storms. These solar events, when interacting with the Earth's magnetosphere, operate mainly in the polar cap and at the subauroral region. Some models indicate that the ice feedback albedo mechanism due to the acceleration of ice melting can be clearly seen at the poles and also at higher latitudes. After each transition of the solar dynamo, the abrupt changes in intensity and frequency that solar storms suffer would be accompanied by abrupt changes in the latitudinal temperature gradient and in the global wind regime.

Furthermore, according to Duhau (2013), this process would explain the cycles of 1–2 ky that took place during the Holocene (the Bond events), and the episodes that took place along the last glaciation (the Dansgaard/Oeschger events). Then, the amplification mechanism of solar irradiance due to the impact of solar storms on the dynamics of the ice could be of great importance during the polarity reversals of the terrestrial magnetic field, during the time when the dipole is not at higher latitudes.

The Sun's magnetic field has two components: a polar component (aligned with the Sun's rotation axis) and the toroidal component (perpendicular to the axis and generated by a current that forms a revolution toroid). Solar activity shows an average periodicity of 11 years (the Schwabe cycle). Both components reverse their polarity at the end of each Schwabe cycle; the complete magnetic cycle (the Hale cycle) corresponds to two Schwabe cycles, that is, 22 years.

Solar cycles also exhibit longer term cycles recorded as Solar Minima and Solar Maxima (or Grand Maxima and Grand Minima). The Solar Minima of the last 600 years seem to correlate fairly well with colder phases or the “Little Ice Age” (e.g. Mörner 2010).

Solar variability is observed taking into account the number of sunspots which are darker zones on the Sun’s surface due to the effect of huge magnetic fields that restrict the emission of radiation. The relative sunspot number (R) is an index of the activity of the entire visible disk of the Sun. It is determined every day without reference to preceding days. Schove (1955) reconstructed the succession of sunspot maxima of every 11-year cycle.

The maxima in the relative sunspot number (R) is the measure of the toroidal magnetic field strength (Beer et al. 1998). Therefore, the long-term variations in R may be used as a proxy for long-term variations in the toroidal Sun’s magnetic field strength (Duhau 2002; Duhau and Chen 2002). Analyzing the last 350 years, we may conclude that the average strength of the toroidal field is constant during the quasi-harmonic states that are suddenly interrupted by the chaotic transitions during which this strength strongly changes (Duhau 2005).

The variations of intensity of the toroidal field influence the frequency of the Sun flares and plasma ejections, which produce high energy particles that reach the Earth’s atmosphere. These particles modify the total content of atmospheric ozone and the intensity of the electric currents circulating between the higher atmosphere layers and the surface.

The polar component is represented by the geomagnetic index aa, that comes from data of GF variations recorded in observatories located near the equator and distributed around the Earth.

The polar field influences the intensity of the so-called solar wind, a plasma supersonic flux.

The cyclic variation in the flux of incoming galactic cosmic rays (GCR) is recorded in the production of ^{14}C in the atmosphere, in the infall of ^{10}Be at the Earth’s surface. These cosmogenic nuclides are formed by the interaction of cosmic radiation (mainly galactic protons) with the molecules of the atmosphere and their production is modulated by the intensity of the magnetic field of the solar wind (Lal and Peters 1967). Another effect of the interaction between the solar wind and the Earth’s magnetic field seems to be that it affects the Earth’s rotational speed where Solar Minima leads to accelerations and Solar Maxima to decelerations.

The massive burst of solar wind and magnetic fields rising above the solar corona or being released into space is named the “coronal mass ejection” (CMEs) and the *geomagnetic* index *Storm Sudden Commencements (SSC)* is a proxy of CMEs strength (Duhau 2003a). CMEs have a huge impact on the Earth’s environment since they are the main source of energetic particle by converting magnetohydrodynamic (MHD) energy in kinetic energy not only in the heliosphere but also in the magnetosphere itself. When the solar activity is maximum the CMEs driven shock waves produce an increase of energetic particles of several orders of magnitude in the heliosphere (Lario and Simnett 2004). They have an important impact on the Earth's environment by increasing the geosynchronous energetic particles flux within the magnetosphere (González et al. 1999, among others). Also, CMEs lead to strong changes in the magnetosphere-ionosphere-ground electrical circuit, producing changes in ozone concentration and surface temperature (Bucha and Bucha 1998; Tinsley and Yu 2004; Jackman and McPeters 2004).

A relationship between SSC and global surface temperature and ozone depletion has been found by Duhau (2003b) and Duhau and Martinez (1995), respectively. Energetic proton events (Jackman and McPeters 2004) play an important role in ozone depletion at the polar cap. Relativistic electrons penetrate at all latitudes, being the cause of 70 % of the ozone depletion at middle latitudes (Callis et al. 1991). The reduction of total ozone content in turn produces changes in the atmospheric circulation (Jadin 1999). Energetic electrons events strongly increase the global electric field and current circuit by increasing the electron content of the ionosphere.

The Sun has an obvious effect on climate since its radiation is the main energy source for the outer envelopes of our planet. But there is a long-standing controversy on whether solar variability can significantly generate climate change or not, and how this might occur.

Since the Earth's surface global temperature rise has been synchronic with increases in the solar dynamo magnetic field strength as well as the solar activity; the solar radiative flux has been considered the principal source of heating of the Earth's atmosphere. The strength of the solar radiative output is measured by solar total irradiance index (see Lean and Rind 1999; Fröhlich and Lean 2004; Soon 2005).

In addition, the irradiance variations are not spectrally homogeneous and the amplitude of the UV variability is, relatively speaking, an order of magnitude larger than the variability of TSI (Lean et al. 1995; Lean 2005). This enhances the stratospheric ozone formation through photochemical reactions (Haigh 1994, 2003) leading to further heating of the stratosphere through absorption of the excess UV radiation by ozone. Modeling studies (Shindell et al. 2001; Palmer et al. 2004) indicated that this mechanism amplifies the global average warming due to the increase in irradiance by about 15–20 %. Therefore, the most suitable mechanism, for driving the climate change, is the influence of UV irradiance on the stratosphere and dynamical coupling to the surface. Solanki et al. (2013) provided an overview of the current state of knowledge, as well as of the main open questions about solar variability and its link with the climate from the present time to the early Holocene.

De Jager and Duhau (2009) found out that there is a relationship between variations of the tropospheric temperature and the Sun's magnetic field. They also concluded that the amplitude of present global warming is not meaningfully different from other relative warming events that occurred several centuries ago. They assumed that the warming recorded in the second half of the twentieth century is the result of the overlapping of the slow warming of the Earth caused by the Sun (Maunder minimum) and the quasi-irregular temperature increases.

Bard and Frank (2006) provided a review of investigations about Sun–climate relationships by considering changes on different timescales, from the last million years up to recent decades. The different studied records also illustrate the multidisciplinary nature of this complex problem, requiring knowledge on several fields such as astronomy and astrophysics, atmospheric dynamics and microphysics, isotope geochemistry and geochronology, as well as geophysics, paleoceanography, and glaciology.

The number of sunspots and the cosmogenetic isotopes (^{10}Be and ^{14}C) allow reconstructing the Holocene solar variability (Vonmoos et al. 2006). The variations depicted 11-year periodicities (the Schwabe cycle), the complete magnetic 22-year cycle (the Hale cycle), Gleissberg in the band ~80 to 90 years, Viers in the band ~170 to 200 years (Ogurtsov et al. 2002), and a possible 2000-year cycle (Sonnett and Finney 1990). Solar variability also exhibits longer period with lack or a few number of sunspots recorded as Solar Minima or Grand Minima. During the last millennium, the Oort, Wolf, Spörer, Maunder, and Dalton Minima have been recorded. They correlate fairly well with colder climate episodes (Vaquero et al. 2002a, b; Usoskin et al. 2005, among others). The Medieval Anomaly Period (900–1400 A.D.) was followed by the Little Ice Age (1500–1800 A.D.) that involved the last three solar minima (Eddy 1976; Bard et al. 2000; Usoskin et al. 2005).

In their review, Gray et al. (2010) specified the different mechanisms by which solar variability affects the climate system.

1.1.2 Orbital Changes—Pleistocene

One of the most relevant forcings suggested to account for the Quaternary glaciations is the variation of the Earth's orbital parameters. Milankovitch (1920) suggested that the periodicity of the Quaternary glaciations rhythm is influenced by orbital cycles. Due to the gravitational influence of the Solar System planets, many astronomic parameters of the movement of the Earth change cyclically. These are: (a) the precession of the equinoxes (relationship between the moments of the equinoxes and solstices and the moment of greater or lesser distance between the Earth and the Sun), (b) the eccentricity of the terrestrial orbit, and (c) the tilt of the Earth's rotation axis (obliquity) with respect to the ecliptic. When the three variation cycles combine with the different times and intensities, complex variations are produced in the amount of solar radiation that reaches each terrestrial latitude.

The variations of the summer solar insolation in higher latitudes sparked off the formation or the melting of the big ice sheets, mainly the Laurentide and Fennoscandian ice sheets. Abundant precipitations of winter snow were necessary for snow to accumulate on those sheets and above all, less insolation so that it would not melt in summer. For the purpose of these variations in insolation other climatic forcings may be added, such as variations of the shielding of the GF.

This topic will not be developed any further in this contribution since the relevant bibliography is very extensive.

1.1.3 Galactic Cosmic Rays (GCR)

It is worth noticing that, in the first part of the twentieth century, the cosmic ray mechanism is related to cosmic ray induced ionization (CRII) of the atmosphere (Ney 1959). However, a great advance related to the mechanisms occurred in the present century. Energetic cosmic rays (CR) initiate a nucleonic–electromagnetic cascade in the atmosphere, affecting its physical and chemical properties, particularly the ion balance (Dorman 2004). This is a dominant source of ionization of the troposphere. Two approaches have been developed to compute the CRII. One model (O'Brien 2005) is based on an analytical approximation of the atmospheric cascade, whereas the other is based on a Monte Carlo simulation of the atmospheric cascade (Usoskin et al. 2004; Desorgher et al. 2005).

GCR could act upon the climate in three ways: (1) through changes in the concentration of cloud condensation nuclei, (2) thunderstorm electrification, or (3) ice formation in cyclones. Svensmark (1998) and Svensmark and Friis-Christensen (1997) have given more support to the hypothesis about the influence of GCR on cloud formation. Briefly, the intensity of the GCR flux depends on the GF and solar wind. A minimum Sun activity is connected to the increase in the cosmic radiation on Earth producing the raise/increase of the number of condensation nuclei and ultimately an increase in cloud cover. This theory received a lot of attention in 1997, when a positive correlation was first presented linking cloudiness to the intensity of cosmic radiation modulated by the Sun over the 1984–1991 period.

However, the GCR flux–cloud correlation has been criticized. The Svensmark's hypothesis was subsequently modified by the same group (Marsh and Svensmark 2000), who proposed that solar influence was limited to low-altitude clouds. Despite these criticisms, the GCR–climate relationship is now being accepted because physicochemical mechanisms are emerging. Recently, the CLOUD international project results published by Kirkby et al. (2011) and 62 coauthors show strong evidence of the influence of the GCR in the nucleation and as a consequence, in the formation of clouds from an experiment. Such results confirmed the hypothesis regarding the processes involved between the GCR input and cloud formation. In turn, Kerminen et al. (2012) presented a revision of the new results of the nucleation processes produced by the GCR.

Particularly, using data recorded in the southern Pacific Ocean clouds, effects on the net radiative flux in the atmosphere were related to the intensity of the Earth's magnetic field. This would be affected by the lower atmosphere cosmic ray effects (Vieira and Da Silva 2006). In the inner region of the Southern Hemisphere Magnetic Anomaly (SHMA) a cooling effect can be observed whereas in the outer region a heating effect has been detected. The variability in the inner region of the SHMA of the net radiative flux is correlated to the GCR flux observed in Huancayo, Peru. According to these authors, the geomagnetic modulation of cloud effects in the net radiative flux in the atmosphere in the SHMA is due to cosmic rays.

During the late Cenozoic, the GCR–climate hypothesis has been tested over a longer time interval using proxy data. For instance, Wagner et al. (2000) have compared the ^{10}Be and ^{36}Cl records (GCR proxies) with a climate proxy record ($\delta^{18}O$) between 20 and 50 ka B.P. It may not be possible to use these cosmonuclides in a robust way as proxies of climate on a timescale from 10^3 to 10^5 years. Besides, the fact that the peak amplitude of ^{36}Cl at 32 ka B.P. is comparable to the one recorded of 39 ka B.P. suggests that the intensity of the dipolar GF during both Mono Lake and Laschamp events was similar and close to zero (Wagner et al. 2000).

These important GF changes in direction and intensity connected to the axial dipole, given the causal relationship produced through the modulation of the cosmic rays lead to consequent modifications on the low cloud cover that would produce climate changes.

We could speculate that maximum paleointensities (correlated to maximum shielding) could produce minima/maximum in the cosmic ray entrance with a consequential low cover of lower clouds, which would imply an increase in global temperature. This process could be possible for any lapse.

Several studies have attempted to extract solar components over periods of ten thousand to hundreds of thousand solar years by subtracting the nuclide production component linked to geomagnetic variations as modeled from paleomagnetic data (Solanki et al. 2004). The currently available reconstructions of GF intensity and cosmogenic nuclide production are still not precise enough to extract a meaningful solar component. To apply this approach, more reliable and comprehensible records are needed of both cosmogenic nuclide production and GF intensity from the past (Bard and Frank 2006).

Recently, Kitaba et al. (2013) suggested that reversals of GF induced the increase of GCR flux that produced enlarged cloud formation with the consequent strong climatic cooling. They presented paleoclimate and paleoenvironmental records of five interglacial periods that include two geomagnetic polarity reversals. Marine oxygen isotope stages 19 and 31 contain both anomalous cooling intervals during the sea-level highstands and the Brunhes–Matuyama and Lower Jaramillo reversals, respectively.

1.1.4 Length of Day Variations (LOD)

Other sources linked climate changes to changes in the Earth's rotation rate which is measured by the excess of length of day variations (LOD) (Duhau 2005, among others).

Golovkov (1983) plotted Earth's rate of rotation (spin rate) against sunspot numbers and found that high spin rates correlated with low sunspot numbers and low spin rates with high sunspot numbers. Mörner (2010) plotted LOD against sunspot numbers for the period 1831–1995 and found a linear relationship where low LOD values (high spin rate) correlated with low sunspot numbers and high LOD values with high sunspot numbers. Consequently, the Earth's rotation accelerates at lower solar activity and decelerates at higher solar activity. Le Mouël et al. (2010) have shown that there is an excellent correlation among the last four sunspot cycles, the changes in LOD, and the atmospheric flux of cosmic rays.

On short timescales (annual or shorter) the nontidal component is dominated by the atmosphere, with small contributions from the ocean and hydrological system. On decadal timescales, millennial (Dumberry and Bloxham 2006), or longer geological times (Greff-Lefftz 2011), the dominant contribution is from angular momentum exchange between the solid mantle and fluid outer core (Holme and Viron 2013). Furthermore, these authors suggested that the nature of leaps in day length leads to a fundamental change in the type of phenomena that may give rise to the jerks, and provides a strong constraint on electrical conductivity of the lower mantle, which can, in turn, constrain its structure and composition. The coupling of the core and mantle allows the torsional oscillations to affect the length of day of the Earth via angular momentum conservation (Teed et al. 2014).

Mörner (2013) suggested that the interchange of angular momentum between solid Earth and the ocean circulation system is of special interest because it implies the redistribution of heat stored in the ocean water (recorded in paleoclimatic data) and the redistribution of ocean water volumes (recorded from sea-level changes).

1.1.5 Volcanic Activity

Volcanic activity occurs naturally and contributes to the total natural variability of the climate system (AR4, IPCC: Solomon 2007). Trace amounts of SO_2 have an important influence on climate. The major historic volcanic eruptions have formed sulfuric acid aerosols in the lower stratosphere. This aerosol cooled the Earth's surface by $\sim$0.5 °C for around 3 years. Geological records show frequency of occurrence between a few to a dozen years. During these periods, the Earth was gradually cooled into major ice ages. However, large frequently erupted volumes of SO_2 appear to overdrive the oxidizing capacity of the atmosphere resulting in very rapid warming (Ward 2009).

It is difficult to connect ancient paleoclimatic change to volcanism because of the challenge that lies in obtaining reliable data regarding the intensity, time, and extension of volcanic activity.

1.1.6 Anthropogenic Influence

The influence on climatic change produced by anthropogenic activity has been evident in the last two centuries. Reports of anthropogenic greenhouse gases in the atmosphere and other effects on climate as well as their impact and mechanisms have been widely described in the various IPCC reports (available on the IPCC web page: http://www.ipcc.ch/). Due to the fact that this subject is out of the scope of the present contribution, this anthropogenic effect is only mentioned as one of the external forcing of the climate system that is quite relevant at present time.

2 Record of GF Reversals in the Past and Their Link to Paleoclimate

2.1 GF Events During the Brunhes Chron

Variations in GF may be large or small. The external field may have undergone rapid variations linked to solar activity. Regarding the internal field, it may have minor fluctuations of intensity and field direction (paleosecular variations), which are slow and progressive. There are also substantive variations of these parameters that lead to GF polarity reversal.

One of the first chronologies of magnetic events of the GF during the Brunhes Chron was proposed by Langereis et al. (1997). A piston core collected from the Calabrian Ridge in the Ionian Sea (Italy) with astronomical calibration and oxygen isotope data as an additional constraint on their age model allowed these authors to obtain the magnetic record of practically the complete Brunhes and the Upper Matuyama chron (Table 1). In the Brunhes Chron, four short reversal excursions (CRO-3) were found and dated (CRO 261, CR1 318, CR2 515, CR3 573 ka). According to these authors, CRO could correspond to one of the Fram Strait, CR3 corresponds to the Emperor, which showed to be equivalent to the Big Lost reversal excursion. CR1 and CR2 were named as Calabrian Ridge 1 and Calabrian Ridge 2.

Lourens (2004) presented new data from the Calabrian Ridge (Ionian Sea) piston cores. New ages for the Calabrian Ridge 2 and 3 magnetic events in the Brunhes were proposed in concordance with minima in the global Sint800 composite record, derived from worldwide deep-sea records of relative paleointensity and have been attributed to the Big Lost and La Palma excursions, respectively (Table 1).

Channell (2006) offered a magnetic record from ODP Site 919 (Irminger Basin) indicating the existence of four intervals of negative inclination in the upper Brunhes Chronozone. According to his age model, these "excursional" intervals occurred in sediments deposited during the following time intervals: 32–34, 39–41, 180–188, and 205–225 ka. These time intervals correspond to polarity excursions detected in other locations, known as Mono Lake, Laschamp, Iceland Basin, and Pringle Falls.

Laj and Channell (2007) published a review of a previous contribution related to magnetic events. According to these authors, the well-documented excursions with acceptable age control occurring during the Brunhes Chron are the following: Mono Lake (33 ka), Laschamp (41 ka), Blake (120 ka), Iceland Basin (188 ka), Pringle falls (211 ka), Big Lost (560–580 ka), and Stage 17 (670 ka).

Sediments and eight lava flows of the Albuquerque volcanoes (Oregon, USA) display excursional paleomagnetic data with virtual geomagnetic poles (VGPs) in the southern hemisphere together with the statistically indistinguishable $^{40}Ar/^{39}Ar$ dates, and established that both sites record the Pringle Falls excursion ($^{40}Ar/^{39}Ar$ isochron ages of 218 ± 14 and 211 ± 13 ka) (Singer et al. 2008). This excursion is also recorded by the 227 ± 8 ka Mamaku Ignimbrite, New Zealand, and by higher deposition rate sediments at ODP site 919 in the North Atlantic Ocean that are dated

Table 1 Brunhes and Matuyama magnetic events and excursion

Langereis et al. (1997)	Lourens (2004)	Macri et al. (2005)	Laj and Channel (2007) Ar/Ar	Singer et al. (2008)	Channell et al. (2012)
		Mono Lake (24–25 Ka)(l)	**Mono Lake** (33 Ka)(l)	**Mono Lake** (32 Ka)(1) Negrini et al (2014) (32–34 Ka)	
Laschamp (45–40 Ka)(2)		**Laschamp** (35–42 Ka)(2)	**Laschamp (41** Ka)(2)	**Laschamp (40,4 ± 1,1** Ka)(2)	**Laschamp (40,8** Ka) duration 0,5 Ka(2)
Norwegian-Greenland sea (80–70 Ka) **55–66 Ka** (O'Regan et al. 2008)(3)					
Blake (110–120 Ka)(4)		**Blake** (110–**120 Ka)(4)**	**Blake (120 Ka) (4)**	**Blake (120 Ka)(4)**	**Blake (115.5** Ka) dur 0,9(4)
Biwal,II yIII 110–200–250????					
Alburquerque 155–165?(5)		Alburquerque 155(5) **High PRI**	**Iceland Basin (188** Ka)(6)	**Iceland Basin (188** Ka)(6)	**Iceland Basin (189,7** Ka) dur 1,4 190 Ka (Channell et al 2014)(6)
Jamaica (205–215 Ka)7)		**Jamaica (210–217** Ka)(7)	**Pringle Falls** (211 Ka)(7)	**Pringle Falls (211 ± 13** ka)(7)	**Pringle Falls (238,8** Ka) dur 0,9(8)
Fram Strait—CRO (265–255 Ka)(9)	CRO 260 Ka (9)	Fram Strait—CRO (253 Ka) (9)			
Calabrian Ridge 1318 ± 3 Ka (10)	CR1 319 Ka (10)	CR1			**Calabrian Ridge I (318.1** Ka) dur 0.6(10)
Levantine (370–360 Ka?)(11)		**Levantine (365** Ka)(ll)			
Unnamed 420–400 Ka)(12)					**Bermuda (412.4 Ka) dur 0.7(12)**
Calabrian Ridge 2.515 ± 3 Ka(13)	CR2 543 Ka (15)	**CR2 (Big lost) 543 Ka(15)**			**Calabrian ridge 2 (529.6** Ka) dur 1.7(14)
Emperor (Big lost) **(570–560** Ka)(16)	CR 3 593 Ka	**CR3 (La palma) 584 Ka**	**Big lost (560–580 Ka)(16)**		**Emperor?** (Big lost) **(557.3** Ka) dur 1.1 540 Ka (Channell et al 2014)(16)
		Delta 693 Ka (17)	**Stage 17 (670 Ka)(17)**		

astrochronologically at 209–207 ka. According to that, Singer et al. (2008) proposed that the names “Albuquerque” and “Jamaica” excursion be abandoned and that a radio isotopic age of 211 ± 13 ka be adopted for the Pringle Falls excursion, which is one of the five globally expressed, well-documented excursions in marine sediment cores, dated by $^{40}Ar/^{39}Ar$ methods, that took place from 220 to 30 ka. Together with at least five other well-dated excursions between 730 and 520 ka, some ten excursions define the Geomagnetic Instability Time Scale (GITS) for the Bruhnes.

Channell et al. (2012) constructed an age model for the Brunhes Chron of Ocean Drilling Program (ODP) Site 1063 (Bermuda Rise) using tandem correlation of oxygen isotope and relative paleointensity data to calibrate reference templates. Four intervals in the Brunhes Chron were correlated with the following magnetic excursions: Laschamp (41 ka), Blake (116 ka), Iceland Basin (190 ka), Pringle Falls (239 ka). The age in brackets corresponds to the ages recorded in Site 1063. These ages are consistent with current age estimates for three of these excursions, but not for the “Pringle Falls” event, which is apparently older than stated in previous contributions. For each of these excursions (termed Category 1 excursions), VGPs reach high southerly latitudes implying paired polarity reversals of the Earth’s main dipole field. In addition, several intervals of low paleomagnetic inclination (low and negative in one case) are observed at 318 ka (MIS 9), 412 ka (MIS 11), and in the 500–600 ka interval. These “Category 2” excursions may constitute inadequately recorded excursions.

Bourne et al. (2012) presented a higher resolution record of the Blake geomagnetic excursion (125 ka) measured in three cores from ODP Site 1062 on the Blake–Bahama Outer Ridge. Paleomagnetic measurements of the cores reveal rapid transitions (500 year) between the contemporary stable normal polarity and a completely reversed state of long duration which spans a stratigraphic interval of 0.7 m. These authors dated the directional excursion as falling between 129 and 122 ka with an estimated duration for the deviation of 6.5 ± 1.3 ka.

Negrini et al. (2014) showed a new record of the Mono Lake excursion (MLE) from the Summer Lake Basin (Oregon, USA). This new record correlates well with other coeval but lower resolution records from western North America including records from the Wilson Creek Formation exposed around Mono Lake. These results confirmed the data recorded at Mono Lake, California, it is not the Laschamp Excursion but rather another one that is several thousand years younger.

Channell et al. (2014) exposed an age model on the basis of planktonic oxygen isotope ($\delta^{18}O$) and relative paleointensity (RPI) data, for the last 1 Myr from the Integrated Ocean Drilling Program (IODP) Site U1306 drilled on the crest of the Eirik Drift (SW Greenland). According to them, the age of Big Lost and Iceland Basin excursions is 540 and 190 ka, respectively.

Finally, Macrì et al. (2005) informed about a composite magnetic record obtained in the Wilkes Land Basin (Antarctica, Southern Hemisphere). It is a high-resolution paleomagnetic record obtained from six Late Pleistocene piston cores recovered on the continental rise from East Antarctica. Paleomagnetic inclinations fluctuate around the expected value (of ca. −77°) for such high latitude sites and always indicate normal magnetic polarity. Short period oscillations to anomalously shallow paleomagnetic inclinations (up to −20°) were identified at different levels in the sampled sequences; positive (reverse) inclination values were however, not observed. This record provided the first experimental data documenting the dynamics and amplitude of the GF variations at high southern latitudes during the Brunhes Chron. Furthermore, the stacking of the individual ChRM inclination records indicates that the recurrent swings toward shallow paleomagnetic inclinations may be correlated to the main known geomagnetic excursions, such as the Mono Lake, Laschamp, Blake, Albuquerque, Jamaica, CR0, CR1, Levantine, Big Lost-CR2, La Palma-CR3, and Delta events.

2.1.1 Results of the Behavior During the Transitions

Although no consensus has been reached so far regarding the properties of the GF during reversals or the main features that might reveal its dynamics, a main characteristic of the reversing field has been observed. That is, a large decrease in the axial dipole and the dominant role of nondipole components. Since the first reversal records (Cox et al. 1963; Dagley and Lawley 1974) were published, the proposition of a dominating nondipolar field was formulated. Other features strongly depend on whether the records derive from sediments or volcanic material. Valet and Plenier (2008) summarized a different hypothesis about the GF behavior during reversals and excursions.

Two main hypotheses have been developed: (1) a recurring field behavior during successive reversals, over a long period of time, which suggests that the Earth's mantle was involved in the reversal process; (2) a time-varying nondipolar transitional field with similar characteristics and time constants to the present field. In this last case, one does not expect a simple morphology of the VGP paths, due to the complex and rapidly changing geometry of the nondipolar field.

The first hypothesis was supported by Laj et al. (1991) (Fig. 2) and Love (1998), who found a bimodal distribution of the longitudes of the transitional poles in sedimentary records of reversals going back into the Miocene and in volcanic records, respectively; although the VGPs did not appear to be exactly located within the same longitudinal band for both records.

Hoffman (1992), using a small volcanic database, noticed the existence of clusters of VGPs over Australia and South America within the same two longitudinal bands as the VGPs from sediments, and proposed that the sedimentary VGP paths showed a smeared picture of the volcanic VGPs. There is also a dominance of transitional VGPs from 11 records of the last reversal within two distinct pairs of clusters near Western Australia and within Siberia and in the southwestern Atlantic and northeastern Canada (Hoffman 1991).

Mazaud et al. (2009) studied sediments collected during the IODP in the North Atlantic, which recorded polarity transitions and geomagnetic excursions at a high resolution. They especially investigated polarity transitions at the upper and lower boundaries of the Jaramillo subchronozone. The records of the lower Jaramillo

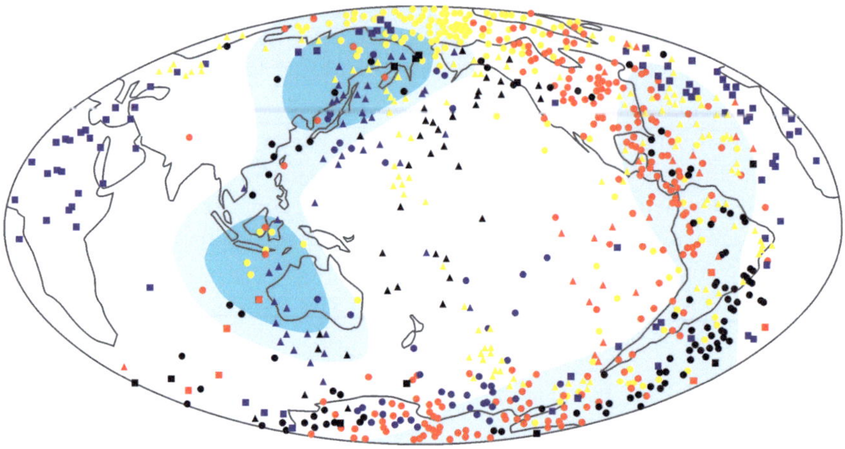

Fig. 2 VGP path during late pleistocene reversions (adapted from Valet and Plenier 2008)

reversal exhibit a marked cluster of VGPs over southern South America, and a secondary accumulation in the region of NE Asia/North Pacific. Records of the upper Jaramillo polarity transition document a VGP loop over the Americas, followed by north to south motion including a secondary VGP accumulation near India. These results are reminiscent of reversal records previously obtained at other IODP, and also of some volcanic records (Channell and Lehman 1997; Mazaud and Channell 1999; Channell et al. 2004; Prévot et al. 1985, Mankinen et al. 1985). Results suggest that a transversal, possibly dipolar, field component fluctuated during these polarity reversals, and that these fluctuations combined with a reduced axial dipole component yielded the observed field on the Earth's surface during the polarity transitions.

Valet et al. (2012) studied the individual trajectories of the north VGPs (Fig. 3) derived from the best volcanic records of reversals, which seem to be rather complex. These authors did not find consistency between the data from these volcanoes, which have different eruption rates. In an attempt to improve the consistency between the individual records, they rescaled the lengths of the R–N and N–R records to match them, by selecting tie points, and interpolating linearly. After this transformation, the entire data set showed a common and consistent dynamical pattern. Many records seem to be punctuated by clusters of nonpolar directions that can reflect uncertainties in reversal angles, very rapid field changes, or both. Valet et al. (2012) inferred that the reversal process has remained unchanged, with the same time constants and durations, for at least 180 million years. These authors proposed that the reversing field is characterized by three successive phases: (a) a precursory event, (b) a 180° polarity switch, and (c) a rebound. The first and third phases reflect the emergence of the nondipole field with large-amplitude secular variation. The actual transit between the two polarities does not last longer than 1000 years and might therefore result from mechanisms other than those governing

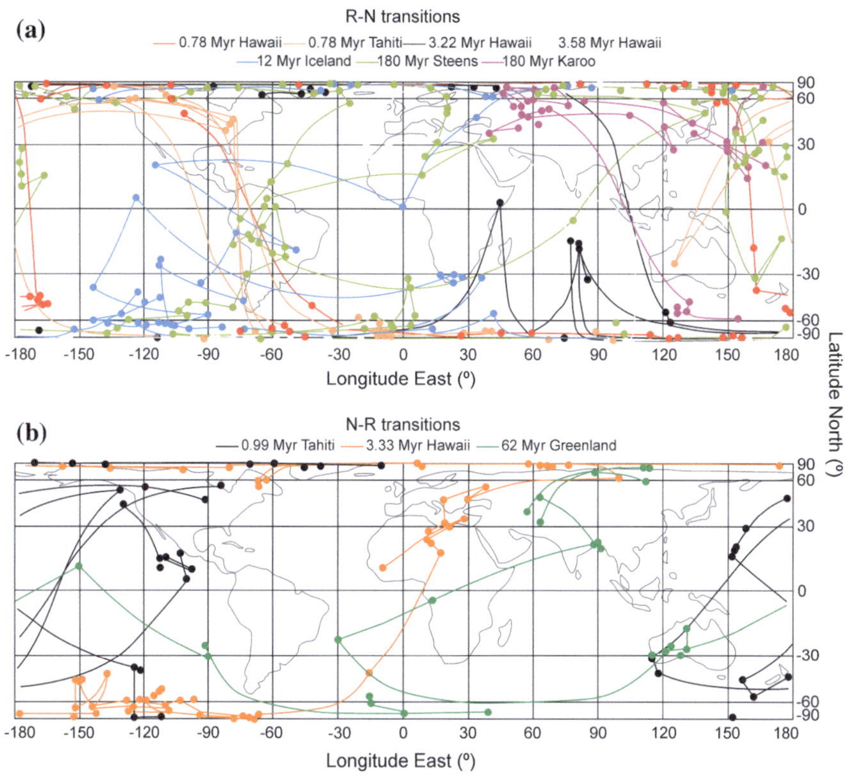

Fig. 3 Paths of VGPs. VGP paths from the most detailed volcanic reversal records of R–N (**a**) and N–R (**b**) transitions, plotted on cylindrical, equal area projections (adapted from Valet et al. 2012)

normal secular variation. Such changes are too brief to be accurately recorded by most sedimentary sequences.

These results support a reversal model which involves control of the lower mantle on the dynamo processes prevailing during the transition, and assumes that flux patches remained essentially stationary through time. Stationary flux lobes at high northern and southern latitudes, but somehow different from the previous ones, were also revealed by downward continuation of the recent and historical field measurements on or above the Earth's surface toward the core–mantle boundary (CMB) (Jackson et al. 2000). The primary locations of the clusters of VGPs near Western Australia and in the southwestern Atlantic Ocean correspond to locations of the present nondipolar field at Earth's surface with the largest vertical field component (Hoffman and Singer 2004).

On the basis of the VGP paths of the most detailed sedimentary records, which are relatively confined within the same large loop, Laj et al. (2006) suggested that the field geometry remains dominantly controlled by the dipole and therefore, that both the dipolar and nondipolar fields decay during these periods. A study on modeling patterns (Lanci et al. 2008) of the same high-resolution records of the Iceland Basin Excursion favors a large reduction of the dipole but remaining higher than the nondipole field throughout the excursional process, except for a very short period.

Some models (Korte and Constable 2005) indicated the persistence of flux lobes in the northern hemisphere during the past thousand years and the presence of more vague negative flux lobes over the Indian Ocean (Fig. 4). This persistence over longer time intervals (about 5 Myr) is argued by other authors (Kelly and Gubbins 1997), giving additional support to some control of the field by the lower mantle.

Brown et al. (2007) used the model CALS7K.2 (Korte and Constable 2005) to explore the possible influence of the time-varying nondipole components of the GF during field reversals and excursions. They suggested that nondipole components could add significant structure to the field during the reversal and excursion processes. The model generates variable reversal paths; however, there is a longitudinal preference of both spatially and, more weakly, temporally. Directional reversal features are not globally synchronous: some polarity changes finished before they started elsewhere. Global intensity variations, however, appear to be more consistent. Large excursions appear naturally when the axial dipole has been reduced to 20 % for the whole time period; however, they are not globally synchronous or uniform.

In order to answer the question of whether VGPs recorded during reversals and excursions show a longitudinal preference, Kutzner and Christensen (2004) studied the heterogeneity of heat flux at the CMB as one possible mechanism for such VGP clustering. They used 3-D convection-driven numerical dynamo models with imposed nonuniform CMB heat flow that show stochastic reversals of the dipole field. They calculated transitional VGPs for a large number of token sites on the Earth's surface. In a model with a simple heat flux variation, they show that the VGP density maps for individual reversals differ substantially from each other, but the VGPs have a tendency to fall around a longitude of high heat flow. The mean VGP density, for many reversals and excursions, shows a statistically significant correlation with the heat flow. They found that regions of high heat flow are centers of magnetic activity where intense magnetic flux bundles are generated, and they contribute to the equatorial dipole component and affects its orientation in longitude. During reversals the equatorial dipole part is not necessarily dominant at the Earth's surface, but it is strong enough to explain the longitudinal preference of VGPs as seen from different sites.

According to the second hypothesis, there is no reproducible field pattern in the secular variation. Thus, no evidence for preferred VGP clusters or for preferred longitudes during reversals (Valet et al. 1992; Prévot et al. 1993). This assertion is justified by the fact that the geographical distribution of the sites contained in the archeomagnetic and paleomagnetic databases is not appropriate for constraining spherical harmonic analyses beyond degree 2 (Carlut and Courtillot 1998; Valet and Plenier 2008). Accordingly, the presence of long-term standing nonzonal features emerging from global field models would be rather fortuitous.

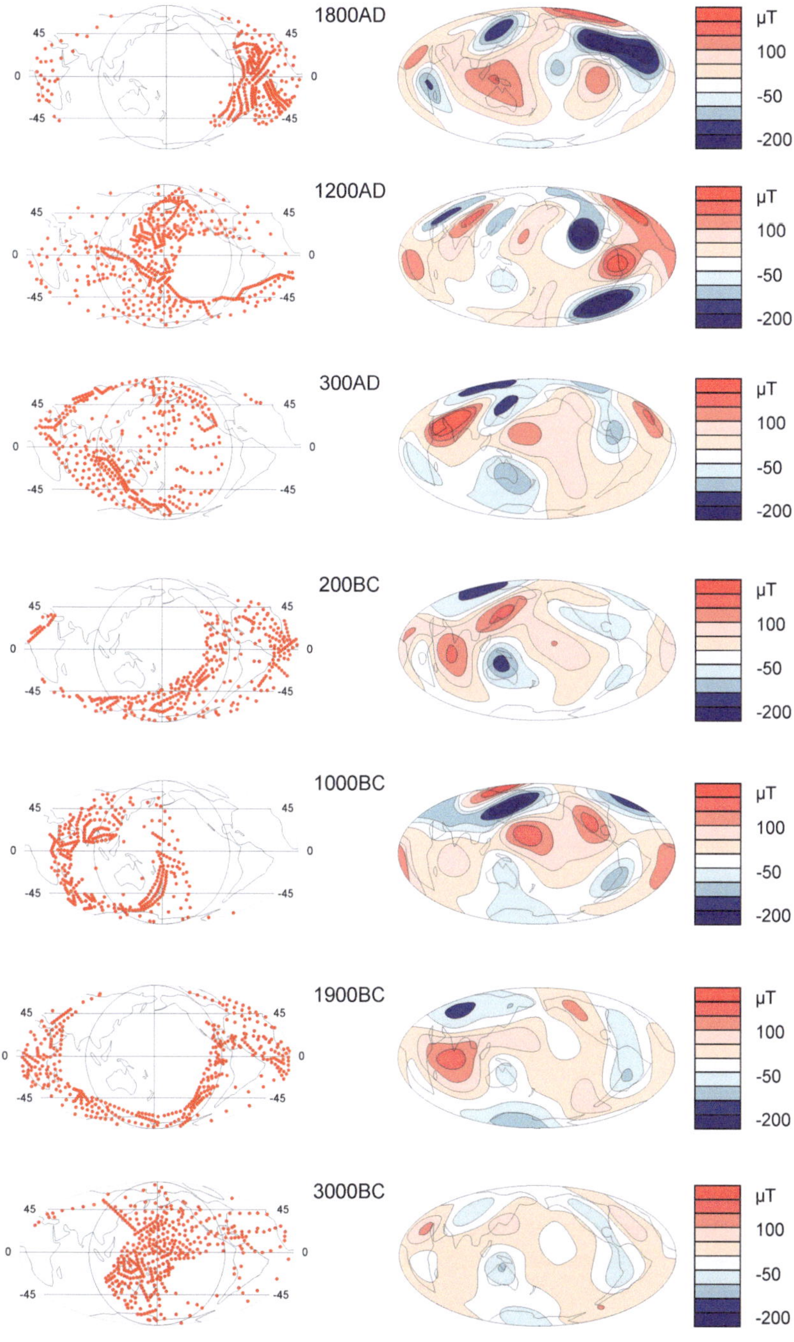

Fig. 4 *Left* snapshots of the VGP positions corresponding to the nonaxial dipole fields at different periods during the past 5 kyr. *Right* snapshots of the radial component of the nonaxial dipole contribution at the CMB (redrawn from Valet and Plenier 2008)

Several other characteristic features of the reversing field have been suggested, most being related to dynamical aspects of the transition. Apparent rapid changes alternating with periods of standstill are frequent in volcanic (Hoffman 1991; Coe and Prévot 1989; Coe et al. 1995) and in sedimentary records obtained with very high resolution (Channell and Lehman 1997; Valet and Herrero-Bervera 2003). Hoffman (2000) also noticed the absence of VGPs between the equator and 30°N in a compilation of records of the last reversal. These dynamic aspects are important, as they can indicate whether magnetic flux is controlled by processes inherent to the core or it is partly linked to the influence of the lower mantle.

Goguitchaichvili et al. (2001) carried out a Thellier paleointensity study of a ~3.6 million year Pliocene geomagnetic excursion recorded in a lava flow succession from southern Georgia (in the lesser Caucasus). They found a stable polarity paleointensity significantly lower than the average Pliocene geomagnetic dipole moment and postintermediate dipole moments recorded in volcanic sequences in Hawaii (~4 Ma) and Steens Mountain (~16.2 Ma), but quite similar to the postintermediate field recorded in Iceland during the Gauss–Matuyama reversal. These results allowed the authors to conclude that the regime of the geodynamo following reversals or excursions may vary significantly from one case to the next without any apparent systematic features.

The GF excursions are very short episodes, with significant deviation of field directions from the range of standard secular variation, even reaching the opposite polarity before returning to its initial state, have been recorded by high-resolution studies. The origin and succession of these excursions reveal field instabilities and may drastically change our concept of a GF characterized by long periods of stable polarity.

Guyodo and Valet (1999) found that the intensity of the Earth's dipole field in the last 800 kyr had undergone large-amplitude variations over this time period with pronounced intensity minima coinciding with known excursions in field direction, reflecting the emergence of nondipole components. They found no stable periodicity in the composite record and therefore, the data set does not support the hypothesis that the Earth's orbital parameters have a direct and strong influence on the geodynamo.

The frequency of these excursions, in numerical (Dormy et al. 2000) as well as in experimental dynamos (Berhanu et al. 2007) is consistent with the assumption that field excursions interrupt the history of the GF.

The available paleomagnetic study of several cores from the Arctic region indicates that the magnetic field has been strongly variable during at least the last 300 ka (Nowaczyk and Antonow 1997; Nowaczyk and Frederichs 1999), with geomagnetic excursions more frequent and of longer duration than anywhere else.

Macrì et al. (2010) provided a collection of well-defined paleomagnetic data at high resolution throughout an Antarctic sequence recovered on the continental rise of the Wilkes Land Basin (East Antarctica), and the consequent compilation of detailed ChRM directions and RPI records brought important contributions to the reconstruction of the GF variability from high latitudes in the southern hemisphere, and therefore, to the understanding of the key general features of the Earth's magnetic field within the polar regions. The sediments showed a detailed record of the Matuyama–Brunhes (M–B) transition and of a sharp oscillation in paleomagnetic directions that may correlate to the M–B precursor event. The whole paleomagnetic record of the Brunhes Chron does not indicate a substantially larger variability of the GF with respect to coeval records from intermediate and low latitudes with comparable sedimentation rates.

In addition, orbital periodicities (100 and 41 ka) found for paleomagnetic inclination records, are similar to those reported in studies of other sedimentary cores at low latitudes (Yamazaki and Oda 2002, 2004). These data support the model of Yamazaki and Oda (2002), which suggests a connection between the geodynamo and the orbital eccentricity, indicating that long-term geomagnetic secular variation in inclination is controlled by changes in the relative strength of the geocentric axial dipole and persistent nondipole components.

Many simulations have been carried out (Valet and Plenier 2008; Leonhardt and Fabian 2007) using VGP for testing simple field geometries. These simulations revealed characteristics of the VGP paths during reversals and excursions.

Valet and Plenier (2008) dealt first with the present nonaxial dipole (NAD) field and noticed that the largest flux concentrations did not induce significant anomalies, at least not large enough to concentrate the pole positions, except for sites lying in their close vicinity. Clusters of VGPs derived from the present NAD field were actually linked to the equatorial dipole. Therefore, they disclaimed the existence of a systematic and direct link between persisting patches of magnetic flux and VGP locations, unless they were considerably more intense than at present during field reversals. Another surprising observation is that the superposition of VGPs from multiple reversals shows a marked and persistent preference for some longitudinal bands. However, individual VGP paths exhibit very different configurations, some with longitudinal preference while some others are quite scattered all over the globe. These results disclaim any long-term persisting pattern of the field and thus fail to see any link to the lower mantle.

Simulations of field excursions using the same nondipole model account for the major features of the paleomagnetic records with large loopings of the pole in alternance with more complex episodes (Fig. 5). In contrast, an excursion field governed by the equatorial dipole (thus assuming a concomitant decrease of the axial dipole and nondipole terms) produces a large loop of the VGPs but fails to reproduce the scatter of the VGPs positions inherent to the volcanic records. The loop of the sedimentary VGP paths is notably amplified by the smearing of the signal in sedimentary records which reinforces low degree harmonics and can thus generate distortion of the initial signal. Finally, paleomagnetic directions of opposite polarity are extremely unlikely without a short period of opposite dipole with strength of at least 15–20 % of initial intensity.

Leonhardt and Fabian (2007) developed a Bayesian inversion method to reconstruct the spherical harmonic expansion of the transitional field corresponding to the Matuyama/Brunhes reversal, from paleomagnetic data. They found that radial magnetic flux patches formed at the equator and moved polewards during the transition. The reconstruction given by the model also offers new answers to the

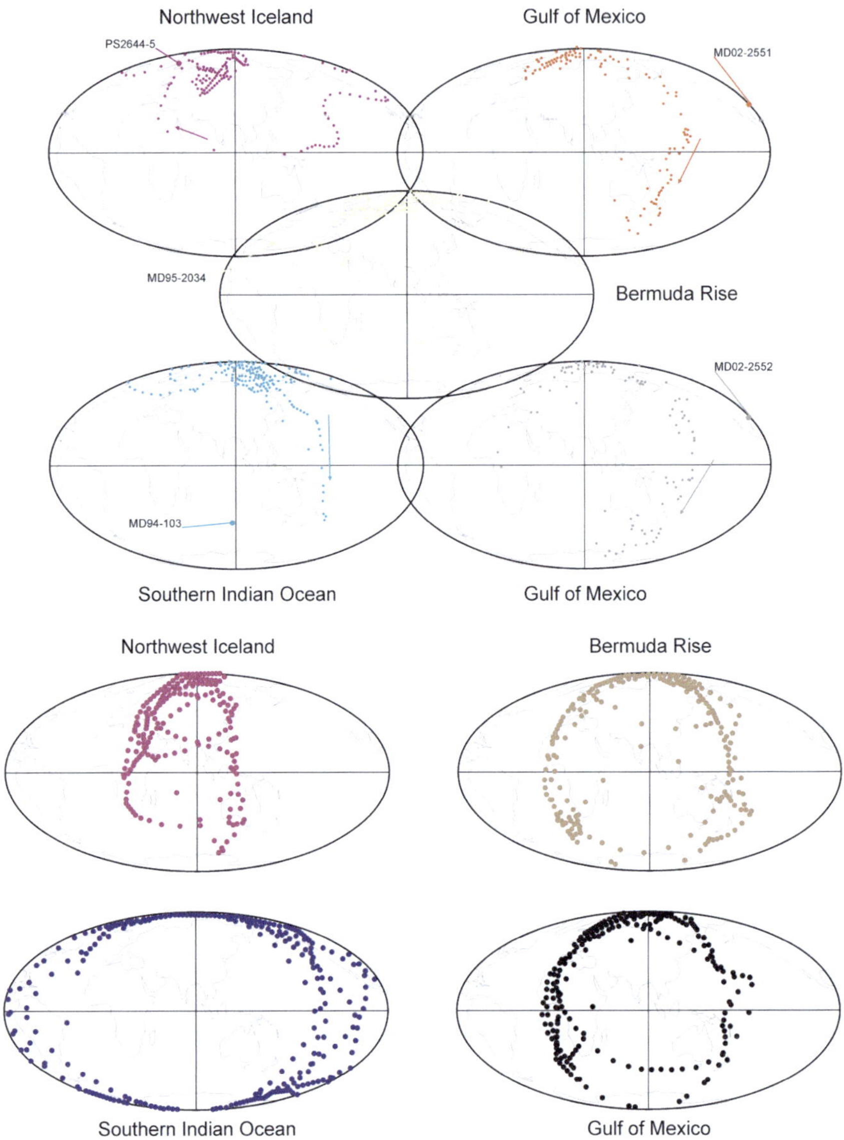

Fig. 5 **a** VGP paths of different records of excursions; **b** simulations of four distinct excursions recorded at the same sites (redrawn from Valet and Plenier 2008)

question of existence of preferred longitudinal bands during the transition and to the problem of reversal duration. Different types of directional variations of the surface GF, continuous or abrupt, are found during the transition. Two preferred longitudinal bands along the Americas and East Asia are not predicted for uniformly distributed sampling locations on the globe. A preference of transitional VGPs for the Pacific hemisphere is found, which is similar to geodynamo models with CMB heat flux derived from present day lower mantle heterogeneities. The paleomagnetic duration of reversals shows not only a latitudinal, but also a longitudinal variation. Even the paleomagnetically determined age of the reversal varies significantly between different sites on the globe.

2.1.2 Long-Term Geomagnetic Changes and Core Convection

Since the geodynamo processes are closely connected with the core–mantle thermal regime, convection in the mantle, and, in particular, the thermal boundary layer (D″) (Loper 1992), it would be logical to expect a modulation of these processes expressed in the existence of periodicities in the number of inversions, field intensity, etc. Long-term variations in the behavior of the GF should thus reflect variations in core convection. It is expected, and geodynamo simulations confirm it, that the amount and pattern of heat that the mantle allows to flow out of the core would affect both the intensity and stability of the field by controlling the robustness and pattern of convection.

Specially, the reversal frequency may be modulated by mantle convection; the overturn time of core fluid is close to the duration time of polarity transition and much shorter than the time interval of stable polarity, which may imply that the geomagnetic polarity reversal is controlled by the dynamo of the mantle; particularly by the core–mantle transition zone (D″ layer). The disturbance of heat in the lower mantle is possibly related to the instability of the heat boundary in the D" layer and can control the convection in the outer core. This will result in stable or quick variations in reversal frequency (Prevot et al. 1990; Shi and Zhu 2002).

Loper and McCartney (1986) provided evidence for a 30 Myr periodicity in the mantle convection, due to instability of D″ at the base of the mantle. They have also provided evidence for periodic thickening and thinning of the D″ layer which eventually affects the activity of the core processes and hence the periodic geomagnetic reversals.

Biggin et al. (2012) presented a synthesis of the latest results from a variety of disciplines, to examine possible causal relationships between geomagnetic behavior and mantle processes on the 10–100 Myr timescale. They considered two measures of geomagnetic behavior: reversal frequency and dipole moment.

Two periods in the last 200 Myr seem to represent examples of the most extreme geomagnetic behavior: the Middle–Late Jurassic (around 150–170 Myr ago) with high reversal frequency (Tominaga et al. 2008; Ogg 2004) and the Cretaceous Normal Superchron (CNS; 84–121 Myr ago) with field of almost single polarity for a period spanning nearly 40 Myr (Gee and Kent 2007; He et al. 2008). Different studies suggest that the average dipole moment during the Cretaceous and the Cenozoic was lower than average for at least part of the Jurassic period.

Also, two earlier Phanerozoic superchrons have been claimed from continental magneto stratigraphic records: the Permian-Carboniferous Reversed Superchron (PCRS; ~265–310 Myr ago, Langereis et al. 2010) and the Ordovician Reversed

Superchron (ORS; ~460–490 Myr ago, Pavlov and Galet 2005). In these cases there is magnetostratigraphic evidence that reversal frequency was very high just before the ORS in the Middle Cambrian, and preliminary measurements of the virtual dipole moment indicate that in the Devonian and Silurian periods this reversal frequency was lower than average. Therefore, the CNS, PCRS, and ORS may all have been preceded by a period of reversal hyperactivity.

Biggin et al. (2012) indicate that simulations of the geodynamo suggest that transitions from periods of rapid polarity reversals to periods of prolonged stability—such as the ones that occurred between the Middle Jurassic and Middle Cretaceous periods—may have been triggered by a decrease in CMB heat either globally or in equatorial regions.

The enhancing convection in dynamo simulations by increasing CMB flow tends to destabilize the dipole generation process, making reversals more likely (Aubert et al. 2009; Olson 2007). Then, reversal frequency and CMB heat flow seem to be positively correlated. Driscoll and Olson (2009) suggested, using a simple numerical dynamo model, that a twofold increase in the CMB heat low could reduce the dipole moment by half. Assuming that the geodynamo lies close to such a transition (Driscoll and Olson 2009; Christensen and Aubert 2006), periods of high reversal frequency/low dynamo activity (superchron), caused by high/low CMB heat low, may be associated with a low/high dipole moment.

CMB heat flow is proportional to the temperature contrast across D″ at the base of the mantle and the thermal conductivity of the lowermost mantle, and inversely proportional to the D″ thickness.

Then, changes in temperature accompanying lower mantle dynamics will affect the GF. Besides, the fluid motion in the outer core can affect the mantle convection first and then, the plate motion and the eruption of ocean basalts. Therefore, the GF is linked not only to the fluid motion in the outer core, but also to many geological events (such as mantle convection, mantle plume activity, global heat flux, true polar wandering, climatic changes, seamount chains, continental basalts, etc.).

2.2 *GF as Climate Forcing*

2.2.1 Holocene

Variations of the terrestrial magnetic field at regular intervals taken place during the last millennia seem to correlate to significant climatic events in the east of the North Atlantic region. Among others, Snowball and Sangren (2004) presented relative palaeointensity estimated from one Holocene lake sediment in central southern Sweden. In Fennoscandia significant peaks in GF intensity occurred at ~8400, ~6400, 3900, and 2800 cal year B.P. The maximum field intensity at 2800 cal year B.P. was associated with the most rapid change in the direction of the geomagnetic vector. High-resolution sedimentary data show that significant century scale increases and decreases in relative field intensity between 4000 and

2000 cal year B.P. and these variations were associated with abrupt changes in the direction of the geomagnetic vector; archaeomagnetic jerks younger than 2000 years were not reproduced in this record of lake sediments. The paleointensity reconstruction displays century to millennial scale trends between 5000 and 450 cal year B.P., and it is coherent with archaeomagnetic data sets from Western Europe and Central Asia. According to these authors, Fennoscandian paleosecular variation and paleointensity records for the earlier Holocene do not show the same rapid style of vector movement; it could suggest that northern Europe has been affected by a more turbulent geodynamo since ~4000 cal year B.P.

In the same time frame, Gallet et al. (2005) presented archeointensity results obtained from French faience potsherds dated from seventeenth to nineteenth century. Their results showed the occurrence of sharp changes in GF secular variation in Western Europe during the studied period. The intensity variation shows several maxima whose rising parts appear to coincide in time with cooling periods documented in the region by natural and historical data.

This perspective seems to change in the Pleistocene as the following examples will show.

2.2.2 Pleistocene. Relation Between GF Magnetic Events and Climate Changes

Studies of Pleistocene glacier deposits from the Scandinavian ice sheet (Houmark-Nielsen 2010) suggested that the ice dynamics during the marine isotopic stage MIS3 had a glacier expansion similar to the post-LGM advances. OSL and ^{14}C available data indicate that glaciers advanced at least twice during the Middle Wiechselian (c. 75–25 ka B.P.), probably related to Dansgaard–Oeschger (D-O) 14–13 (54–46 ka B.P.) and 8–5 (35–30 ka B.P.) events.

Additionally, during the present isotope time frame two magnetic events took place: the Mono Lake event (or excursion, 1, Table 1) and the Laschamp event (2). The Laschamp event was a short reversal of the magnetic field, around 39–41 ka B. P., originally discovered in lava flows. The Mono Lake event (or 'Excursional' Interval) took place approximately between 32 and 34 ka B.P. (Channell 2006). Due to their proximity, the relation between the magnetic phenomena and the climatic events mentioned here could be that of causativeness.

As it was previously mentioned, Kitaba et al. (2013) found new evidence of climatic effects of cloud formation induced by galactic cosmic rays (CRs) and variations in GF intensity. They presented paleoclimate and paleoenvironment records of five interglacial periods that include two geomagnetic polarity reversals. Marine oxygen isotope stages 19 and 31 contain both anomalous cooling intervals during the sea-level highstands and the Matuyama–Brunhes and Lower Jaramillo reversals, respectively.

At present, there is a lack of a consistent study of the relationship between the geomagnetic events and climate changes of the paleoclimatic record.

This contribution intends to validate the hypothetical relationship climate versus GF. To analyze the influence of the reversals to climate, the present authors decided to compare global variations of temperature and known magnetic events in various locations in the world during the Brunhes Chron (Table 1)

Due to the fact that benthic $\delta^{18}O$ is often used as a stratigraphic tool to place marine records on a common age model and as a proxy for the timing of ice volume/sea-level change, these available data have been used. The paleoclimate inferences made (Fig. 2) arise from a series of data of both oceans presented by Lisiecki and Raymo (2012). They presented Atlantic and Pacific 800 ka benthic $\delta^{18}O$ stacks.

Figure 6 shows the $\delta^{18}O$ variations recorded in both oceans and the GF reversals. As it can be seen, there is a good match between most of the events and a marked cooling of the oceanic waters triggered at the beginning or during the GF reversal.

The process should be interpreted as cooling by the formation of clouds connected to the reduction of the intensity of GF during polarity transition due to the GCR flow increment. The cooling would be more or less intense according to the characteristics of GF during the reversal. After the reversal, the climate cooling continues due to the feedback processes triggered within the system as a consequence of ice increase, decrease of greenhouse effect gases concentration, increase of the albedo, and the conveyor belt intensity, among other factors (Rial et al. 2004, among others).

It is also necessary to consider the effect of the TSI changes resulting from orbital variations within the time frame of the observed cooling (Ruddiman 2006, among others).

As it was summarized before in 2.1, the GF does not keep the same behavior during its polarity reversal. The decreases of intensity are not regular (Macrì et al. 2005, among others) or the geometry during the event is not always the same, since a nondipolar behavior has been inferred during those lapses (Valet and Plenier 2008,

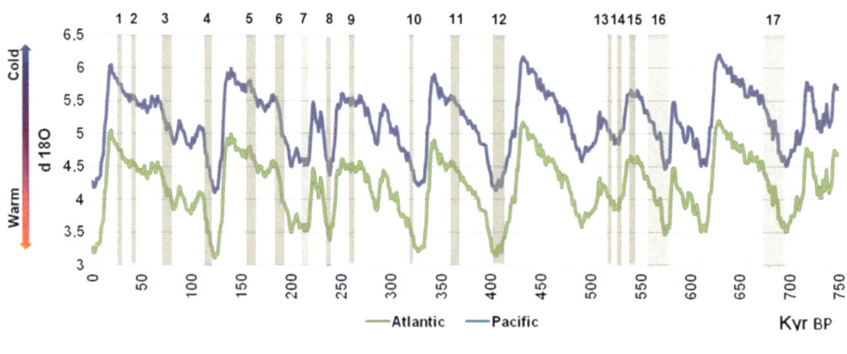

Fig. 6 Stacked (averaged) benthic foraminifera $\delta^{18}O$ records from 20 Atlantic Ocean sites and 14 Pacific Ocean sites, covering the past 800,000 years from Lisiecki and Raymo (2012) and the reversal GF (see Table 1). The above numbers correspond to each reversal GP

among others). Consequently, the input of GCR and their climatic effects were not always the same for different polarity transitions of the GF.

It should be noticed that in some cases the synchronicity is evident. In Fig. 6 there are some relevant examples such as the MIS5 and the Blake event (number 4: around 120–110 ka), the MIS2, and the Laschamp (number 2 around 40 ka) and Mono Lake events (number 1: around 30 ka). It must be pointed out that the Bermuda excursion (number 12: around 400–429 ka) might not have been global since it was recorded only in some locations. However, even during the MIS 11 the climatic signal (well documented in marine and ice sheet isotopic records as well as in terrestrial sequences; Ashton et al. 2008, among others), comprises at least two major warmer episodes with an intervening cooler phase (Koutsodendris et al. 2012; Candy et al. 2014). Therefore, the Bermuda event could have induced this colder phase. Besides, during this event the MIS 10 started, which also seems to have been promoted by the Bermuda event, and then reinforced by the posterior Levantide (N°11, around 365 ka).

Evidently, climate changes are complex and multiple-caused phenomena but evidence could be pointing to a relevant incidence of cosmic rays (and the subsequent cloud formation) in “windows in the GF shielding” during the transitions. To confirm these assumptions it is necessary to perform a detailed study of each particular event, variations in morphology and paleointensity of the field during the transition. Especially, the beginning of MIS12 and MIS 14 should be carefully revised based on other external forcings, since no GF reversal coinciding event has been recorded.

Considering that during the different transitions the GF has not behaved in the same way, the same climatic effect cannot be expected. On the other hand, the simultaneous action of other forcings (orbital parameters, etc.) should be considered. This generates an interference whose magnitude is very difficult to assess. The effects caused by different forcings could be synergetic or not.

Finally, we should bear in mind that the disparity in the ages attributed to the events according to the location where they were recorded make it difficult to carry out comparisons. This disparity has two probable origins: the own nature of the GF reversals that makes these events look of different ages on different sites of the globe, and the different errors in the dating methods used.

3 Conclusions

As proposed by several authors, the forcings affecting the climate are varied and complex, such as the time frame in study, the effects produced by variations in solar activity, and variations in solar irradiance due to orbital cycles, among others.

The influence of GF during polarity reversals seems to be important, at least to the time frame corresponding to the Bruhnes Chron. The process would be linked to a decrease in the intensity of the geomagnetic dipole that would favor the input of cosmic rays which would promote the increase of low clouds and the consequent climate cooling.

In the current climate context affected by global warming, studies develop in a complex scenario. On one hand, the influence of anthropogenic activity over climate warming seems to rise with the increase of greenhouse gases and the cutting down of native forests and desertification. On the other hand, solar activity has started to diminish since the beginning of the twenty-first century and various authors have forecasted the beginning of a new Grand Solar Minima similar to the Dalton or even to the Maünder minima. At the same time, the GF is weakening. Both factors are related to the increment of GCR input. Also, if the GF reversed, would this be superior to the anthropogenic forcing in effect that would prevent natural cooling?

References

Ashton N, Lewis SG, Parfitt SA, Penkman KEH, Coope GR (2008) New evidence for complex climate change in MIS 11 from Hoxne, Suffolk, UK. Quat Sci Rev 27:652–668

Aubert J, Labrosse S, Poitou C (2009) Modelling the palaeo-evolution of the geodynamo. Geophys J Int 179:1414–1428

Bard E, Frank M (2006) Climate change and solar variability: what's new under the sun? Earth and Planet Sci Lett 248:1–14

Bard E, Raisbeck G, Yiou F, Jouzel J (2000) Solar irradiance during the last 1200 years based on cosmogenic nuclides. Tellus B 5(2):985–992

Beer JS, Tobias S, Weiss N (1998) An active sun trough the maunder minimum. Solar Phys 181:237–249

Berhanu M, Moncaux R, Fauve S, Mordant N, Pétrélis F, Chiffaudel A, Daviaud F et al (2007) Magnetic field reversals in an experimental turbulent dynamo. Europhy Lett 77(590):01

Biggin J, Steinberger B, Aubert J, Suttie N, Holme R, Torsvik TH, van der Meer DG, van Hinsbergen DJJ (2012) Possible links between long-term geomagnetic variations and whole-mantle convection processes. Nat Geosci. doi:10.1038/NGEO1521

Bourne M, Niocaill CM, Thomas AL, Knudsen MF, Henderson GM (2012) Rapid directional changes associated with a 6.5 kyr-long Blake geomagnetic excursion at the Blake-Bahama Outer Ridge. Earth and Planet Sci Lett 333–334:21–34. doi:10.1016/j.epsl.2012.04.017

Brown MC, Holme R, Bargery A (2007) Exploring the influence of the non-dipole field on magnetic records for field reversals and excursions. Geophys J Int 168:541–550

Bucha V, Bucha V Jr (1998) Geomagnetic forcing of changes in climate and in the atmospheric circulation. J Atmos Solar-Terrestrial Phys 60(1):145–169

Callis LB, Boughner RE, Natarajan M, Lambeth JB, Baker DN, Blake JB (1991) Ozone depletion in the high latitude lower stratosphere: 1979–1990. J Geophys Res 96:2921–2937

Candy I, Schreve DC, Sherriff J, Tye GJ (2014) Marine isotope stage 11: palaeoclimates, palaeoenvironments and its role as an analogue for the current interglacial. Earth-Sci Rev 128:18–51

Carlut J, Courtillot V (1998) How complex is the time-averaged field over the past 5 Myr? Geophys J Int 134:527–544

Channell JET (2006) Late Brunhes polarity excursions (Mono Lake, Laschamp, Iceland Basin and Pringle Falls) recorded at ODP Site 919 (Irminger Basin). Earth Planet Sci Lett 244:378–393

Channell JET, Lehman B (1997) The last two geomagnetic polarity reversals recorded in high deposition-rate sediments drifts. Nature 389:712–715

Channell JET, Curtis JH, Flower BP (2004) The Matuyama-Brunhes boundary interval (500–900 ka) in North Atlantic drift sediments. Geophys J Int 158(4):89–505
Chanell JET, Hodell DA, Curtis JH (2012) ODP site 1063 (Bermuda Rise) revisited: oxygen isotopes, excursions and paleointensity in the Brunhes Chron. Geochem Geophys Geosyst 13 (2):27
Channell JET, Wright JD, Mazaud A, Stoner JS (2014) Age through tandem correlation of quaternary relative paleointensity (RPI) and oxygen isotope data at IODP Site U1306 (Eirik Drift, SW Greenland). Quat Sci Rev 88:135–146
Christensen UR, Aubert J (2006) Scaling properties of convection-driven dynamos in rotating spherical shells and application to planetary magnetic fields. Geophys J Int 166(1):97–114
Coe RS, Prévot M (1989) Evidence suggesting extremely rapid field variation during a geomagnetic reversal. Earth Planet Sci Lett 92:292–298
Coe RS, Prévot M, Camps P (1995) New evidence for extraordinarily rapid change of the geomagnetic field during a reversal. Nature 374:687–692
Courtillot V, Gallet Y, Le Mouel JL, Fluteau F, Genevey A (2007) Are there connections between the earth's magnetic field and climate? Earth Planet Sci Lett 253:328–339
Cox A, Doell DR, Dalrymple GB (1963) Geomagnetic polarity epochs and pleistocene geochronometry. Nature 198:1049–1952
Dagley P, Lawley E (1974) Paleomagnetic evidence for the transitional behavior of the geomagnetic field. Geophys J R Astron Soc 36:577–598
De Jager C, Duhau S (2009) The variable solar dynamo and the forecast of solar activity. Influence in terrestrial surface temperature, In: Cossia J (ed) Global warming of the 21th century, NOVA Science Publishers, available November 2011 in http://www.cdejager.com/wp-content/uploads/2008/09/2010-Variable-solar-dynamo3.pdf
Desorgher L, Flückiger EO, Gurtner M, Moser MR, Bütikofer R (2005) Atmocosmics: a Geant4 code for computing the interaction of cosmic rays with the earth's atmosphere. Int J Mod Phys A 20:6802–6804
Dorman LI (2004) Cosmic rays in the earth's atmosphere and underground, Chap. 12. Springer, New York, 855 p
Dormy E, Valet JP, Courtillot V (2000) Numerical models of the geodynamo and observational constraints. Geochem Geophys Geosyst 1, 2000GC00062
Driscoll P, Olson P (2009) Effects of buoyancy and rotation on the polarity reversal frequency of gravitationally driven numerical dynamos. Geophys J Int 178:1337–1350
Duhau S (2002) Two coupled 88 and 22 year oscillators in the modulation of sunspot cycles. Anales AFA 13:37 (Buenos Aires)
Duhau S (2003a) Solar variability as an input to the earth's environment, proceedings of ISCS 2003, Tatranska Lomnica. Slovak Republic. In: A Wilson (ed) ESA SP-535, p. 91
Duhau S (2003b) An early prediction of maximum sunspot cycle 24. Solar Phys 213:203–212
Duhau S (2005) Long term variations in solar magnetic field, geomagnetic field and climate. In: Procceding of 9th Asian-pacific regional IAU meeting
Duhau S (2013) Un Nuevo mecanismo de amplificación de la irradiancia solar en la modulación de las grandes fluctuaciones climáticas. Abstract "Simposio Multidisciplinario: El Estadio Isotópico 3 en la Argentina y el sur de América del Sur: 60.000 a 25.000 años atrás", 27 and 28 June 2013, La Plata, Argentina, p 9
Duhau S, Chen A (2002) The sudden increases of solar and geomagnetic activity after 1923 as a manifestation of a non-linear solar dynamo. Geophys Res Lett 13:1–6
Duhau S, Martinez EA (1995) On the origin of the fluctuations in the length of day and in the geomagnetic field on a decadal time scale. Geophys Res Lett 22(23):3283–3286
Duhau S, Martínez EA (2012) Solar dynamo transitions as drivers of sudden climate changes. Global warming-impacts and future perspectives, INTECH Open Science/Open Minds, pp 185–204
Dumberry M, Bloxham J (2006) Azimuthal flows in the Earth's core and changes in length of day at millennial timescales. Geophys J Int 165(1):32–46

Fröhlich C, Lean J (2004) Solar radiative output and its variability: evidence and mechanisms. Astron Astrophys Rev 12:273–320
Gallet Y, Genevey A, Fluteau F (2005) Does earth's magnetic secular variation control centennial climate change? Earth Planet Sci Lett 236:339–347
Gee JS, Kent DV (2007) Source of oceanic magnetic anomalies and the geomagnetic polarity timescale. In: Kono M (ed) Treatise Geophysics, vol 5 Geomagnetism, pp 455–507
Goguitchaichvili A, Camps P, Urrutia-Fucugauchi J (2001) On the features of the geodynamo following reversals or excursions: by absolute geomagnetic paleointensity data. Phys Earth Planet Int 124:81–93
Gonzalez WD, Tsurutani BT, de Gonzalez ALC (1999) Interplanetary origin of geomagnetic storms. Space Sci Rev 88:529–562
Gray LJ, Beer J, Geller M, Haigh JD, Lockwood M, Matthes K, Cubasch U, Fleitmann D, Harrison G, Hood L, Luterbacher J, Meehl GA, Shindell D, van Geel B, White W (2010) Solar influences on climate. Rev Geophys 48:RG4001. doi:10.1029/2009RG000282
Greff-Lefftz M (2011) Length of day variations due to mantle dynamics at geological timescale. Geophys J Int 187(2):595–612
Golovkov VP (1983) Dynamics of the geomagnetic field and the internal structure of the earth. In: Bucha V (ed) Magnetic field and processes in the earth's interior. Czeck Acad Sci, Pragúe, pp 395–501
Guyodo Y, Valet JP (1999) Global changes in intensity of the earth's magnetic field during the past 800 kyr. Nature 399:249–252
Haigh JD (1994) The role of stratospheric ozone in modulating the solar radiative forcing of climate. Nature 370:544–546
Haigh JD (2003) The effects of solar variability on the earth's climate. Philos Trans R Soc Lond A 361:95–111
He H, Pan Y, Tauxe L, Qin H, Zhu R (2008) Toward age determination of the M0r (Barremian–Aptian boundary) of the early Cretaceous. Phys Earth Planet Inter 169:41–48
Hoffman KA (1991) Long-lived transitional states of the geomagnetic field and the two dynamo families. Nature 354:273–277
Hoffman KA (1992) Dipolar reversal states of the geomagnetic field and core–mantle dynamics. Nature 359:789–794
Hoffman KA (2000) Temporal aspects o f the last reversal of earth's magnetic field. Phil Trans R Soc Lond A 358:1181–1190
Hoffman KA, Singer BS (2004) Regionally recurrent paleomagnetic transitional fields and mantle processes. AGU Geophys Monogr 145:233–243
Holme R, de Viron O (2013) Characterization and implications of intradecadal variations in length of day. Nature 499(7457):202–204
Houmark-Nielsen M (2010) Extent, age and dynamics of marine isotope stage 3 glaciations in the southwestern Baltic Basin. Boreas. doi:10.1111/j.1502-3885.2009.00136.x
Jackman CH, McPeters H (2004) The effect of solar proton events on ozone and other constituents, In: Pap JM, Fox P (eds), Solar variability and its effect on climate, AGU Geophys Monogr 141, pp. 305–319
Jackson A, Jonkers A, Walker M (2000) Four centuries of geomagnetic secular variation from historical records. Phil Trans R Soc Lond 358:957–990. doi:10.10.98/rsta.0569
Jadin EA (1999) Interannual variation of total ozone and stratospheric angular momentum. Int J Geomag Aeronomy 1(2):169–180
Kelly P, Gubbins D (1997) The geomagnetic field over the past 5 million years. Geophys J Int 128 (2):315–330
Kirkby J, Curtius J, Almeida J, Dunne E, Duplissy J, Ehrhart S, Franchin A, Gagne S, Ickes L, Kurten A, Kupc A, Metzger A, Riccobono F, Rondo L, Schobesberger S, Tsagkogeorgas G, Wimmer D, Amorim A, Bianchi F, Breitenlechner M, David A, Dommen J, Downard A, Ehn M, Flagan RC, Haider S, Hansel A, Hauser D, Jud W, Junninen H, Kreissl F, Kvashin A, Laaksonen A, Lehtipalo K, Lima J, Lovejoy ER, Makhmutov V, Mathot S, Mikkila J, Minginette P, Mogo S, Nieminen T, Onnela A, Pereira P, Petaja T, Schnitzhofer R,

Seinfeld JH, Sipila M, Stozhkov Y, Stratmann F, Tome A, Vanhanen J, Viisanen Y, Vrtala A, Wagner PE, Walther H, Weingartner E, Wex H, Winkler PM, Carslaw KS, Worsnop DR, Baltensperger U, Kulmala M (2011) Role of sulphuric acid, ammonia and galactic cosmic rays in atmospheric aerosol nucleation. Nature 476(7361):429–433

Kitaba I, Hyodo M, Katoh S, Dettman DL, Sato H (2013) Midlatitude cooling caused by geomagnetic field minimum during polarity reversal. Proc Natl Acad Sci 110(4):1215–1220

Kerminen VM, Paramonov M, Anttila T, Riipinen I, Fountoukis C, Korhonen H, Asmi E, Laakso L, Lihavainen H, Swietlicki E, Svenningsson B, Asmi A, Pandis SN, Kulmala M, Petäjä T (2012) Cloud condensation nuclei production associated with atmospheric nucleation: a synthesis based on existing literature and new results. Atmos Chem Phys 12(24):12037–12059

Korte M, Constable CG (2005) Continuous geomagnetic models for the past 7 millennia II: CALS7K. Geochem Geophys Geosyst 6(1). doi:10.1029/2004GC000801

Koutsodendris A, Pross J, Müller UC, Brauer A, Fletcher WJ, Kühl N, Kirilova EP, Verhagen FTM, Lücke A, Lotter AF (2012) A short-term climate oscillation during the Holsteinian interglacial (MIS 11c): An analogy to the 8.2 ka climatic event? Global Planet Change 92–93:224–235. doi:10.1016/j.gloplacha.2012.05.011

Kutzner C, Christensen UR (2004) Simulated geomagnetic reversals and preferred virtual geomagnetic pole paths. Geophys J Int 157:1105–1118

Laj C, Channell JET (2007) Geomagnetic excursions. In: Kono M (ed) Treatise on geophysics 5, geomagnetism. Elsevier, Amsterdam, pp 373–416

Laj C, Mazaud A, Weeks R, Fuller M, Herrero-Bervera E (1991) Geomagnetic reversal paths. Nature 351:447–448

Laj C, Kissel C, Roberts AP (2006) Geomagnetic field behaviour during the Iceland Basin and Laschamp geomagnetic excursions: a simple transitional field geometry? Geochem Geophys Geosyst 7(3). doi:10.1029/2005GC001122

Lal L, Peters B (1967) Cosmic ray produced radioactivity on the earth, Handbuch der Physik, XLVI/2. Springer, Berlin, pp 551–612

Lanci L, Kissel C, Leonhardt R, Laj C (2008) Morphology of the Iceland Basin excursion from a spherical harmonics analysis and an iterative Bayesian inversion procedure of sedimentary records. Phys Earth Planet Interiors 169:131–139

Langereis CG, Dekkers MJ, De Lange GJ, Paterne M, Van Santvoort PJM (1997) Magnetostratigraphy and astronomical calibration of the last 1.1 Myr from an eastern mediterranean piston core and dating of short events in the Brunhes. Geophys J Int 129 (1):75–94

Langereis CG, Krijgsman W, Muttoni G, Menning M (2010) Magnetostratigraphy—concepts, definitions, and applications. Newslett Stratigr 43:207–233

Lario D, Simmet G (2004) Solar Energetic particle variations. In: Pap JM, Fox P (eds) Solar variability and its effects on climate. AGU Geophys Monogr 141, pp 195–220

Lean J (2005) Living with a variable sun. Phys Today 6:32–38

Lean J, Rind D (1999) Evaluating sun-climate relationships since the little ice age. J Atmosph Solar-Terr Phys 61:25–36

Lean J, Beer J, Bradley R (1995) Reconstruction of solar irradiance since 1610: implications for climate change. Geophys Res Lett 22:3195–3198

Le Mouël JL, Blanter E, Shnirmanm M, Cortillot V (2010) Solar forcing of the semi-annual variation of length-of-day. Geophys Res Lett 37:L15307 (5 p). doi:10.1029/2010GL043185

Leonhardt R, Fabian K (2007) Paleomagnetic reconstruction of the global geomagnetic field evolution during the Matuyama/Brunhes transition: Iterative Bayesian inversion and independent verification. Earth Planet Sci Lett 253:172–195

Lisiecki LE, Raymo ME (2012) Atlantic and Pacific 800 KYr benthic d18O stacks. IGBP PAGES/world data center for paleoclimatology data contribution series# 2012-115. NOAA/NCDC Paleoclimatology Program, Boulder CO, USA

Love JJ (1998) Paleomagnetic volcanic data and geometric regularity of reversals and excursions. J Geophys Res 103(B6):12,435–12,452

Loper DE (1992) On the correlation between mantle plume flux and the frequency of reversals of the geomagnetic field. Geophys Res Lett 19(1):25–28

Loper DE, McCartney K (1986) Mantle plumes and the periodicity of magnetic field reversals. Geophys Res Lett 13:1525–1528

Lourens LJ (2004) Revised tuning of ocean drilling program site 964 and KC01B (Mediterranean) and implications for the δ18O, tephra, calcareous nannofossil, and geomagnetic reversal chronologies of the past 1.1 Myr. Paleoceanography 19(3)

Macrì P, Sagnotti L, Dinarès-Turell J, Caburlotto A (2005) A composite record of late pleistocene relative geomagnetic paleointensity from the Wilkes Land Basin (Antarctica). Phys Earth Planet Int 151(3):223–242

Macrì P, Sagnotti L, Dinarès-Turell J, Caburlotto A (2010) Relative geomagnetic paleointensity of the Brunhes Chron and the Matuyama-Brunhes precursor as recorded in sediment core from Wilkes Land Basin (Antarctica). Phys Earth Planet Int 179:72–86

Mankinen EA, Prévot M, Grommé CS, Coe RS (1985) The steens mountain (Oregon) geomagnetic polarity transition. 1. Directional history, duration of episodes, and rock magnetism. J Geophys Res 90(10393–10):416

Marsh ND, Svensmark H (2000) Low cloud properties influenced by cosmic rays. Phys Rev Lett 85:5004–5007

Mazaud A, Channell JET (1999) The top Olduvai polarity transition at ODP Site 983 (Iceland Basin). Earth Planet Sci Lett 16(6):1–13

Mazaud A, Channell JET, Xuan C, Stoner JS (2009) Upper and lower Jaramillo polarity transitions recorded in IODP Expedition 303 North Atlantic sediments: Implications for transitional field geometry. Phys Earth Planet Sci 172:131–140

Milankovitch M (1920) Théorie Mathématique de phénomènes thermiques produits par la radiation solaire. Gauthiers-Volars, Paris

Mörner N-A (2010) Solar minima, earth's rotation and little ice ages in the past and in the future. The North Atlantic—European case. Glob Planet Change 72:282293

Mörner N (2013) Solar wind, earth's rotation and changes in terrestrial climate. Phys Rev Res Int 3 (2):117–136

Negrini RM, McCuan DT, Horton RA, Lopez JD, Cassata WS, Channell JE, Verosub KL, Knott JR, Coe RS, Liddicoat JC, Lund SP, Benson LV, Sarna-Wojcicki AM (2014) Nongeocentric axial dipole field behavior during the Mono Lake excursion. J Geophys Res Solid Earth 119(4):2567–2581

Ney EP (1959) Cosmic radiation and the weather. Nature 183:451–452

Nowaczyk NR, Antonow M (1997) High resolution magnetostratigraphy of four sediment cores from the Greenland Sea. I. Identification of the Mono Lake excursion, Laschamp and Biwa I/Jamaica geomagnetic polarity events. Geophys J Int 131:310–324

Nowaczyk N, Frederichs T (1999) Geomagnetic events and relative paleointensity variations during the last 300 ka as recorded in Kolbeinsey Ridge sediments, Iceland Sea, indication for a strongly variable geomagnetic field. Int J Earth Sci 88:116–131

O'Brien K (2005) The theory of cosmic-ray and high-energy solar-particle transport in the atmosphere. In: McLaughlin JP, Simopoulos SE, Steinhusler F (eds) Proceedings of the 7th international symposium on the natural radiation environment. Elsevier, New York, pp 29–44

O'Regan M, King J, Backman J, Jakobsson M, Palike H, Moran K, Heil C, Sakamoto T, Cronin TM, Jordan RW (2008) Constraints on the pleistocene chronology of sediments from the lomonosov ridge. Paleoceanography 23, PA1S19. doi:10.1029/2007PA001551

Ogurtsov MG, Nagovitsyn YA, Kocharov GE, Jungner H (2002) Long-period cycles of the sun's activity recorded in direct solar data and proxies. Solar Phys 11:371–394

Ogg JG (2004) In: Gradstein FM, Ogg JG, Smith AG (eds) A geological time scale. Cambridge University Press, Cambridge, pp 307–339

Olson P (2007) Gravitational dynamos and the low-frequency geomagnetic secular variation. Proc Natl Acad Sci USA 105:20159–20166

Palmer MA, Gray LJ, Allen MR, Norton WA (2004) Solar forcing of climate: model results. Adv Space Res 34:343–348

Pavlov V, Gallet Y (2005) A third superchron during the early paleozoic. Episode 28:78–84
Prévot M, Mankinen EA, Grommé CS, Coe R (1985) How the geomagnetic field vector reverses polarity. Nature 316:230–234
Prevot M, Derder ME, McWilliams M (1990) Intensity of the earth's magnetic field: evidence for a Mesozoic dipole low. Earth Planet Sci Lett 97:129–139
Prévot M, Mankinen EA, Gromme CS, Coe RS (1993) Absence of longitudinal confinement of poles in volcanic records of geomagnetic reversals. Nature 366:53–57
Rial JA, Pielke RA Sr, Beniston M, Claussen M, Canadell J, Cox P, Held H, de Noblet-Ducoudre N, Prinn R, Reynolds J, Salas JD (2004) Nonlinearities, feedbacks and critical thresholds within the Earth's climate system. Clim Change 65(1–2):11–38
Ruddiman WF (2006) Orbital changes and climate. Quat Sci Rev 25(23):3092–3112
Schove DJ (1955) The sunspot cycle, 649 BC to AD 2000. J Geophys Res 60(2):127–146
Shi RP, Zhu RX (2002) Possible links between abnormal geological events and geodynamics during Cretaceous. Progress Geophys 17(2):295–300
Shindell DT, Schmidt GA, Mann ME, Rind D, Waple A (2001) A Solar forcing of regional climate change during the Maunder minimum. Science 294:2149–2152
Singer BS, Jicha BR, Kirby BT, Geissman JW, Herrero-Bervera E (2008) 40Ar/39Ar dating links albuquerque volcanoes to the Pringle falls excursion and the geomagnetic instability time scale. Earth Planet Sci Lett 267(3):584–595
Snowball I, Sangren P (2004) Geomagnetic field intensity changes in Sweden between 9000 and 450 cal BP: extending the record of "archaeomagnetic jerks" by means of lake sediments and the pseudo-Thellier technique. Earth Planet Sci Lett 227:361–376
Solanki SK, Usoskin IG, Kromer B, Schüssler M, Beer J (2004) Unusual activity of the sun during recent decades compared to the previous 11,000 years. Nature 431:1–12
Solanki SK, Krivova NA, Haigh JD (2013) Solar irradiance variability and climate. Ann Rev Astron Astrophys 51:311–351
Solomon S (ed) (2007) Climate change 2007-the physical science basis: working group I contribution to the fourth assessment report of the IPCC (vol 4). Cambridge University Press, Cambridge
Sonnett CP, Finney SA (1990) The spectrum of radiocarbon. Philos Trans R Soc Lond 30A: 413–426
Soon W (2005) Variable solar irradiance as a plausible agent for multidecadal variations in the arctic-wide surface air temperature record of the past 130 years. Geophys Res Lett 32:L16712. doi:10.1029/2005GL023429
Stocker TF, Qin D, Plattner GK, Tignor M, Allen SK, Boschung J, Nauels A, Xia Y, Bex V., Midgley PM (eds) (2013) IPCC, 2013: climate change 2013: the physical science basis. Contribution of working group I to the fifth assessment report of the intergovernmental panel on climate change. Cambridge University Press, New York, 1535 p
Svensmark H (1998) Influence of cosmic rays on earth's climate. Phys Rev Lett 81:5027–5030
Svensmark H, Friis-Christensen E (1997) Variation of cosmic ray flux and global cloud coverage—a missing link in solar—climate relationships. J Atmos Sol Terr Phys 59:1225–1232
Teed RJ, Jones CA, Tobias SM (2014) The transition to earth-like torsional oscillations in magnetoconvection simulations, (preprint available: http://www1.maths.leeds.ac.uk/~rjteed/2013GJIpp.pdf)
Tinsley BA, Yu F (2004) Atmosferic ionization and clouds as links between solar activity and climate. In: Pap JM, Fox P (eds) Solar variability and its effect on climate, geophysical monograph series, vol 141, pp 321–339
Tominaga M, Sager WW, Tivey MA, Lee SM (2008) Deep-tow magnetic anomaly study of the Pacific Jurassic Quiet Zone and implications for the geomagnetic polarity reversal timescale and geomagnetic field behavior. J Geophys Res Sol Ea 113:B07110
Usoskin IG, Gladysheva OG, Kovaltsov GA (2004) Cosmic ray induced ionization in the atmosphere: spatial and temporal changes. J Atmos Sol Terr Phys 66(18):1791–1796

Usoskin IG, Schüssler M, Solanki SK, Mursula K (2005) Solar activity, cosmic rays, and earth's temperature: a millennium scale comparison. J Geophys Res 110:A 10102. doi:10.1029/2004JA010946

Valet JP, Herrero-Bervera E (2003) Characteristics of geomagnetic reversals inferred from detailed volcanic records. C R Acad Sci Paris 335:79–90

Valet JP, Tucholka P, Courtillot V, Meynadier L (1992) Paleomagnetic constraints on the geometry of the geomagnetic field during reversals. Nature 356:400–407

Valet JP, Plenier G (2008) Simulations of a time-varying non-dipole field during geomagnetic reversals and excursions. Phys Earth Planet Sc 169:178–193

Valet JP, Fournier A, Courtillot V, Herrero-Bervera E (2012) Dynamical similarity of geomagnetic field reversals. Nature 490:89–93

Vaquero JM, Gallego MC, Garcıa JA (2002a) A 250-year cycle in naked-eye observations of sunspot. Geophys Res Lett 29(20):1997

Vaquero JM, Sanchez-Bajo F, Gallego MC (2002b) On the reliability of the de la rue sunspot area measurements. Solar Phys 209:311–319

Vieira LEA, da Silva LA (2006) Geomagnetic modulation of cloud effect in the southern hemisphere magnetic anomaly through lower atmosphere cosmic ray effects. Geophys Res Lett 33:L14802. doi:10.1029/2006GL026389

Vonmoos M, Beer J, Muscheler R (2006) Large variations in Holocene solar activity: constraints from 10Be in the Greenland ice core project ice core. J Geophys Res 111:A10105. doi:10.1029/2005JA011500

Wagner G, Beer J, Laj C, Kissel C, Masarik J, Muscheler R, Synal HA (2000) Chlorine-36 evidence for the mono lake event in the summit GRIP ice core. Earth Planet Sci Lett 181:1–6

Ward PL (2009) Sulfur dioxide initiates global climate change in four ways. Thin Solid Films 517 (11):3188–3203

Yamazaki T, Oda H (2002) Orbital influence on earth's magnetic field: 100,000-year periodicity in inclination. Science 295:2435–2438

Yamazaki T, Oda H (2004) Intensity-inclination correlation on long-term secular variation of the geomagnetic field and its relevance to persistent non-dipole component. AGU monograph 145 "timescales of the internal geomagnetic field", pp 287–298

Abrupt Climate Changes During the Marine Isotope Stage 3 (MIS 3)

Eduardo Andrés Agosta and Rosa Hilda Compagnucci

Abstract The climate in the North Atlantic Ocean during the Marine Isotope Stage 3 (MIS 3)—roughly between 80,000 years before present (B.P.) and 20,000 years B.P., within the last glacial period—is characterized by great instability, with opposing climate transitions including at least six colder Heinrich (H) events and fourteen warmer Dansgaard–Oeschger (D-O) events. Periodic longer cooling cycles encompassing two D-O events and ending in a colder Heinrich episode occurred lasting about 10 to 15 ky each, known as the Bond cycle. Heinrich events occurred less frequently than D-O events. These were recurrent every 1.5 ky on average, while $\sim$10 ky elapsed between two H events. Neither of the two types of events is strictly periodical, however. After H events abrupt shifted to warmer climate, the D-O events followed immediately. During an H event, abnormally large amounts of rock debris transported by icebergs were deposited as layers at the bottom of the North Atlantic Ocean. The various theories on the causes include factors internal to the dynamics of ice sheets, and external factors such as changes in the solar flux and changes in the Atlantic Meridional Overturning Circulation (AMOC). The latter is the most robust hypothesis. At certain times, these ice sheets released large amounts of freshwater into the North Atlantic Ocean. Heinrich events are an extreme example of this, when the Laurentide ice sheet disgorged excessively large amounts of freshwater into the Labrador Sea in the form of icebergs. These freshwater dumps

E.A. Agosta (✉) · R.H. Compagnucci
Equipo Interdisciplinario para el Estudio de Procesos Atmosféricos en el Cambio Global [PEPACG], Facultad de Ciencias Físicomatemática e Ingeniería, Pontificia Universidad Católica Argentina [UCA], Buenos Aires, Argentina
e-mail: eduardo_agosta@uca.edu.ar

R.H. Compagnucci
e-mail: rosa_compagnucci@uca.edu.ar

E.A. Agosta
Facultad de Ciencias Astronómicas y Geofísica [FACG], Universidad Nacional de La Plata, La Plata, Argentina

E.A. Agosta
Consejo Nacional de Investigaciones Científicas y Técnicas [CONICET], Buenos Aires, Argentina

G.M. Gasparini et al. (eds.), *Marine Isotope Stage 3 in Southern South America, 60 ka B.P.–30 ka B.P.*, Springer Earth System Sciences,
DOI 10.1007/978-3-319-40000-6_5

reduced ocean salinity enough to slow down deep-water formation and AMOC. Since AMOC plays an important role in transporting heat northward, a slowdown would cause the North Atlantic Ocean to cool. Later, as the addition of freshwater decreased, ocean salinity and deep-water formation increased and climate conditions recovered. During the D-O events, the high-latitude warming occurred abruptly (probably in decades to centuries), reaching temperatures close to interglacial conditions. Even though H and D-O events seemed to have been initiated in the North Atlantic Ocean, they had a global footprint. Global climate anomalies were consistent with a slowdown of AMOC and reduced ocean heat transport into the northern high latitudes. The bipolar pattern with warming conditions in the Northern Hemisphere (NH) and cooling in the Southern Hemisphere (SH) is discussed from the information published by various authors who have used the limited data available for the SH, and palaeoclimatic simulations obtained by numerical modelling. Results show that the SH mid-latitude anomalies presented much smaller magnitude than those of the NH.

Keywords MIS 3 · Abrupt climatic change · Dansgaard–Oeschger events · Heinrich events · Ice drift in the North Atlantic Ocean · GISP2 oxygen isotope ($\delta^{18}O$) · Oceanic circulation · Atmospheric circulation

1 Introduction

Proxy data of the Quaternary climate of the Earth have provided abundant evidence for millennial-scale climate oscillations throughout the last glacial period, beyond the restricted range of historical and instrumental data. The suborbital oscillations are related to natural climate changes that were both abrupt and large in amplitude. Paleoclimate records with sufficient resolution are particularly confident from the northern North Atlantic Ocean and Greenland, and have been thoroughly analyzed. Amplitude and timing of regional abrupt climate change were first identified in the Greenland ice core records over timescale of a few years to a few decades at most (Alley et al. 1993; Taylor et al. 1993). These events of rapid warming of 10 ± 5 °C (Severinghaus and Brook 1999; Lang et al. 1999; Huber et al. 2006) in annual average temperature every ~1500 years are known as the Dansgaard–Oeschger (D-O) events (Dansgaard et al. 1993). Such sudden warming events in Greenland were likewise quite synchronized by sea surface temperature (SST) variations, recorded in marine sediment cores from the North Atlantic Ocean (site DSDP 609; Bond et al. 1993).

Each abrupt warming shared a qualitatively comparable temperature progression. It was preceded by a period of about 1000 years of relatively cold stable conditions (stadial climate), ended by a rapid shift of less than 10 years (Landais et al. 2004, 2006; Huber et al. 2006) toward much warmer conditions (interstadial climate) that persisted for ~200 to 400 years, then followed by a more gradual transition (50–200 years) back to the colder conditions (stadial climate) that preceded the warming event (Schwander et al. 1997; Lang et al. 1999; Peterson et al. 2013). This trapezoid-like sequence of climate shifts is known as a D-O cycle (Dokken et al. 2013), and can be deduced from the Greenland Ice Sheet Project (GISP) 2 of ice core oxygen isotope ($\delta^{18}O$) records shown in Fig. 1a. Note that this Greenland temperature variability was further confirmed through virtually complete ice core records provided by the North Greenland Ice Core Project (NGRIP 2004) and the previous Greenland Ice Core Project (GRIP). The $\delta^{18}O$ records show that each D-O cycle began with an abrupt jump in $\delta^{18}O$ ice, occurring in as little as a few years (Steffensen et al. 2008; Thomas et al. 2009), which was associated with a large warming of $\sim 12°$ (Wolff et al. 2010). Other properties of the ice, including electrical conductivity (Taylor et al. 1993), deuterium excess (Dansgaard et al. 1993; Steffensen et al. 2008), dust content (Fuhrer et al. 1999), and methane concentrations (Brook et al. 1996) changed in less than a decade. At the same time, accumulation rates roughly doubled and proportionally more precipitation fell in winter months (Alley et al. 1993; Cuffey and Clow 1997).

The D-O events in Greenland show a periodicity of 1470 years, though this is an average value between D-O #5 and #13 (Schulz 2002; Rahmstorf 2003), and the

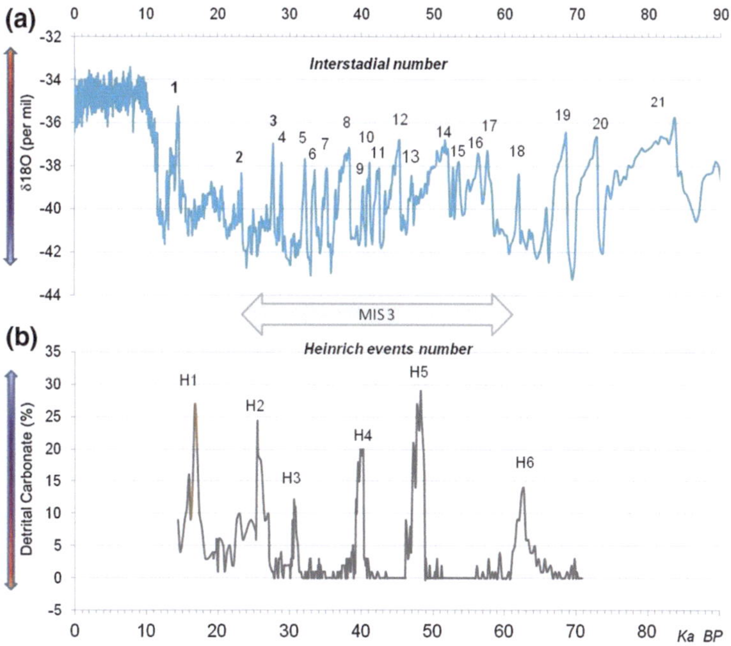

Fig. 1 Climatic variability during the MIS 3; **a** The $\delta^{18}O$ (per mil) record from the GISP2 ice core in Greenland, showing 20 of the 25 observed Dansgaard–Oeschger events (numbered). Data from Grootes et al. (1993) and Grootes and Stuiver (1997); **b** A record of ice-rafted material Detrital Carbonate (percent of 63–150 μm lithic fraction) from a deep-sea core in the North Atlantic Ocean at DSPD 609 site (49° 52.70 N, 24° 14.30 W; 3884 m below sea level) and Heinrich events (numbered) [data from Bond and Lotti (1995) and recalibration from Obrochta et al. (2012)]

recurrence interval can vary from event to event between 1.1 thousand years (ky) and 8 ky. The debate still persists over whether a 1470-year oscillation exists in D-O time series owing to diverse age models and statistical techniques used in the analysis (Ditlevsen et al. 2007; Petersen et al. 2013). Another important point with respect to the abruptness of climate change in Greenland is that the warming episodes in the ice core record tend to be more abrupt than the cooling events (Seager and Battisti 2007).

During the last ice period, 14 D-O events out of the 25 numbered events (D-O from #4 to #17) were observed during the Marine Isotope Stage 3 (MIS 3), between approximately 60 ka B.P. (thousand years before present) and 30 ka B.P. In turn, successive observational studies confirmed that many global climate proxies show D-O-like variability on similar timescales. Thus D-O events represent at least near hemispheric, if not global, scale climate shift (Voelker 2002; Ramhstorf 2003).

The six coldest periods of stadial climate associated to D-O cycles are named as Heinrich (H) events (Fig. 1b). They were marked by an enhanced discharge of icebergs into the North Atlantic Ocean, increasing the deposition of discrete layers of ice-rafted debris (IRD) that were found on the sediments of the ocean floor and originated primarily from the areas around Hudson Bay (Fig. 2; Heinrich 1988; Broecker 1994; Hemming 2004). The initial studies by Bond et al. (1992), (1993) were also able to establish the temporal link between the millennial-scale temperature oscillations in Greenland and slower time-varying Heinrich events, whose periodicity ranges between 7 and 10 ky. Further details on the origin, distribution, and timing of Heinrich event layers can be found in the review made by Hemming (2004).

An extra intriguing feature encompassing the relationship between D-O events and Heinrich events, shown also by the seminal paper of Bond et al. (1993), is the so-called

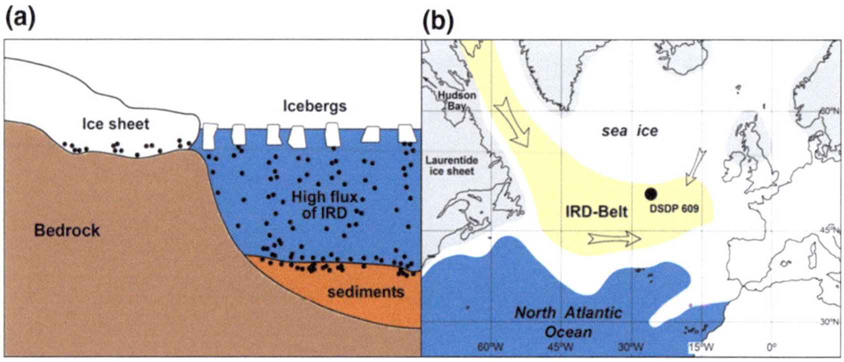

Fig. 2 Schematic of ice-rafted debris (IRD) deposition forming marine layers of detrital carbonate (sediment) from the Laurentide ice sheet along the Hudson Bay. **a** During Heinrich events, armadas of icebergs broke off from glaciers and traversed the North Atlantic Ocean. The icebergs contained rock mass eroded by the glaciers, and as they melted, this matter was dropped onto the sea floor as "ice-rafted debris" (IRD). **b** Iceberg, ice-rafted detritus (IRD) from a sediment core, the IRD belt in the North Atlantic Ocean, and the location of core DSPD 609 (Conkright 2002)

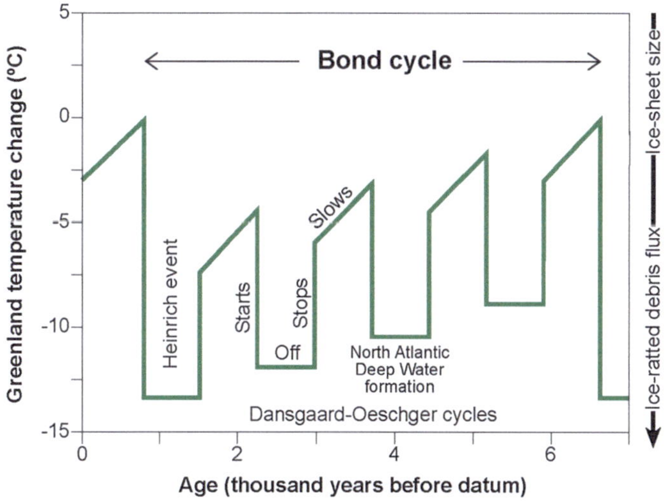

Fig. 3 An idealized bond cycle. Successive D-O oscillations become progressively cooler as the ice sheet grows in the Hudson Bay. Then, the base of the ice thaws and Heinrich events occur when the Laurentide ice sheet disgorged excessively large amounts of freshwater into the Labrador Sea in the form of icebergs

Bond cycle. In the period 80 ka B.P. and 20 ka B.P. (that is, roughly in the MIS 3 period), the bundling of (two or more) D-O cycles, a Heinrich event, and the successive abrupt shift to markedly warm interstadial clearly defines a longer term cooling cycle with a duration of roughly 10–15 ky (see Fig. 1). Such asymmetrical sawtooth shape-like cycle (Fig. 3) initiates with full D-O cycles (successive progression of colder stadial—sharp warming—warmer interstadial—gradual cooling—colder stadial) and culminates in a prolonged cold period (stadial) during which a Heinrich event occurs. Following the stadial, a rapid termination-like shift to a prominent warm interstadial marks the beginning of the next Bond cycle. The progression of the Bond cycles describes progressively cooler interstadials up to the current warmest interstadial of the last ice period, known as the Holocene (Bond et al. 1997).

2 Proposed Mechanisms for the MIS 3 Climatic Abrupt Changes

The most commonly inferred connection between these rapid climate changes is related to the release of cold freshwater by iceberg melting. The intrusion of this freshwater is assumed to slow or shut down the formation of the North Atlantic Deep Water (NADW), thus preventing the penetration of the North Atlantic Drift, the northern branch of the Gulf Stream, into high latitudes. The climatic significance of the Gulf Stream stems from the enormous quantity of heat it transports to northwestern Europe and its enablement to the exchange of moisture between the ocean and atmosphere. A slowdown of the NADW formation is related to D-O cycle whereas its shutdown, which corresponds to a substantial weakening of the Atlantic Meridional Overturning Circulation (AMOC), is related to Heinrich events (McManus et al. 2004; Zhang et al. 2015).

This leading and prosperous hypothesis was first proposed by Birchfield and Broecker (1990) as a mechanism to explain how salinity variations in the North Atlantic Ocean influence the strength of AMOC, causing D-O cycles during the last ice age. The notion is that the trade winds, blowing nearly east to west across the tropical Atlantic Ocean, transport moisture air masses across Central America into the Pacific Ocean. Consequently, a net transport of freshwater out of the tropical North Atlantic Ocean basin occurs making the surface Atlantic water the most saline in the world (Talley 2002). The Gulf Stream transports warm and saline waters from the tropical Atlantic Ocean toward northern subpolar regions, forming the surface branch of AMOC (Fig. 4). Therefore, this oceanic current, primarily driven by surface winds, transfers heat and moisture to the atmosphere through evaporation, warming the North Atlantic Ocean region (Rahmstorf 1996). This evaporation process, together with the turbulent mixing with deep colder waters, causes the surface waters to cool and sink because their density increases as temperature decreases and salinity increases (Colling 2001). The main sinking of sea water occurs to the south of Greenland and in the Norwegian Sea, forming NADW. Along the bottom of the Atlantic Ocean, the NADW flows southward carrying

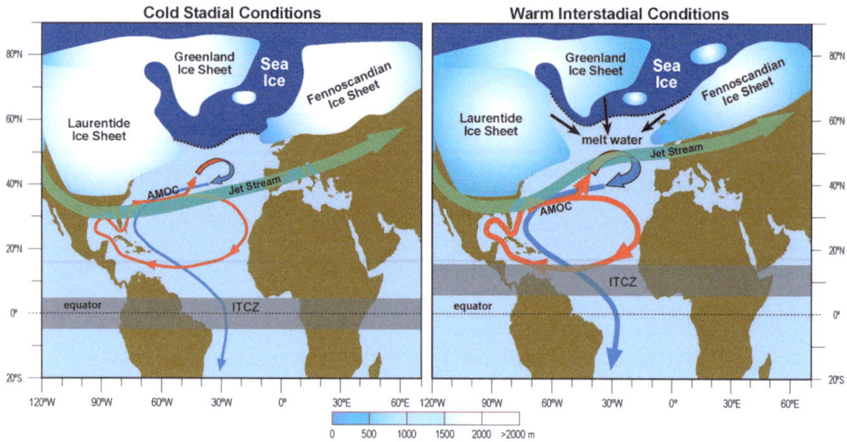

Fig. 4 Schematic of main mechanisms of the ocean, ice, and atmosphere acting to generate a D-O cycle. The presence of massive ice sheets around the margins of the North Atlantic Ocean during the last ice age provided a freshwater source (melt water; *black arrows*) that could rapidly alter surface salinity at sites of deep-water formation, thus perturbing the intensity of the Atlantic Meridional Overturning Circulation (AMOC; *red arrows*, superficial branch; *blue arrows*, deep branch), shifting the meridional position of the Intertropical Convergence Zone (ITCZ), and changing the meandering of the subtropical jet stream of maximum upper level winds (*green arrow*)

saline waters. Note that it is known as AMOC the net transport of northward flowing surface waters and southward flowing deep waters within the North Atlantic Ocean. Given that variations in temperature and salinity are the physical mechanisms involved in driving these Atlantic Ocean flows, the AMOC is also referred to as the oceanic thermohaline circulation, being ultimately integrated as part of the globally extended Great Ocean Conveyor Belt (Rahmstorf 1996, 2006; Stouffer et al. 2006).

If surface waters in the regions of deep-water formation become too fresh and less dense preventing sinking, then AMOC wakens or shuts down. The weakening of AMOC reduces the heat transport from tropical latitudes to higher latitudes and ice sheets are able to grow (Fig. 4a). During the colder stadial in the North Atlantic Ocean, both ice sheets melting and surface salinity increase diminish. Under these conditions the Atlantic Inter Tropical Convergence Zone (ITCZ) shifts southward reducing the amount of freshwater input to the tropical North Atlantic Ocean (Peterson and Huag 2006; Stouffer et al. 2006). Therefore, the tropical hydrologic cycle may act as a negative feedback response that increases North Atlantic Ocean salinity during periods of reduced AMOC (Krebs and Timmermann 2007). Furthermore, the presence of sea ice barriers at high latitudes produces the salt accumulation in the North Atlantic Ocean during periods of reduced NADW formation. Therefore, surface waters at key sites of deep-water formation become saline and dense enough again to sink. Both effects ultimately cause AMOC to strengthen, leading to an abrupt warming in the high-latitude North Atlantic Ocean, shifting the climate to an interstadial warm phase (Fig. 4b). Hence, a warm interstadial phase (D-O event) occurs when AMOS is more vigorous, leading to enhanced northward oceanic heat transport toward higher latitudes in the North Atlantic Ocean (Fig. 4b). Warmer conditions cause the ices sheets in the North Atlantic Ocean to melt, gradually reducing water salinity. Simultaneously, a more northerly position of the ITCZ during the interstadial phase enhanced freshwater precipitation in the tropical North Atlantic Ocean, and consequently reducing the surface salinity (Wang and Mysak 2006; Stouffer et al. 2006). Eventually, surface salinity is reduced enough to weaken AMOC, shifting the climatc again into a colder stadial phase of the D-O cycle (Fig. 4a).

Several modelling studies have suggested that it only takes a small reduction in sea surface salinity to alter the rate of NADW formation, and that large inputs of freshwater from melting ice sheets to the North Atlantic Ocean could even cause a complete collapse of AMOC (Clark et al. 2007; Stouffer et al. 2006).

The previous hypothesis of the "thermohaline oceanic circulation" is complemented with large-scale changes in upper tropospheric westerly winds over the North Atlantic Ocean, linked to D-O variability (Romanova et al. 2006). Most of the winter interannual variability in the present climate of the Northern Hemisphere is related to stationary waves caused by high mountain ranges and thermal contrast between land and sea (Held et al. 2002). These standing tropospheric waves modulate the meandering position of the upper level strong winds known as jet streams. A jet stream is responsible of driving baroclinic weather systems in the middle latitudes around the globe, acting as a pathway for the storms. Hence, the storm tracks change from a more zonal (west-to-east) to a more meridional (south-to-north) path according to the modulation induced by the jet stream wave. In turn, during the last ice age, large ice sheets covered much of North America and northern Europe, namely the Laurentide and Fennoscandian ice sheets, respectively (Romanova et al. 2006). Like mountain ranges, the ice sheets were large-scale obstacles that influenced stationary waves and changed the path of the jet streams. Therefore, the altered orography in the Northern Hemisphere and different oceanic heat transports resulted in a changed hydrological cycle, a reduction of the Hadley circulation, and a southward shift of the ITCZ in the boreal winter during glacial times (Romanova et al. 2006). In consequence, during the interstadial phase of the D-O cycle, the storm tracks over the North Atlantic Ocean were more meridian (Fig. 4a) and during the stadial phase they were more zonal (Fig. 4b). A productive point of this hypothesis is that wind fields are efficient generators of oceanic circulation change, capable of great volatility and very rapid global-scale teleconnections (Wunsch 2006).

The triggering mechanism for these coupled atmospheric and oceanic changes are still under discussion. Both data analysis and modelling suggested the possibility of stochastic resonance, leading to threshold transition of the climate system forced by a weak external forcing, such as the solar cycle, and random variability (Ramhstorf 2006; Braun et al. 2007; Braun and Kurths 2010; Saha 2015). Thus, the observed 1470-year cycle in the Greenland data could have originated from solar variability, despite the lack of a 1470-year spectral contribution in records of solar proxies, in combination with atmospheric–oceanic system variability. This could also arise from an internal non-periodic oscillation of the ocean–atmosphere system alone (Ditlevsen et al. 2007). In turn, the millennial frequency band is also consistent with dual 1000 and 2000-year forcing, similar to the variability inferred in ^{14}C and ^{10}Be estimates of solar variability (Obrochta et al. 2012).

Additional mechanisms have been proposed to explain the Heinrich anomalous ice discharge events, since what causes the meltwater pulses is still under debate. The most classical theory considers these to be internal instabilities of the Laurentide ice sheet associated with alterations of basal conditions (MacAyeal 1993; Calov et al. 2002; Hulbe et al. 2004). A sudden break-up of ice shelves was also concerned through atmospheric warming (Hulbe et al. 2004) or tidal effects (Arbic et al. 2004). Thus oscillations of the Laurentide ice sheet were assumed to lead to important disruptions of the AMOC and NADW formation, preventing the interhemispheric exchange of heat, and the concomitant northern latitude cooling. However, most recent paleoclimate data revealed that most of these events probably occurred after AMOC had already slowed down or/and NADW largely collapsed, within about one thousand years (Hall et al. 2006; Hemming 2004; Jonkers et al. 2010; Roche et al. 2004). This implies that the initial AMOC reduction could not have been caused by the Heinrich events themselves. Hence, the effects of oceanic circulation on ice sheets dynamics have been recently proposed as triggering mechanism.

In summary, though the mechanisms responsible for the D-O cycles are not fully understood, they are widely thought to involve abrupt changes in AMOC owing to freshwater perturbations. Both Heinrich events and D-O variability during MIS 3 are thought to be caused by Northern Hemisphere ice sheet calving, freshwater discharges, and high-latitude storm tracks variations, which subsequently influenced the strength of the AMOC, poleward heat transport, and eventually global climate (Dokken et al. 2013; Petersen et al. 2013). A complete understanding of the mechanisms of abrupt change requires inclusion of processes at both lower and higher latitudes, as well as the potential for feedback between them (Clement and Peterson 2008). In the leading studies of the MIS 3 climate variability, these exploratory hypotheses are still subject to test using coupled atmospheric–oceanic General Circulation Models (GCM) in order to combine theory and paleoclimate proxies.

3 Abrupt Climatic Changes Beyond the North Atlantic Ocean-Northern Europe Borders

There is plenty of evidence for global imprints of abruptness in climate beyond Greenland and the North Atlantic Ocean sector. Warming in Greenland and the North Atlantic Ocean is coincident with warmer and wetter conditions in Europe (Genty et al. 2003; Moseley et al. 2014), an enhanced summer monsoon in the northwest Indian Ocean (Schulz et al. 1998; Pausata et al. 2011), a northward shift of precipitation belts in the Cariaco Basin (Peterson et al. 2000), drier and warmer conditions in the southwestern United States (Wagner et al. 2001; Oster et al. 2014), changes in ocean ventilation off the shore of Santa Barbara, California (Hendy et al. 2002). Moreover, whereas the effect of warming expanded mostly across the Northern Hemisphere high latitudes, a near-simultaneous cooling occurred in Antarctica resulting in a North–South seesaw pattern (EPICA Members 2006; Wolff et al. 2010). Though quantitatively less explored, antiphase changes in precipitation and hydrology have also been observed in tropical and subtropical latitudes of the Northern and Southern Hemispheres. These changes have already been linked to North Atlantic Ocean colder stadials through a southward displacement of the ITCZ (Baker et al. 2001; Huang et al. 2000; Wang et al. 2007; Muller et al. 2006; Clement and Peterson 2008; Margalef et al. 2015). Far away, the ITCZ changes over the Western Tropical Atlantic Ocean (Cariaco Basin record, Peterson et al. 2000, vs. Northeastern Brazil speleothem record, Wang et al. 2004), the dust record from the tropical East Atlantic Ocean (Jullien et al. 2007) and monsoon variations over China (e.g., the Hulu cave speleothem record of Wang et al. 2001), India (Leuschner and Sirocko 2000), and Africa (Mulitza et al. 2008; Niedermeyer et al. 2009) have been correlated to the abrupt events recorded in Greenland and/or the North Atlantic Ocean. Since the evidence of abrupt, millennial timescale changes appears in other regions throughout the globe and in quantities of the atmosphere, oceans, and ice sheets, it is perhaps more appropriate to think about these processes as being components of a global, coupled feedback (Clement and Peterson 2008), making the Earth climate during MIS 3 the most variable in the last glacial period.

3.1 Bipolar Seesaw Between Northern and Southern Hemispheres

The "bipolar seesaw" is an expression to denominate the asynchronous relationship between millennial-scale temperature changes over Greenland and Antarctica during the last glacial period which acts to redistribute heat depending on the state of AMOC within the entire basin of the Atlantic Ocean (Broecker 1998; Barker et al. 2009). The D-O cycle observed in the Greenland ice core records have counterparts in the Antarctic records, though their shape is different and phase is shifted (Stocker and Johnsen 2003). Thus, large-scale features of Antarctic $\delta^{18}O$ variations are consistent with an overall warming during the Greenland colder phases, ending when Greenland is abruptly warming (Fig. 5). This overall seesaw pattern is associated with changes in the Atlantic Ocean circulation modulated by the inertia of the Southern Ocean (Landais et al. 2015). The simplest thermodynamical model to connect this interhemispheric variability at millennial scale was proposed by Stocker and Jonhsen (2003). The bipolar seesaw pattern was explained by changes in the heat and freshwater flux connected to AMOC throughout the entire Atlantic Ocean basin, where a stronger AMOC leads to increased drainage of heat from the Southern Ocean heat reservoir. Thus, North Atlantic Ocean temperature changes are associated with those of the same magnitude but of opposite sign in the South. The heat of the Southern Ocean, by which Antarctic temperatures are influenced, then drifts

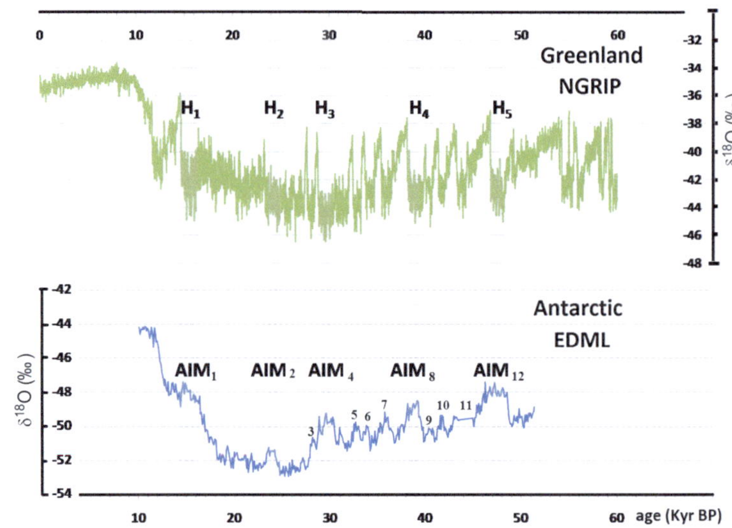

Fig. 5 Greenland (NGRIP, *green*, data from EPICA Community Members (2006)) and Antarctic (EDML, *blue*, data from EPICA Community Members (2004) and Buiron et al. (2011)) climate records. Both *curves* are $\delta^{18}O$ values representing temperatures. Stacks of Greenland and Antarctic water isotopic records nicely illustrate a seesaw pattern with the abrupt warming in Greenland being concomitant with the beginning of the cooling in Antarctica at the Antarctic Isotopic Maximum (AIM; Landais et al. 2015). Data retrieved from http://www.iceandclimate.nbi.ku.dk/

toward the South Atlantic Ocean. The response of the Southern Ocean is slow compared to the Atlantic temperature seesaw, reflecting the relatively strong departure of the Southern Ocean and Antarctic part of the climate system from the remaining global circulation system. This simple seesaw model can, to a broader extent, account for both the phase change and the shape of the events.

In principle, an interhemispheric climate coupling by the bipolar seesaw should also apply for all the shorter D-O events. However, to what extent this concept is also able to explain the higher frequency climate variability in Antarctic ice cores remains uncertain. The Antarctic temperature variability from high resolution data also depicts submillennial features occurring without a Greenland counterpart. This suggests possible teleconnections between changes in low latitude tropospheric circulation and Antarctic temperature without any Greenland temperature fingerprint (Ding et al. 2011). The bipolar seesaw concept therefore does not correctly reflect the full complexity of processes at play (Landais et al. 2015).

3.2 Stadial and Interstadial Imprints from Global Records

Based on the database of the global distribution of centennial-scale records for MIS 3 collected by Voelker et al. (2002), Fig. 6 shows the general climate conditions during stadials and interstadials. In general, climate conditions in the Northern Hemisphere were cooler and drier during stadials (Fig. 6a), warmer and wetter during interstadials (Fig. 6b), especially over the North Atlantic Ocean sector. As expected from the coupling between oceanic and tropospheric circulation, climate conditions on land agree well with those reconstructed in the neighbouring ocean basin. According to the bipolar seesaw model, the Southern Hemisphere should be warmer and wetter during stadials and cooler and drier during interstadials. However, some exceptions appear related to temperature in the Southern Ocean and eastern Antarctica, and to precipitation in the South Pacific Ocean. Thus, the climate pattern of proxy data reveals a lack of zonal uniformity in the Southern Hemisphere. While the South Atlantic Ocean shows signals opposite to the North Atlantic, as expected by the bipolar seesaw model, the South Pacific Ocean and southern South America are unexpectedly consistent with the North Atlantic Ocean signals. The climate pattern, thus, tends to be more zonally uniform in the Northern Hemisphere whereas, in contrast, more zonally asymmetric in the Southern Hemisphere.

In spite of the fact that different proxies are likely to record even the same event differently (Hemming 2004) and that the ability to date records precisely is limited (Clements and Petersen 2008), it is possible to draw a big picture of the mean climate conditions in each D-O phase in order to compare GCM results and assess their discrepancies.

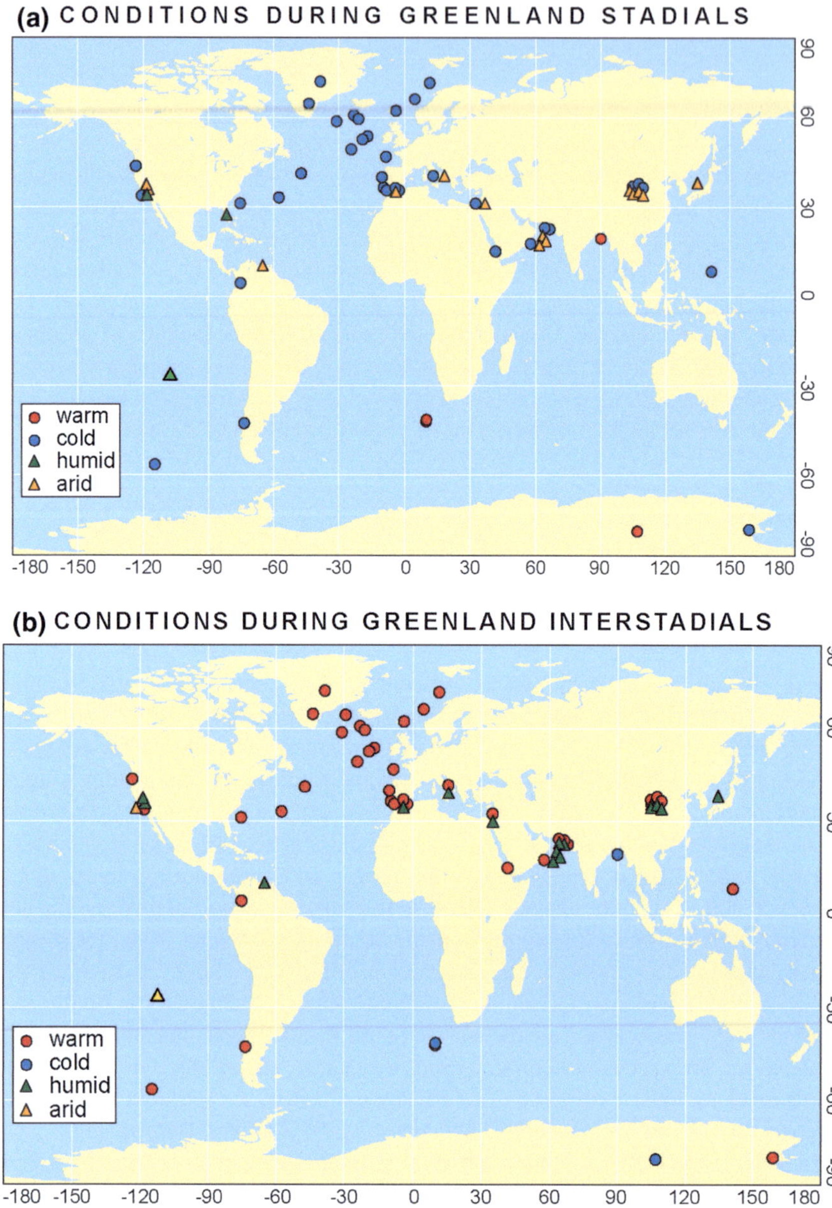

Fig. 6 Spatial distribution of sites of paleclimatic proxies with centennial-scale resolution able to detect **a** stadial and **b** interstadial phases of the D-O cycle during MIS 3, according to the database collected by Voelker et el. (2002). Data is retrieved from Margalef et al. (2015) for Easter Island (Chile, 27° 07′ S, 109° 22′ W). *Red dot* represents warmer paleoclimate conditions; *blue dot*, cold conditions; *green triangle*, humid conditions; and *orange triangle*, arid conditions

4 Scenarios of Climatic Variability During D-O and Heinrich Events

Several simulations were performed using coupled atmospheric–oceanic General Circulation Models (AOGCMs) to identify the mechanisms responsible for the climate behaviour observed during millennial-scale abrupt changes. Heinrich or colder D-O cycle phase events are represented by prescribing an abrupt freshwater release in the North Atlantic Ocean region, indicating a slowing down of AMOC (Stouffer et al. 2006). In general, when the freshwater forcing is stopped, the AMOC recovers fully after 120–260 years. These first modelling studies of this nature were performed in the present or pre-industrial contexts (modern boundary conditions) in order to highlight the climate signature of different climate states (e.g., Manabe and Stouffer 1988). However, the initial state and boundary conditions may have an important effect on the response and recovery to freshwater forcing together with the spatial distribution and intensity of the concomitant climatic anomalies. Glacial boundary conditions lead to a slower recovery of AMOC due to a more extensive sea ice and a more stable density below the surface layer in the North Atlantic Ocean (Bitz et al. 2007). In addition, with the Bering Strait closed during glacial conditions, the export of North Atlantic Ocean meltwater takes longer, having only an outlet through the southern end of the North Atlantic Ocean, and thus bringing about a slower recovery of AMOC (Hu et al. 2008).

Most recent experiments were performed under Last Glacial Maximum (Kageyama et al. 2010; Otto-Bliesner and Brady 2010) or even Marine Isotopic Stage 3 (Merkel et al. 2010; Buiron et al. 2012) boundary conditions. All these models have confirmed the robustness of the bipolar seesaw signature of the climate response to AMOC weakening. Common climatic responses in these simulations include North Atlantic Ocean cooling and a tendency for a southward shift of the Atlantic ITCZ (Dahl et al. 2005; Broccoli et al. 2006; Krebs and Timmermann 2007; Chiang et al. 2008; Swingedouw et al. 2009). Some models also simulate a northward shift of the Nordic Seas convection sites (locations where NADW form), which coincides with the warming of the north of the hosing area (Stouffer et al. 2006; Kageyama et al. 2013).

An example is the Otto-Bliesner and Brady (2010) scenario (Fig. 7) that illustrates the conventional bipolar seesaw pattern with uniform warming in the Southern Hemisphere and generalized cooling in the Northern Hemisphere, obtained by reproducing a Heinrich simulation under the Last Glacial Maximum boundary conditions. Among all the experiments performed with different amount of freshwater forcing only, the one shown in Fig. 7 was run long enough to allow AMOC to recover to the pre-hosing state of 15 Sv. The recovery time was 500 years. Therefore, according to this type of simulation, while there is a strong cooling (from $\sim 8°$ to 15 °C) in the North Atlantic Ocean, in the Argentine Patagonia there is a warming of about 1°–4 °C, being cooler further southward. In general in this kind of simulations, the symmetric adjustments in the northern extratropics are a consequence of advection of the North Atlantic Ocean cooling by the mean westerlies, which propagate the cooling into the tropics through the

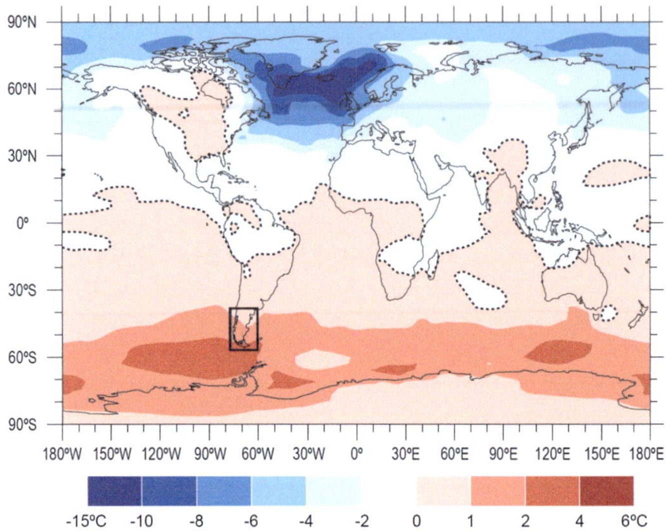

Fig. 7 Annual surface air temperature anomalies (°C) averaged over the last 20 years of freshwater forcing for the experiment NATL_1ex: freshwater in the North Atlantic Ocean, with an amount of 0.1 Sv during 500 years of hosing and 4.5 m equivalent sea-level change. Adapted from Otto-Bliesner and Brady (2010)

wind-evaporation-SST feedback in all basins, in concert with a southward shift of ITZC in all basins as well (Chiang et al. 2008).

However, there are other simulations showing regional differences in the thermal response of the Southern Oceans (Clement and Peterson 2008; Timmermann et al. 2010). Likewise, some models simulate contrasted patterns with a cooling in the West Pacific Ocean and a warming in the South Atlantic and Indian oceans (Kageyama et al. 2010, 2013). In some of these simulations the major changes in the North Atlantic Ocean, that propagate into the tropical Atlantic Ocean by atmospheric and ocean processes, then would trigger contrasted responses in the other tropical basins by modifications of the Walker cells or by fast oceanic teleconnections related to wave propagation due to the sea-level height (Clement and Petersen 2008; Kageyama et al. 2010, 2013). More recently, the zonal asymmetries are partly attributed to tropospheric teleconnections induced by the sea surface temperature changes in the tropical oceans.

To study the background state dependence to freshwater perturbations, Menviel et al. (2008) conducted water hosing experiments under pre-industrial conditions (FC) and Last Glacial Maximum (LMG) conditions (FL). To mimic a typical Heinrich event, anomalous freshwater was injected into the northern North Atlantic Ocean (55° W–10° W, 50° N–65° N). Noticeable differences in the wind responses were obtained for the Southern Hemisphere around southern South America (Fig. 8). In the scenario under pre-industrial (PIN) conditions the zonal westerlies were perturbed by a circulation dipole located in high latitude of an anticyclone in the South Pacific Ocean, and a cyclone in the South Atlantic Ocean. The consequent

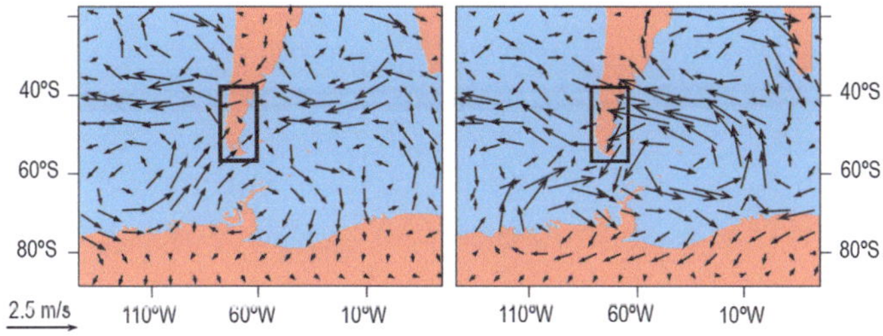

Fig. 8 Annual mean wind velocity anomalies (m s^{-1}) at 800 mbar (*left*) FC-PIN and (*right*) FL-LGM, (where PIN is pre-industrial conditions, FC the state after forcing the PIN by anomalous freshwater and LGM is the Last Glacial Maximum conditions and FL is the state after forcing LGM with freshwater). Adapted from Menviel et al. (2008)

anomaly of wind is from the southwest over Patagonia. On the other hand, under the LGM conditions the westerlies weaken and the wind anomalies are strong from the east, suggesting the possibility of easterly wind occurrence. These results clearly depict the strong dependence of the resulting anomalies to the background state.

Under a more realistic scenario, Buiron et al. (2012) performed simulations using LGM climate boundary conditions (PMIP2 protocol, Braconnot et al. 2007). The ICE-5G ice sheet reconstruction (Peltier 2004), and greenhouse gases atmospheric mixing ratios of 185 ppm, 350 ppb, and 200 ppb for CO_2, CH_4, and N_2O, respectively (Flückiger et al. 1999; Dällenbach et al. 2000; Monnin et al. 2001), were prescribed, as well as 21 ky orbital parameters (Berger 1978). The vegetation cover remained similar to modern conditions over ice-free areas but river pathways were adapted for LGM conditions (Alkama et al. 2007), with a stable AMOC at around 15 Sv, as reference. To simulate a rapid cooling event in the North Atlantic Ocean through a collapse of AMOC, they imposed an additional 0.1 Sv freshwater flux in the North Atlantic Ocean and the Arctic (Kageyama et al. 2010). Figure 9 shows the mean annual surface air temperature (SAT) anomaly between the LGM reference and the LGM after 100 years of the freshwater flux input. Zonal asymmetries appear in both the Northern and Southern Hemispheres, reproducing a global pattern of cooling/warming that is consistent with the paleoclimate proxies described in Fig. 6a. While the seesaw of cooling/warming is present between the North and South Atlantic Ocean, there is a cooling in the eastern South Pacific Ocean in concert with southern South America that includes the Patagonian sector. Buiron et al. (2012) suggested that the zonal hemispherical asymmetry of the bipolar mechanism can be locally offset by faster tropospheric teleconnections originating from the tropics, even though the precise location of this fast response was uncertain.

Even keeping similar boundary and forcing conditions, the simulations obtained with different AOGCMs differ from each other. In order to analyze the stability of the bipolar response in SAT between the North and South Atlantic oceans, Timmermann et al. (2010) used four different models (CCSM2, ECHAM5-OM1, GFDL CM2.1, and HadCM3). As expected, they showed that a shutdown of AMOC led to an overall cooling of the Northern Hemisphere, a warming of the

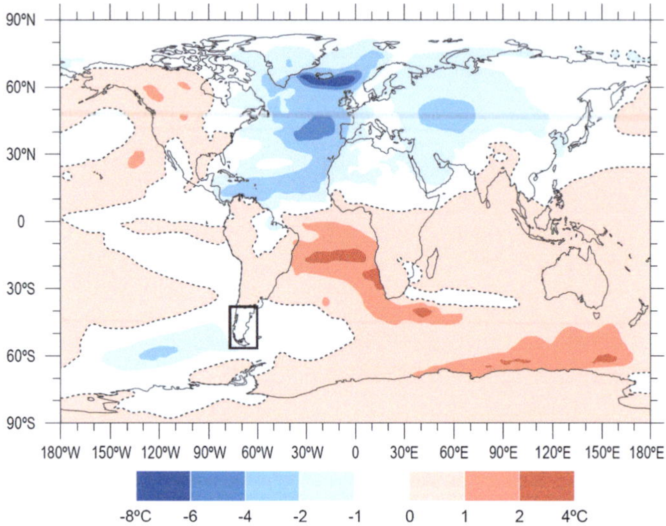

Fig. 9 Mean annual SAT anomaly (in °C) between LGM reference and LGM forced by freshwater. Adapted from Buiron et al. (2012)

southeastern Atlantic Ocean, an intensification of the north-easterly trade winds in the tropical North Atlantic and north-eastern tropical Pacific oceans, and a southward shift of the Atlantic ITCZ. However, the main differences among the model simulations appeared in the Southern Hemisphere, between 150°W and 0°E (Fig. 10). The main simulation difference can be observed in GFDLCM2.1 in comparison with CCSM2 and GFDL CM2.1. These latter two simulations showed negative SST anomalies in the South Pacific Ocean while the HadCM3 simulation showed positive anomalies. These differences were reflected also in the low-level wind anomalies. The westerlies seemed to be strengthened in all the model simulations but that of ECHAM5-OM1.

Timmermann et al. (2010) attributed such dipole-like surface air temperature to an intensification of the negative phase of the low-frequency tropospheric variability mode, which connects the South Pacific Ocean and South America (PSA) in the present climate (Mo 2000). Negative SST anomalies in the western equatorial Pacific Ocean led to a reduction of convection, tropospheric cooling, and the generation of an austral tropospheric atmospheric Rossby wave response that encompasses the Southern Hemisphere. Simulated anomalies in the meridional heat advection produced an anomalous cyclonic circulation in the Amundsen–Bellingshausen Sea. Therefore, such anomalous behaviour of the South Pacific Ocean could have been due to SST anomalies from the Tropical Pacific Ocean. Ding et al. (2011) later explained a possible link between the central equatorial Pacific Ocean and the Pacific sector of East Antarctica climate through atmospheric Rossby waves propagating from the tropics to Antarctica. Atmospheric teleconnections have therefore the

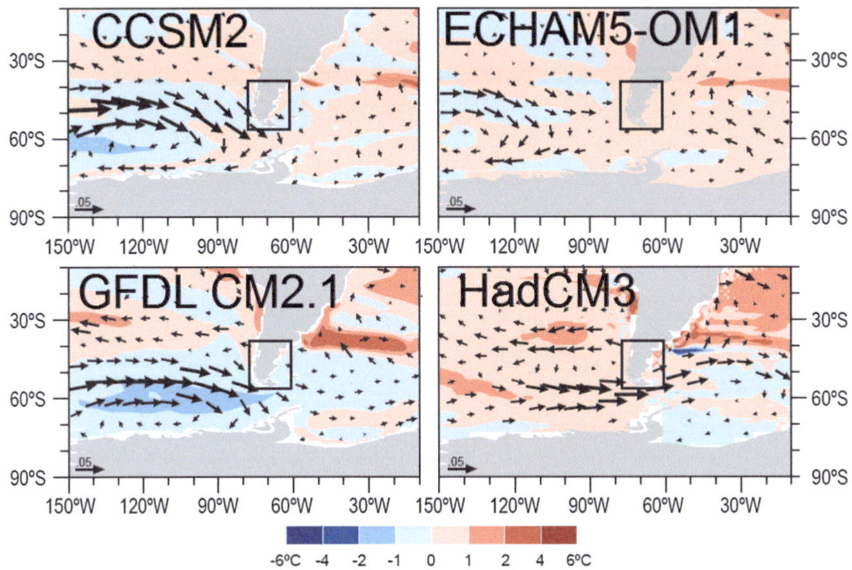

Fig. 10 Ocean–atmosphere anomalies averaged over the austral winter season (June to August) simulated by four different CGCMs: CCSM2, ECHAM5-OM1, GFDL CM2.1, and HadCM3, in the 1-Sv freshwater perturbation experiments under present-day or pre-industrial conditions. SST (shading; °C) and surface wind stress (vectors; N/m^2). Adapted from Timmerman et al. (2010)]

potential to cause zonal asymmetries in Patagonia and the circum-Antarctic response to AMOC changes (Kageyama et al. 2013).

In order to simulate the millennial-scale variability associated with a D-O cycle, Menviel et al. (2014a, b) set out an intermediate complexity global climate model. The model was forced by time-varying external boundary conditions (greenhouse gases, orbital forcing, and ice sheet orography and albedo) and anomalous North Atlantic Ocean freshwater fluxes. Figure 11 shows their first EOF pattern of air temperature field obtained from this model, which assumed that the continuum of MIS 3 climate variability on centennial to millennial timescales can be generated by a suitable North Atlantic Ocean freshwater forcing and the associated AMOC response. The polarity of the pattern corresponds to a D-O event or an interstadial phase of the D-O cycle. Thus, the interstadial conditions are characterized by a Northern Hemisphere warming with strongest amplitudes over the Greenland-Iceland-Norway Sea and the Arctic Ocean. The warming extends into Northern Africa, Asia, and the western North Pacific. Simulated Southern Hemisphere cooling during interstadials is consistent with the presence of a bipolar temperature seesaw strong. The absence of zonal hemispheric asymmetries could be an artefact due to retaining only the first principal component of the EOF analysis. In turn, the first EOF pattern of precipitation field is shown in Fig. 12. It shows precipitation increases in the region of positive temperature anomalies. The greater values are observed in the northern tropics of Africa and South America. In other areas of

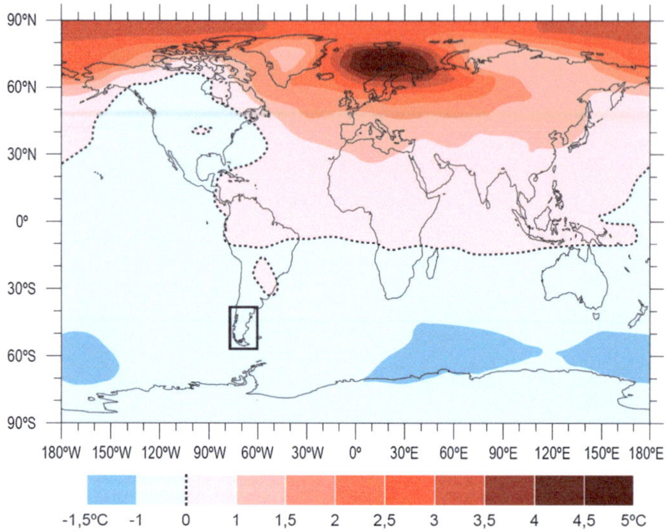

Fig. 11 Pattern of detrended (linear trend filtered out) 2 m air temperature anomalies (°C) shown by the first EOF, which explains 64 % of the variance. Adapted from Menviel et al. (2014a, b)

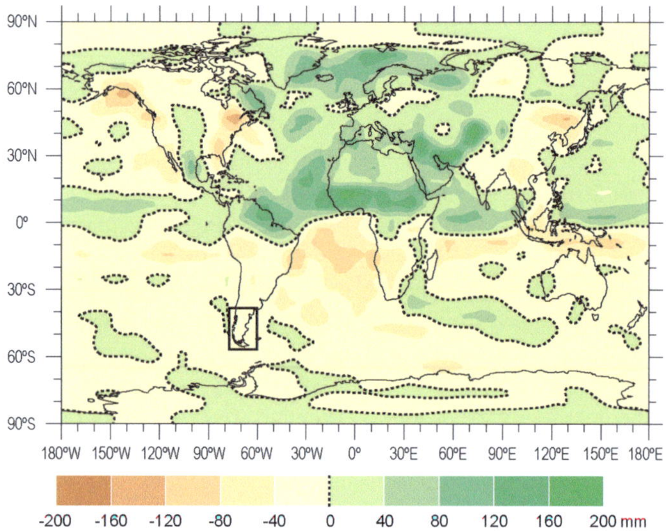

Fig. 12 Pattern of detrended precipitation anomalies (mm year^{-1}) shown by the 1st EOF, which explains 16 % of the variance. Adapted from Menviel et al. (2014a, b)

South America, negative precipitation anomalies are observed, especially in the Patagonian region, in concert with negative temperature anomalies there. This global temperature/global patterns are consistent with a direct thermodynamical relationship, i.e., with the fact to have lower (higher) temperature over the oceans in the Southern (Northern) Hemisphere oceans, involving less (more) precipitation there. The inverse pattern in both temperature (Fig. 11) and precipitation (Fig. 12) must be expected for the colder stadial D-O cycle phase.

5 Concluding Remarks

In recent years, progress in the paleoclimate modelling and proxy reconstruction has been quantitatively improved, nonetheless, the nature of abrupt climate changes related to millennial-scale oscillations during the MIS 3 still remains elusive. The effects of abruptness of these climate shifts are undoubtedly seen in vast regions of the Northern Hemisphere middle and higher latitudes and in the tropics, as exposed by abundance of paleoclimate records. Evidence shows that the bipolar seesaw pattern can be well documented from climatic events in Antarctic and sub-Arctic areas, being almost coeval. By contrast, the signature of abruptness is poorly recorded, or could be potentially absent, in middle and higher latitudes of the Southern Hemisphere, since only the territories of New Zealand and Patagonia extend farther south of 40° S. Questions arise whether the seesaw effect diminishes with distance from the poles (Williams et al. 2015). Marine paleoclimate proxies of the Southern Hemisphere have no temporal resolution good enough to properly mirror rapid climatic fluctuations. The sophisticated climate models are of great aid for understanding the extent and intensity of the past climate anomalies. Most, if not all, models reproduce the seesaw mechanism under a freshwater perturbation forcing in the North Atlantic Ocean. Even without prescribed freshwater hosing, a model simulation of gradual changes in the height of the Northern Hemisphere ice sheets has recently shown how these can alter the coupled atmosphere–ocean system and cause rapid glacial climate shifts closely resembling D-O events with the concomitant bipolar pattern (Zhang et al. 2015). However, as shown in the precedent sections, model results are highly dependent on the background state and the boundary conditions of the experiment. Different models and experiments can lead to very diverse interpretations of the past climate conditions. The major differences in the simulated anomalies among the experiments are found to be in the Southern Hemisphere, especially over southern South America and the adjacent oceans. There, the simulated temperature anomalies could be either of warming or cooling during the colder stadial D-O cycle phase or Heinrich events; somewhat a similar disparity is found for the wind anomalies. Under similar LGM conditions some models showed that westerlies are markedly weakened in Patagonia while other models showed the opposite. The inconsistent signals raise the question whether westerlies were weakened or localized easterlies more frequently in the region under glacial conditions. These issues can only be undertaken by a substantial improvement in the number of proxies that still are scarce in southern South America to evaluate the uncertainties of simulated past climate conditions.

A major understanding of the physical mechanisms involved in the MIS3 abrupt climate transitions, as well as the spatial reach of temperature and precipitation anomalies, assists to set out a better assessment of the current context of global climate change. The period encompassing the late past century and early present century is characterized by a negative trend in sea ice extent in the Northern Hemisphere and, by contrast, a positive trend in the circum-Antarctic areas (Parkinson and Cavalieri 2012; also at http://nsidc.org/data/seaice_index/). The latter is simultaneous with an apparently divergent long-term warming of the Antarctic Peninsula (Mulvaney et al. 2012; also at http://earthobservatory.nasa.gov/IOTD/view.php?id=6502). Some experts would argue that the bipolar-like temperature pattern, currently observed globally, could be similar to the one associated with an interstadial D-O cycle phase, suggesting an imminent collapse or weakening of AMOC (Vellinga and Wood 2008). Recent observations found that until now AMOC seems not to be reducing its intensity, and its recent variations to a less vigorous extent are part of the oscillatory nature (Rahmstorf 2006; Rossby et al. 2014; Parker and Ollier 2016). Other studies claim the opposite (McCarthy et al. 2012; Andres et al. 2013; Yin and Goddard 2013; Ezer et al. 2013). At this time the debate has just started.

Acknowledgments We thank the National Agency of Science and Technique Promotion of Argentina (ANPCyT) for supporting the project PICT-2013-0043. Many thanks to the Carmelite Order.

References

Alkama R, Kageyama M, Ramstein G, Marti O, Ribstein P, Swingedouw D (2007) Impact of a realistic river routing in coupled ocean atmosphere simulations of the Last Glacial Maximum climate. Clim Dyn 30:855–869

Alley RB, Meese DA, Shuman CA, Gow AJ, Taylor KC, Grootes PM, Zielinski GA (1993) Abrupt increase in Greenland snow accumulation at the end of the Younger Dryas event. Nat London 362:527–527

Andres M, Gawarkiewicz GG, Toole JM (2013) Interannual sea level variability in the western North Atlantic: regional forcing and remote response. Geophys Res Lett 40:5915–5919. doi:10.1002/2013GL058013

Arbic BK, MacAyeal DR, Mitrovica JX, Milne GA (2004) Paleoclimate: Ocean tides and Heinrich events. Nature 432:460. doi:10.1038/432460a

Baker PA, Rigsby CA, Seltzer GO, Fritz SC, Lowenstein TK, Bacher NP, Veliz C (2001) Tropical climate changes at millennial and orbital time scales on the Bolivian Altiplano. Nature 409:698–701

Barker S, Diz P, Vautravers MJ, Pike J, Knorr G, Hall IR, Broecker WS (2009) Interhemispheric Atlantic seesaw response during the last deglaciation. Nature 457(7233):1097–1102

Berger A (1978) Long term variations of daily insolation and quaternary climatic changes. J Atmos Sci 35(2):2362–2367

Birchfield GE, Broecker WS (1990) A salt oscillator in the glacial Atlantic? 2. A "scale analysis" model. Paleoceanography 5(6):835–843

Bitz CM, Chiang JCH, Cheng W, Barsugli JJ (2007) Rates of thermohaline recovery from freshwater pulses in modern, Last Glacial Maximum, and greenhouse warming climates. Geophys Res Lett 34(7). doi:10.1029/2006GL029237
Bond G, Lotti R (1995) Iceberg discharges into the North Atlantic on millennial time scales during the last glaciation. Science 267:1005–1010
Bond G, Heinrich H, Broecker W, Labeyrie L, McManus J, Andrews J, Huon S, Jantschik R, Clasen S, Simet C, Tedesco K, Klas M, Bonani G, Ivy S (1992) Evidence for massive discharges of icebergs into the North Atlantic Ocean during the last glacial period. Nature 360:245–249
Bond G, Broecker W, Johnsen S, McManus J, Labeyrie L, Jouzel J, Bonani G (1993) Correlations between climate records from North Atlantic sediments and Greenland ice. Nature 365:143–147
Bond GC, Showers W, Cheseby M, Lotti R, Almasi P, deMenocal P, Priore P, Cullen H, Hajdas I, Bonani G (1997) A pervasive millennial-scale cycle in North Atlantic Holocene and glacial climates. Science 278(5341):1257–1266
Braconnot P, Otto-Bliesner B, Harrison S, Joussaume S, Peterchmitt J-Y, Abe- Ouchi A, Crucifix M, Driesschaert E, Fichefet T, Hewitt CD, Kageyama M, Kitoh A, Laîné A, Loutre M-F, Marti O, Merkel U, Ramstein G, Valdes P, Weber SL, Yu Y, Zhao Y (2007) Results of PMIP2 coupled simulations of the mid-Holocene and Last Glacial Maximum e part 1: experiments and large-scale features. Clim Past 3:261–277
Braun H, Kurths J (2010) Were Dansgaard-Oeschger events forced by the Sun? Eur Phys J-Spec Top 191(1):117–129
Braun H, Ganopolski A, Christl M, Chialvo DR (2007) A simple conceptual model of abrupt glacial climate events. Nonlin Processes Geophys 14:709–721
Broccoli AJ, Dahl KA, Stouffer RJ (2006) Response of the ITCZ to northern hemisphere cooling. Geophys Res Lett 33:L01702. doi:10.1029/2005GL024546
Broecker WS (1994) Massive iceberg discharges as triggers for global climate change. Nature 372:421–424
Broecker WS (1998) Paleocean circulation during the last deglaciation: a bipolar seesaw? Paleoceanography 13:119–121
Brook EJ, Sowers T, Orchardo J (1996) Rapid variations in atmospheric methane concentration during the past 110,000 years. Science 273(5278):1087–1091
Buiron D, Chappellaz J, Stenni B, Frezzotti M, Baumgartner M, Capron E, Landais A, Lemieux-Dudon B, Masson-Delmotte V, Montagnat M, Parrenin F, Schilt A (2011) TALDICE-1 age scale of the Talos Dome deep ice core, East Antarctica. Clim Past 7:1–16. doi:10.5194/cp-7-1-2011
Buiron D, Stenni B, Chappellaz J, Landais A, Baumgartner M, Bonazza M, Capron E, Frezzotti M, Kageyama M, Lemieux-Dudon B, Masson-Delmotte V, Parrenin F, Schilt A, Selmo E, Severi M, Swingedouw D, Udisti R (2012) Regional imprints of millennial variability during the MIS 3 period around Antarctica. Quat Sci Rev 48:99–112
Calov R, Ganopolski A, Petoukhov V, Claussen M, Greve R (2002) Large-scale instabilities of the Laurentide ice sheet simulated in a fully coupled climate-system model. Geophys Res Lett 29(24)
Chiang JCH, Cheng W, Bitz CM (2008) Fast teleconnections to the tropical Atlantic sector from Atlantic thermohaline adjustment. Geophys Res Lett 35:L07704. doi:10.1029/2008GL033292
Clark PU, Hostetler SW, Pisias NG, Schmittner A, Meissner KJ (2007) Mechanisms for an ~7-Kyr climate and sea-level oscillation during marine isotope stage 3. Ocean circulation: mechanisms and impacts-past and future changes of meridional overturning, pp 209–246
Clement AC, Peterson LC (2008) Mechanisms of abrupt climate change of the last glacial period. Rev Geophys 46. doi:10.1029/2006RG000204
Colling A (2001) Ocean circulation, vol 3. Open University, Oceanography Course Team Butterworth-Heinemann, Science, Butterworth-Heinemann, Oxford
Conkright ME (2002) World ocean atlas. Objective analyses, data statistics, and figures. Silver Springs, MD: CD-ROM Documentation

Cuffey KM, Clow GD (1997) Temperature, accumulation, and ice sheet elevation in central Greenland through the last deglacial transition. J Geophys Res Oceans (1978–2012) 102 (C12):26383–26396

Dahl KA, Broccoli AJ, Stouffer RJ (2005) Assessing the role of North Atlantic freshwater forcing in millennial scale climate variability: a tropical Atlantic perspective. Clim Dyn 24:325–346

Dällenbach A, Blunier T, Flückiger J, Stauffer B, Chappellaz J, Raynaud D (2000) Changes in the atmospheric CH4 gradient between Greenland and Antarctica during the last glacial and the transition to the Holocene. Geophys Res Lett 27:1005–1008

Dansgaard W, Johnsen SJ, Clausen HB, Dahl-Jensen D, Gundestrup NS, Hammer CU, Bond G (1993) Evidence for general instability of past climate from a 250-kyr ice-core record. Nature 364(6434):218–220

Ding Q, Steig EJ, Battisti DS, Küttel M (2011) Winter warming in West Antarctica caused by central tropical Pacific warming. Nat Geosci 4:398–403. doi:10.1038/ngeo1129

Ditlevsen PD, Andersen KK, Svensson A (2007) The DO-climate events are probably noise induced: statistical investigation of the claimed 1470 years cycle. Clim Past 3(1):129–134

Dokken TM, Nisancioglu KH, Li C, Battisti DS, Kissel C (2013) Dansgaard-Oeschger cycles: interactions between ocean and sea ice intrinsic to the Nordic seas. Paleoceanography 28 (3):491–502

EPICA Community Members (2004) Eight glacial cycles from an Antarctic ice core. Nature 429:623–628

EPICA Community Members (2006) One-to-one coupling of glacial climate variability in Greenland and Antarctica. Nature 444:195–198. doi:10.1038/nature05301

Ezer T, Atkinson LP, Corlett WB, Blanco JL (2013) Gulf stream's induced sea level rise and variability along the U.S. mid-Atlantic coast. J Geophys Res Oceans 118:685–697. doi:10.1002/jgrc.20091

Flückiger J, Dällenbach A, Blunier T, Stauffer B, Stocker TF, Raynaud D, Barnola J-M (1999) Variations in atmospheric N_2O concentration during abrupt climatic changes. Science 285:227–230

Fuhrer K, Wolff EW, Johnsen SJ (1999) Timescales for dust variability in the Greenland Ice Core Project (GRIP) ice core in the last 100,000 years. J Geophys Res Atmos (1984–2012) 104 (D24):31043–31052

Genty D, Blamart D, Ouahdi R, Gilmour M, Baker A, Jouzel J, Van-Exter S (2003) Precise dating of Dansgaard-Oeschger climate oscillations in western Europe from stalagmite data. Nature 421:833–937

Grootes PM, Stuiver M (1997) Oxygen 18/16 variability in Greenland snow and ice with 103-to 105-year time resolution. J Geophys Res Oceans (1978–2012) 102(C12):26455–26470

Grootes PM, Stuiver M, White JWC, Johnsen SJ, Jouzel J (1993) Comparison of oxygen isotope records from the GISP2 and GRIP Greenland ice cores. Nature 366:552–554

Hall IR, Moran SB, Zahn R, Knutz PC, Shen CC, Edwards RL (2006) Accelerated drawdown of meridional overturning in the late-glacial Atlantic triggered 10 by transient pre-H event freshwater perturbation. Geophys Res Lett 33:L16616. doi:10.1029/2006GL026239

Heinrich H (1988) Origin and consequences of cyclic ice rafting in the northeast Atlantic Ocean during the past 130,000 years. Quat Res 29:142–152

Held IM, Ting M, Wang H (2002) Northern Winter stationary waves: theory and modeling. J Climate 15:2125–2144. doi:10.1175/1520-0442(2002)015<2125:NWSWTA>2.0.CO;2

Hendy IL, Kennett JP, Roark EB, Ingram BL (2002) Apparent synchroneity of submillennial scale climate events between Greenland and Santa Barbara Basin, California from 30–10ka. Quaternary Sci Rev 21(10):1167–1184

Hemming S (2004) Heinrich events: massive late Pleistocene detritus layers of the North Atlantic and their global climate imprint. Rev Geophys 42: RG1005. doi:10.1029/2003RG000128

Hu A, Otto-Bliesner BL, Meehl GA, Han W, Morrill C, Brady EC, Briegleb B (2008) Response of thermohaline circulation to freshwater forcing under present-day and LGM conditions. J Climate 21(10):2239–2258

Hulbe CL, MacAyeal DR, Denton GH, Kleman J, Lowell TV (2004) Catastrophic ice shelf breakup as the source of Heinrich event icebergs. Paleoceanography 19(1). doi:10.1029/2003PA000890

Huang RX, Cane MA, Naik N, Goodman P (2000) Global adjustment of the thermocline in response to deepwater formation. Geophys Res Lett 27:759–762

Huber C, Leuenberger M, Spahni R, Flückiger J, Schwander J, Stocker TF, Jouzel J (2006) Isotope calibrated Greenland temperature record over Marine Isotope Stage 3 and its relation to CH 4. Earth Planet Sci Lett 243(3):504–519

Jonkers L, Moros M, Prins M, Dokken T, Dahl C, Dijkstra N, Perner K, Brummer G (2010) A reconstruction of sea surface warming in the northern North Atlantic during MIS 3 ice-rafting events. Quaternary Sci Rev 29:1791–1800

Jullien E, Grousset F, Malaize B, Duprat J, Sanchez-Goni MF, Eynaud F, Charlier K, Schneider R, Bory A, Bout V, Flores JA (2007) Low-latitude "dusty events" vs. high-latitude "icy Heinrich events". Quat Res 68:379–386

Kageyama M, Paul A, Roche DM, van Meerbeeck CJ (2010) Modelling glacial climatic millennial-scale variability related to changes in the Atlantic meridional overturning circulation: a review. Quat Sci Rev 29:2931–2956

Kageyama M, Merkel U, Otto-Bliesner B, Prange M, Abe-Ouchi A, Lohmann G, Ohgaito R, Roche DM, Singarayer J, Swingedouw D, Zhang X (2013) Climatic impacts of fresh water hosing under Last Glacial Maximum conditions: a multi-model study. Clim Past 9:935–953

Krebs U, Timmermann A (2007) Tropical air-sea interactions accelerate the recovery of the Atlantic meridional overturning circulation after a major shutdown. J Clim 20:4940–4956

Landais A, Caillon N, Goujon C, Grachev AM, Barnola JM, Chappellaz J, Jouzel J, Masson-Delmotte V, Leuenberger M (2004) Quantification of rapid temperature change during DO event 12 and phasing with methane inferred from air isotopic measurements. Earth Planet Sci Lett 225:221–232

Landais A, Masson-Delmotte V, Jouzel J, Raynaud D, Johnsen S, Huber C, Minster B (2006) The glacial inception as recorded in the North GRIP Greenland ice core: timing, structure and associated abrupt temperature changes. Clim Dyn 26(2–3):273–284

Landais A, Masson-Delmotte V, Stenni B, Selmo E, Roche DM, Jouzel J, Popp T (2015) A review of the bipolar see–saw from synchronized and high resolution ice core water stable isotope records from Greenland and East Antarctica. Quat Sci Rev 114:18–32

Lang C, Leuenberger M, Schwander J (1999) 16 °C rapid temperature variation in central Greenland 70,000 years ago. Science 286:934–937

Lebreiro SM, Moreno JC, McCave IN, Weaver PPE (1996) Evidence for Heinrich layers off Portugal (Tore Seamount: 39° N, 12° W). Mar Geol 131:47–56

Leuschner DC, Sirocko F (2000) The low-latitude monsoon climate during Dansgaard-Oeschger cycles and Heinrich Events. Quat Sci Rev 22:925–941

MacAyeal DR (1993) Binge/purge oscillations of the Laurentide ice sheet as a cause of the North Atlantic's Heinrich events. Paleoceanography 8(6):775–784

Manabe S, Stouffer RJ (1988) Two stable equilibria of a coupled ocean-atmosphere model. J Clim 1:841–866

Margalef O, Cacho I, Pla-Rabes S, Cañellas-Boltà N, Pueyo JJ, Sáez A, Pena LD, Valero-Garcés BL, Rull V, Giralt S (2015) Millennial-scale precipitation variability over Easter Island (South Pacific) during MIS 3: inter-hemispheric teleconnections with North Atlantic abrupt cold events. Clim Past Discuss 11:1407–1435. doi:10.5194/cpd-11-1407-2015

McCarthy G, Frejka-Williams E, Johns WE, Baringer MO, Meinen CS, Bryden HL, Rayner D, Duchez A, Roberts C, Cunningham SA (2012) Observed interannual variability of the Atlantic Meridional overturning circulation at 26.5 N. Geophys Res Lett 39:L19609. doi:10.1029/2012GL052933

McManus FJ, Francois R, Gherardi JM, Keigwin LD, Brown-Leger S (2004) Collapse and rapid resumption of Atlantic meridional circulation linked to deglacial climate changes. Nature 428:834–837

Menviel L, Timmermann A, Mouchet A, Timm O (2008) Meridional reorganizations of marine and terrestrial productivity during Heinrich events. Paleoceanography 23:PA1203. doi:10.1029/2007PA001445

Menviel L, England MH, Meissner KJ, Mouchet A, Yu J (2014a) Atlantic-Pacific seesaw and its role in outgassing CO_2 during Heinrich events. Paleoceanography 29. doi:10.1002/2013PA002542

Menviel L, Timmermann A, Friedrich T, England MH (2014b) Hindcasting the continuum of Dansgaard-Oeschger variability: mechanisms, patterns and timing. Clim Past 10(1):63–77

Merkel U, Prange M, Schulz M (2010) ENSO variability and teleconnections during glacial climates. Quat Sci Rev 29:86–100

Mo KC (2000) Relationships between interdecadal variability in the Southern Hemisphere and sea surface temperature anomalies. J Clim 13:3599–3610

Monnin E, Indermühle A, Dällenbach A, Flückiger J, Stauffer B, Stocker TF, Raynaud D, Barnola J-M (2001) Atmospheric CO_2 concentrations over the last glacial termination. Science 291:112–114

Moseley GE, Spötl C, Svensson A, Cheng H, Brandstätter S, Edwards RL (2014) Multi-speleothem record reveals tightly coupled climate between central Europe and Greenland during Marine Isotope Stage 3. Geology 42(12):1043–1046

Mulitza S, Prange M, Stuut J, Zabel M, von Dobeneck T, Itambi AC, Nizou J, Schulz M, Wefer G (2008) Sahel megadroughts triggered by glacial slowdowns of Atlantic meridional overturning, Paleoceanography 23:PA420. doi:10.1029/2008PA001637

Muller J, Wüst RAJ, Weiss DJ, Hu Y (2006) Geochemical and stratigraphic evidence of environmental change at Lynch's Crater, Queensland, Australia. Glob Planet Change 53: 269–277

Mulvaney R, Abram NJ, Hindmarsh RCA, Arrowsmith C, Fleet L, Triest J, Sime LC, Alemany O, Foord S (2012) Recent Antarctic Peninsula warming relative to Holocene climate and ice-shelf history. Nature 489(7414):141–144

Niedermeyer EM, Prange M, Mulitza S, Mollenhauer G, Schefuss E, Schulz M (2009) Extratropical forcing of Sahel aridity during Heinrich stadials. Geophys Res Lett 36:L20707. doi:10.1029/2009GL039687

North Greenland Ice Core Project Members. Andersen KK, Azuma N, Barnola JM, Bigler M, Biscaye P, Caillon N, White JWC (2004) High-resolution record of Northern Hemisphere climate extending into the last interglacial period. Nature 431:147–151

Obrochta SP, Miyahara H, Yokoyama Y, Crowley TJ (2012) A re-examination of evidence for the North Atlantic "1500-year cycle" at Site 609. Quat Sci Rev 55:23–33

Otto-Bliesner BL, Brady EC (2010) The sensitivity of the climate response to the magnitude and location of freshwater forcing: Last Glacial Maximum experiments. Quat Sci Rev 29:56–73

Oster JL, Montañez IP, Mertz-Kraus R, Sharp WD, Stock GM, Spero HJ, John Tinsley, James C Zachos (2014) Millennial-scale variationsin western Sierra Nevada precipitation during the last glacial cycle MIS 4/3 transition. Quaternary Res 82(1):236–248

Parker A, Ollier CD (2016) There is no real evidence for a diminishing trend of the Atlantic meridional overturning circulation. J Ocean Eng Sci 1(1):30–35

Parkinson CL, Cavalieri DJ (2012) Antarctic sea ice variability and trends, 1979–2010. Cryosphere 6(4):871–880

Pausata FSR, Battisti DS, Nisancioglu KH, Bitz CM (2011) Chinese stalagmite 18O controlled by changes in the Indian monsoon during a simulated Heinrich event. Nat Geosci 4:474–480

Peltier WR (2004) Global glacial isostasy and the surface of the ice-age Earth: The ICE-5G (VM2) model and GRACE. Annu Rev Earth Planet Sci 32:111–149

Peterson LC, Haug GH (2006) Variability in the mean latitude of the Atlantic Intertropical Convergence Zone as recorded by riverine input of sediments to the Cariaco Basin (Venezuela). Palaeogeogr Palaeoclimatol Palaeoecol 234(1):97–113

Peterson LC, Haug GH, Hughen KA, Röhl U (2000) Rapid changes in the hydrologic cycle of the tropical North Atlantic during the last glacial. Science 290:1947–1951

Petersen SV, Schrag DP, Clark PU (2013) A new mechanism for Dansgaard-Oeschger cycles. Paleoceanography 28(1):24–30
Rahmstorf S (1996) On the freshwater forcing and transport of the Atlantic thermohaline circulation. Clim Dyn 12(12):799–811
Rahmstorf S (2003) Timing of abrupt climate change: a precise clock. Geophys Res Lett 30 (10):1510
Rahmstorf S (2006) Thermohaline ocean circulation. In: Elias SA (ed) Encyclopedia of quaternary sciences. Elsevier, Amsterdam
Roche D, Paillard D, Cortijo E (2004) Constraints on the duration and freshwater release of Heinrich event 4 through isotope modelling. Nature 432:379–382
Romanova V, Lohmann G, Grosfeld K, Butzin M (2006) The relative role of oceanic heat transport and orography on glacial climate. Quat Sci Rev 25(7):832–845
Rossby T, Flagg CN, Donohue K, Sanchez-Franks A, Lillibridge J (2014) On the long-term stability of Gulf stream transport based on 20 years of direct measurements. Geophys Res Lett 41(1):114–120
Saha R (2015) Millennial-scale stable oscillations between sea ice and convective deep water formation. Preprint arXiv:1503.03494
Schulz M (2002) On the 1470-year pacing of Dansgaard-Oeschger warm events. Paleoceanography 17(2):1–4
Schulz H, von Rad U, Erlenkeuser H (1998) Correlation between Arabian Sea and Greenland climate oscillations of the past 110,000 years. Nature 393:54–57
Schwander J, Sowers T, Barnola J-M, Blunier T, Fuchs A, Malaize B (1997) Age scale of the air in the summit ice: implication for glacial-interglacial temperature change. J Geophys Res 102:19483–19494
Seager R, Battisti DS (2007) Challenges to our understanding of the general circulation: abrupt climate change. Glob Circ Atmos 331–371
Severinghaus JP, Brook EJ (1999) Abrupt climate change at the end of the last glacial period inferred from trapped air in polar ice. Science 286(5441):930–934
Steffensen JP, Andersen KK, Bigler M, Clausen HB, Dahl-Jensen D, Fischer H, Goto-Azuma K, Hansson M, Johnsen SJ, Jouzel J, Masson-Delmotte V, Popp T, Rasmussen SO, Röthlisberger R, Ruth U, Stauffer B, Siggaard-Andersen M, Sveinbjörnsdóttir AE, Svensson A, White JCW (2008) High-resolution Greenland ice core data show abrupt climate change happens in few years. Science 321(5889), 680–684
Stocker TF, Johnsen SJ (2003) A minimum thermodynamic model for the bipolar seesaw. Paleoceanography 18(4):1087. doi:10.1029/2003PA000920
Stouffer RJ, Yin J, Gregory JM, Dixon KW, Spelman MJ, Hurlin W, Weaver AJ, Eby M, Flato GM, Hasumi H, Hu A, Jungclaus JH, Kamenkovich IV, Levermann A, Montoya M, Murakami S, Nawrath S, Oka A, Peltier WR, Robitaille DY, Sokolov A, Vettoretti G, Weber SL (2006) Investigating the causes of the response of the thermohaline circulation to past and future climate changes. J Clim 19:1365–1387. doi:10.1175/JCLI3689.1
Swingedouw D, Mignot J, Braconnot P, Mosquet E, Kageyama M, Alkama R (2009) Impact of freshwater release in the North Atlantic under different climate conditions in an OAGCM. J Clim 22:6377–6403
Talley LD (2002) Salinity patterns in the ocean. Encyclopedia of global change. In: MacCracken MC, Perry JS (eds) Volume: the earth system: physical and chemical dimensions of global environmental change, pp 629–640
Taylor KC, Lamorey GW, Doyle GA, Alley RB, Grootes PM, Mayewski PA, White JWC, Barlow LK (1993), The flickering switch of late Pleistocene climate change. Nature 361:432–436
Timmermann A, Menviel L, Okumura Y, Schilla A, Merkel U, Timm O, Hu A, Otto-Bliesner B, Schulz M (2010) Towards a quantitative understanding of millennial-scale Antarctic Warming events. Quat Sci Rev 29:74–85

Thomas ER, Wolff EW, Mulvaney R, Johnsen SJ, Steffensen JP, Arrowsmith C (2009) Anatomy of a Dansgaard-Oeschger warming transition: High-resolution analysis of the North Greenland Ice Core Project ice core. J Geophys Res-Atmos, 114(D8)

Vellinga M, Wood RA (2008) Impacts of thermohaline circulation shutdown in the twenty-first century. Clim Change 91(1–2):43–63

Voelker AHL (2002) Global distribution of centennial-scale records for marine isotope stage (MIS) 3: a database. Quat Sci Rev 21:1185–1212. doi:10.1016/S0277-3791(01)00139-1

Wang Z, Mysak LA (2006) Glacial abrupt climate changes and Dansgaard-Oeschger oscillations in a coupled climate model. Paleoceanography 21(2)

Wagner G, Beer J, Masarik J, Muscheler R, Kubik PW, Mende W, Yiou F (2001) Presence of the solar de Vries cycle (~205 years) during the last ice age. Geophys Res Lett 28(2):303–306

Wang YJ, Cheng H, Edwards RL, An ZS, Wu JY, Shen C-C, Dorale JA (2001) A high-resolution absolute-dated late Pleistocene monsoon record from Hulu Cave, China. Science 294:2345–2348

Wang X, Auler AS, Edwards RL, Cheng H, Cristalli PS, Smart PL, Richards DA, Shen C-C (2004) Wet periods in northeastern Brazil over the past 210 kyr linked to distant climate anomalies. Nature 432:740–743. doi:10.1038/nature03067

Wang X, Auler AS, Edwards RL, Cheng H, Ito E, Wang Y, Kong X, Solheid M (2007) Millennial-scale precipitation changes in southern Brazil over the past 90,000 years. Geophys Res Lett 34:1–5

Williams PW, McGlone M, Neil H, Zhao JX (2015) A review of New Zealand palaeoclimate from the Last Interglacial to the global Last Glacial Maximum. Quaternary Sci Rev 110:92–106

Wolff EW, Barbante C, Becagli S, Bigler M, Boutron CF, Castellano E, de Angelis M, Federer U, Fischer H, Fundel F, Hansson M, Hutterli M, Jonsell U, Karlin T, Kaufmann P, Lambert F, Littot GC, Mulvaney R, Röthlisberger R, Ruth U, Severi M, Siggaard-Andersen ML, Sime LC, Steffensen JP, Stocker TF, Traversi R, Twarloh B, Udisti R,Wagenbach D, Wegner A (2010) Changes in environment over the last 800,000 years from chemical analysis of the EPICA Dome C ice core. Quat Sci Rev 29:285–295

Wunsch C (2006) Abrupt climate change: an alternative view. Quat Res 65:191–203

Yin J, Goddard PB (2013) Oceanic control of sea level rise patterns along the East Coast of the United States. Geophys Res Lett 40(20):5514–5520. doi:10.1002/2013GL057992

Zhang X, Prange M, Merkel U, Schulz M (2015) Spatial fingerprint and magnitude of changes in the Atlantic meridional overturning circulation during marine isotope stage 3. Geophys Res Lett 42(6):1903–1911

Active Deformation, Uplift and Subsidence in Southern South America Throughout the Quaternary: A General Review About Their Development and Mechanisms

Andrés Folguera, Guido Gianni, Lucía Sagripanti, Emilio Rojas Vera, Bruno Colavitto, Darío Orts and Víctor Alberto Ramos

Abstract A broad range of processes act today and have acted simultaneously during the Quaternary, producing relief from the Chilean coast, where the Pacific Ocean floor is sinking underneath the South American margin, to the Brazilian and Argentine Atlantic Ocean platform area. This picture shows to be complex and responds to a variety of processes which are just started to be considered. These processes involve mountains created in a passive margin setting along vast sections of the Brazilian Atlantic Ocean coast and regions located inland, to "current" orogenic processes along the Andean zone. On one hand, mountains in the passive margin seem to be created in the area where the forearc region eastwardly shifts at a similar rate than the westward advancing continent and, therefore, it can be considered as relatively stationary and dynamically sustained by a perpendicular-to-the-margin asthenospheric flow. On the other hand, the orogenic processes associated with the eastern Andes show to be highly active at two particular areas: the Subandean region, where the trench is stationary and the Pampean flat sub-

A. Folguera (✉) · G. Gianni · L. Sagripanti · E. Rojas Vera · B. Colavitto
D. Orts · V.A. Ramos
Instituto de Estudios Andinos, Laboratorio de Tectónica Andina,
Conicet, Facultad de Ciencias Exactas Y Naturales,
Universidad de Buenos Aires, Buenos Aires, Argentina
e-mail: andresfolguera2@yahoo.com.ar

G. Gianni
e-mail: guidogianni22@gmail.com

L. Sagripanti
e-mail: lusagripanti@gmail.com

E. Rojas Vera
e-mail: erv081@yahoo.com.ar

B. Colavitto
e-mail: bcolavitto@gmail.com

D. Orts
e-mail: dlorts@yahoo.com

V.A. Ramos
e-mail: andes@gl.fcen.uba.ar

G.M. Gasparini et al. (eds.), *Marine Isotope Stage 3 in Southern South America, 60 ka B.P.–30 ka B.P.*, Springer Earth System Sciences,
DOI 10.1007/978-3-319-40000-6_6

duction zone to the south, where a shallower geometry of the Nazca plate creates particular conditions for deformation and rapid propagation of the orogenic front generating a high-amplitude orogen. In the Southern Central and Patagonian Andes, mountain (orogenic) building processes are attenuated, and other mechanisms of regional uplift become dominant, such as the (i) crustal weakening and deformation linked to the impact of mantle plumes originated in the 660 km mantle transition, (ii) the retirement of ice masses from the Andes after the Pleistocene producing an isostatic rebound, (iii) the dynamic topography associated with the opening of asthenospheric windows during the subduction of the Chile ridge and slab tearing processes, (iv) the subduction of oceanic plateaux linked to transform zones and (v) the accretion of oceanic materials beneath the forearc region. Additionally and after recent geodetic studies, (vi) forearc coastal uplift due to co-seismic and post-seismic lithospheric stretching associated with large earthquakes along the subduction zone, also shows to be a factor associated with regional uplift that needs to be further considered as an additional mechanism from the Chilean coast to presumably the arc zone.

Keywords Regional uplift and subsidence · Quaternary · Neotectonics · Andes · Pacific and Atlantic ocean coasts

1 Introduction

Active uplift in the Andes has been generally associated with contraction imposed by the convergence between the Pacific Ocean subducted plates and the South American plate (Ramos et al. 2004; Costa et al. 2006a; Oncken et al. 2006; Schellart et al. 2011). However, in the last years multiple mechanisms have been recognized along the Andes that produce, together with orogenic forces, regional to local uplift and potentially exhumation of the upper crust, from isostatic readjustments due to crustal stretching interrupting Andean constructional stages and delamination, to co-seismic extension, and even deep mantle dynamics. Additionally, segments where exhumation and uplift seem to be governed by thrusting are not clearly delimited and their associated mechanisms are not totally understood.

In general terms a narrow band of thrusts has been described bordering the eastern Altiplano and Pampean regions from southern Perú and Bolivia to northern and central Argentina between 10° and 33° S (Fig. 1). This segment coincides with a broad and high mountain chain associated with important amounts of intracrustal earthquakes that denote active mountain building processes (see Costa et al. 2006a, for a recent revision). South of 33° S, crustal seismicity on the eastern Andes diminishes meaningfully, mountain morphology becoming narrower and lower. Even though orogenic mechanisms are described for these Southern Andes at least discontinuously, other factors have been linked to uplift and exhumation in the last years, in particular for the Patagonian region.

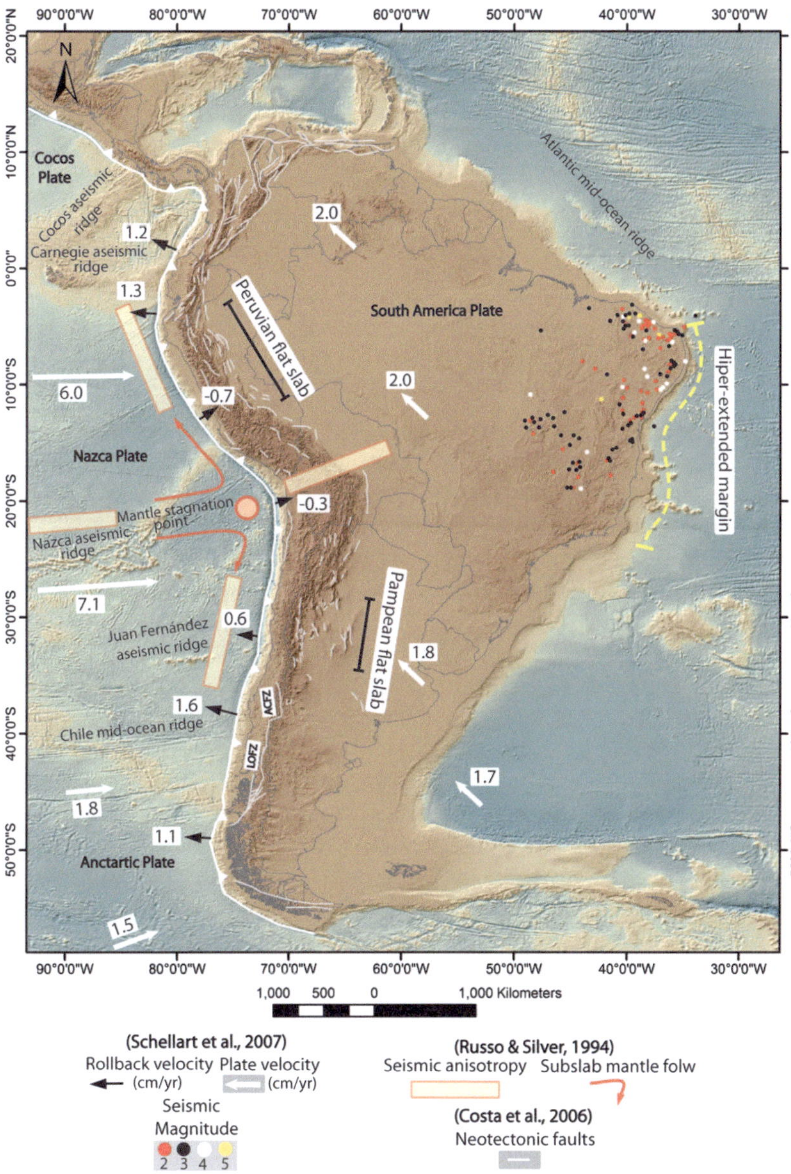

Fig. 1 Uplift and exhumation mechanisms associated with the Central Andes and Atlantic Ocean passive margin. Location of intracrustal seismicity spread over the Atlantic Brazilian coast is after Bezerra et al. (2006), Riccomini and Assumpçao (1999) and Rossetti et al. (2012). Trench retreat rates are after Schellart et al. (2007, 2011). Note maximum amplitude of the Andean orogen corresponding to the Altiplano region at the site of stationary trench associated with neotectonic structures on both mountain flanks (Farías et al. 2005; Ramos et al. 2006; Brooks et al. 2011). Structures with recognized Quaternary activity along the Andean margin are taken from a compilation of Costa et al. (2006a). Directions of maximum velocities, derived from seismic anisotropy in the asthenosphere beneath the Nazca plate, are after Russo and Silver (1994)

This work outlines the main processes that are associated with uplift, regional subsidence and exhumation of vast sectors of the Southern Andes and their foreland area, mainly recognized in the last years since technological innovations have allowed (i) recognizing centimeter to millimeter fluctuations of the landscape, (ii) illuminating the thermal structure of the lower crust and upper mantle, and (iii) analyzing variations in the gravity field through time, from the Altiplano region to Patagonia, passing through the foreland area and even the passive margin where no subduction of ocean lithosphere takes place.

2 Central and Patagonian Andes Tectonic Setting

The Andes are formed over a subduction system consisting of two oceanic plates subducting underneath the South American plate. This configuration shows a noticeable symmetry with the Altiplano at its mid-sector flanked by two flat subduction settings, the Peruvian in the north and the Pampean in the south (Fig. 1) (Gephart 1994). Topography is higher at the mid sector and diminishes steadily toward both edges of the subductive system, where narrow mountain systems are connected to transform boundaries between the South American, Caribbean and Scotia plates respectively (Fig. 1). An active arc is produced at the sites where the subduction configuration is steeper than 20–30° S and oceanic lithosphere subducts with ages older than 5 Ma, and is discontinued through four magmatic gaps along the Andes.

Even though this system has a striking symmetry, from north to south the Nazca and Antarctic plates sink beneath the western border of the South American continent at different rates. While Nazca penetrates beneath the continent with varying relatively high rates between 6 and 7 cm/yr, the Antarctic plate sinks at less than 2 cm/yr. This change has been attributed to the ridge–trench collision and the migration of the triple junction from south to north in the lasts 14 Myr between Nazca, Antarctica and South American plates that provoked the opening of an asthenospheric window beneath Patagonia and mechanical disconnection of Antarctica and Nazca plates at depth, consequently producing a drastic drop in slab pull forces (Cande and Leslie 1986; Ramos and Kay 1992).

3 Active Uplift in the Central Andes and Atlantic Ocean Passive Margin

Anomalously high topography is visualized north and south of the Arica bend region where the Altiplano is developed (Fig. 1). The Altiplano region is elevated by delamination of the lower lithosphere (McQuarrie et al. 2005; Asch et al. 2006; Calixto et al. 2013) producing a high topography partially linked to isostatic readjustments (Isacks 1988; Sobolev and Babeyko 2005; Dávila et al. 2012). This feature is eastwardly flanked by the Eastern Cordillera and Subandean system, with a deep decollement in the Paleozoic basement, frontally inserted in Paleozoic sections that accommodated contraction since the last 10 Myr up to the present (Echavarría et al. 2003; Ramos et al. 2004; Oncken et al. 2006; Brooks et al. 2011). At the Atlantic Ocean Brazilian margin another topographic anomaly is recognized in a passive margin (Fig. 1). This has been associated with active mountain uplifts restricted to the Atlantic Ocean coastal area associated with crustal seismicity and neotectonics (Bezerra et al. 2006; Riccomini and Assumpçao 1999; Rossetti et al. 2012). These uplifts are roughly coincident with the area where the Chilean forearc is relatively static respect to the Brazilian Atlantic Ocean spreading center (Schellart et al. 2007, 2011). While the Brazilian craton displaces north-westwardly some 20 mm/year, the northern Chilean coast moves between 22.9 and 23.5 mm/year eastwardly implying that the area interposed between both is shortened somewhere along a transect perpendicular to the trench (Kendrick et al. 1999). Even though the main part of this contraction is indeed absorbed at the Subandean region (Brooks et al. 2011), a fraction is accommodated at the Brazilian coastal zone (Fig. 1). This situation seems to be a longstanding phenomenon since thermochronological data along the Brazilian coast show Cretaceous, Eocene and Miocene exhumations that represent exactly the known Andean growth stages (Cogné et al. 2012). To the south, the Chilean forearc escapes westwardly away the Argentine craton, accommodating rates lesser than 1 mm/year of elastic and permanent deformation in the Andes. In this context, the interposed area does not experience high amounts of contraction such as to the north (Fig. 1), since the trench moves to the Pacific Ocean at similar rates as the north-westwardly advancing continent does.

4 Uplift of the Southern Central Andes

Active uplift between 27 and 36° S along the Andes is associated with orogenic mechanisms determined by the Pampean flat subduction zone (27–33° S) and a segment to the south where the Nazca plate changes its dip angle smoothly from flat to 30° E (Pesicek et al. 2012) (Fig. 2). This flat to shallow subduction system determines a high-amplitude orogen where the foreland area is fragmented in a series of basement blocks such as the Sierras Pampeanas and the San Rafael Block (Bastías et al. 1993; Costa and Vita-Finzi 1996; Costa et al. 2006a; Vergés et al. 2006). Both flanks of the Andes are affected by active structures: along the Precordilera and Sierras Pampeanas over the Argentine side and over the eastern Chilean flank (Baker et al. 2009; Armijo et al. 2007; Turienzo et al. 2012; Rodríguez et al. 2013; Díaz et al. 2014). The Precordillera to the south loses its topographic expression beyond the southern limit of the Pampean flat slab, although blind structures fold and uplift the piedmont zone revealing a strong gradient in GPS displacements respect to the Sierras Pampeanas to the east, showing that young structures accommodate contraction at the foothills (Brooks et al. 2010).

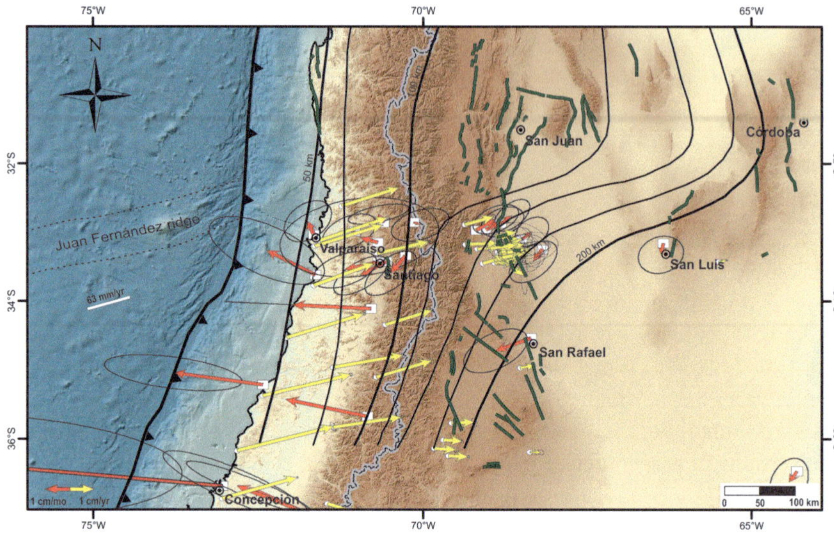

Fig. 2 Uplift mechanisms associated with the area comprised between the Pampean flat subduction zone and the Southern Central Andes. Digital elevation model with iso-depth contours of the subducted Nazca plate (Cahill and Isacks 1992). Note that between 27 and 33° S a flat subduction zone is depicted by interplate seismicity (Pampean flat subduction zone), whereas to the south a normal (30° E) subduction segment is established. *Green lines* describe fault segments with recognized Quaternary activity (after Bastías et al. 1993; Siame et al. 1997; Cortés 1999; Costa and Vita-Finzi 1996; Costa et al. 2006a; Vergés et al. 2006; Baker et al. 2009; Armijo et al. 2007; Turienzo et al. 2012; Rodríguez et al. 2013; Díaz et al. 2014). *Yellow arrows* denote interseismic displacements measured using GPS stations, whereas *red arrows* show co-seismic displacements associated with the Mw 8.8 Maule earthquake (27/2/2010) (Brooks et al. 2003)

5 Active Uplift at the Transition Zone Between the Southern Central Andes and Northern Patagonia

The area interposed between 36 and 38° S shows transitional characteristics in the mechanisms responsible for exhumation between the Southern Central Andes and the Patagonian Andes. Structures affecting Quaternary strata that accommodate contraction are present at the Tromen volcanic plateau in the retroarc zone (Fig. 3) (Galland et al. 2007). Evidence of younger deformation is found in volcanic products of less than 2 Ma, whereas morphometric analyses through the fluvial network allow the recognition of a nonequilibrium state for most of the fluvial channels (Sagripanti et al. 2014).

Thus, this zone of active deformation is considerably retracted to the west respect to the active thrust front that exhumes the San Rafael block to the north (Fig. 2). The arc zone is actively exhumed by a pattern of strike-slip and reverse faults named the Liquiñe-Ofqui fault system and its continuation to the north over the Argentine side of the Andes known as the Guañacos fault system (Lavenu and Cembrano 1999; Folguera et al. 2004, 2006a, b; Rosenau et al. 2006). The retroarc

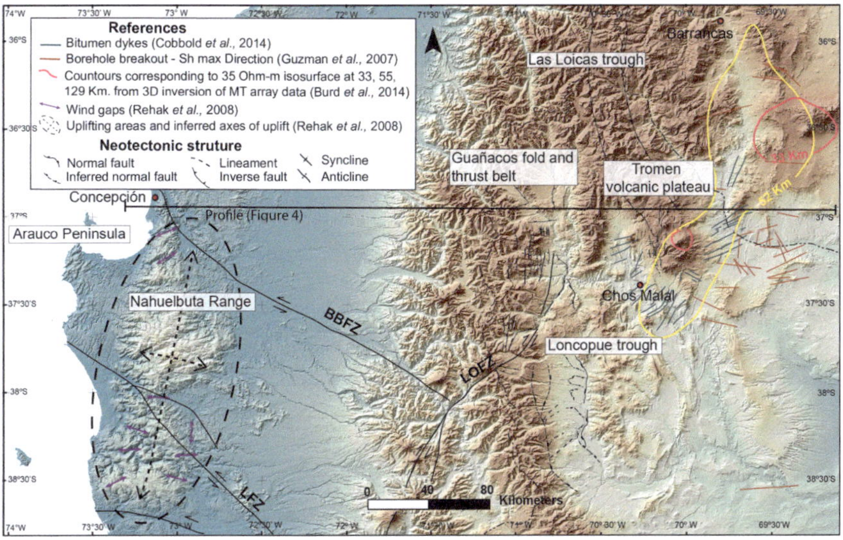

Fig. 3 Uplift mechanisms in the transition zone between the Southern Central and the Northern Patagonian Andes linked to sediment underthrusting and basal accretion to the overriding plate, and development of splay faults along the marine platform, coastal zones and Coastal Cordillera (Melnick et al. 2006; Rehak et al. 2008; Stefer et al. 2009; Geersen et al. 2011), forearc detachment along the arc zone through the northern Liquiñe-Ofqui fault system (Lavenu and Cembrano 1999; Folguera et al. 2004, 2006a, b, Rosenau et al. 2006), and crustal stretching in the retroarc zone linked to isostatic readjustments and volcanism (Rojas Vera et al. 2009, 2010; Folguera et al. 2006a, b, 2007, 2008, 2010; Varekamp et al. 2010; Søager et al. 2013). Oblique compression parallel to the oblique plate convergence is inferred by orientation of bitumen dykes (Cobbold et al. 2014), break-out orientations (Guzmán et al. 2007), and neotectonic structures (Galland et al. 2007; Messager et al. 2010; Sagripanti et al. 2014). Iso-resistivity anomalies depicting geometry of mantle plumes impacting the lower crust in the retroarc zone are after Burd et al. (2014) (*BBFZ* Bío–Bío fault zone, *LFZ* Lanalhue fault zone, *black double arrows* denote locations of wind gaps and *broken ellipsoidal lines* denote uplifting areas in the forearc zone)

zone is also affected by exhumation linked to the development of extensional troughs such as the Loncopué and Las Loicas extensional systems (Rojas Vera et al. 2009, 2010; Folguera et al. 2006a, b, 2007, 2008, 2010; Varekamp et al. 2010; Søager et al. 2013). These systems are spatially linked to a complex pattern of mantle plumes that are impacting the lower crust originated below the 410 km mantle boundary mainly at the retroarc zone (Fig. 4) (Burd et al. 2014). Locally, a main asthenospheric anomaly is branched into a series of minor anomalies that impact the lower crust at the sites of neotectonic activity implying a possible mechanical connection between thermally weakened crust and horizontal crustal yielding (Galland et al. 2007; Messager et al. 2010; Sagripanti et al. 2014). The foreland zone is connected to active uplift affecting broad portions of the landscape at these latitudes (Vogt et al. 2010; Nievière et al. 2013). Mechanics of uplift have been linked there to the flexion of the lithosphere at the foreland zone due to the

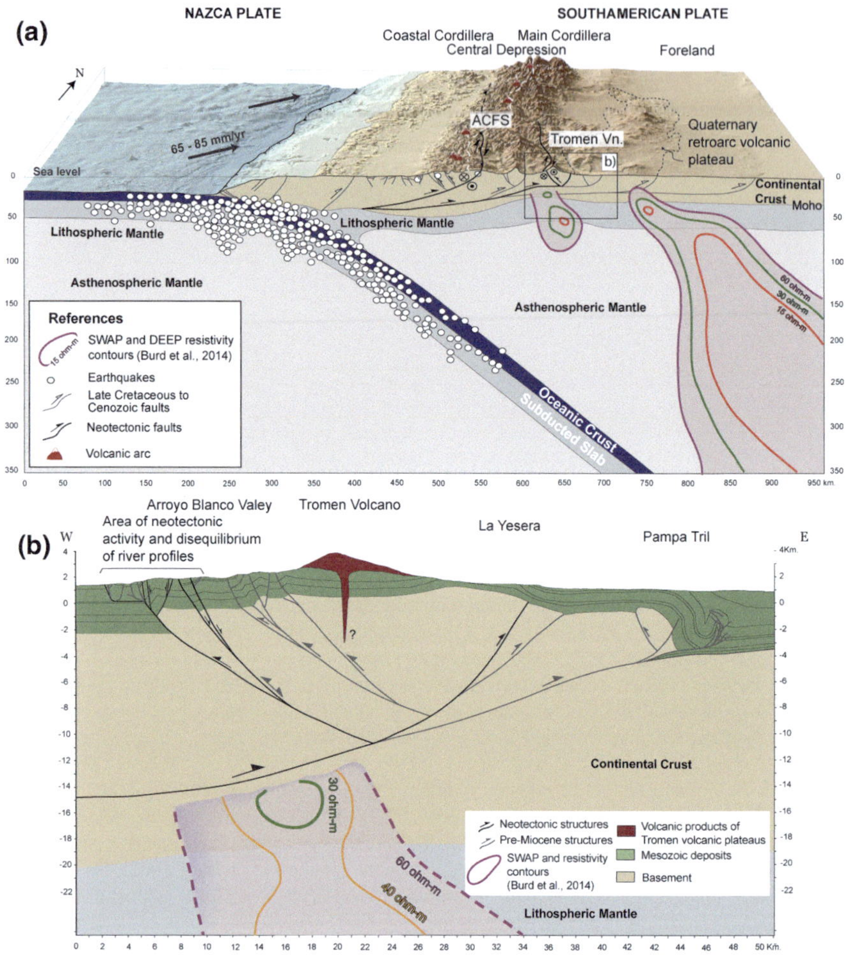

Fig. 4 **a** Digital elevation model with cross section that represents upper crustal structure at 37° S (see location in Fig. 3). Seismicity and geometry of the subduction zone are after Pesicek et al. (2012), while the resistivity anomalies in the mantle and lower crust are a cross section of the 3D model of Burd et al. (2014). Neotectonic structural systems are represented with thicker lines, along the arc front as the Antiñir-Copahue fault zone (ACFS) and the retroarc zone in the Tromen volcanic plateau area (after Messager et al. 2010 and Sagripanti et al. 2014). These neotectonic systems accommodate contractional and right-lateral displacements imposed by the oblique convergence between the Nazca and South American plates. Foreland exhumation has been described for the Pleistocene at these latitudes by Vogt et al. (2010) and Nievière et al. (2013). **b** Detail of the inferred mantle plume from the 3D model of Burd et al. (2014) impacting the lower crust and correlation with neotectonic deformation in the Tromen volcanic plateau (after Sagripanti et al. 2014)

Andean load (peripheral bulge), although a direct spatial connection with the areas where the mantle plumes are impacting, described in Burd et al. (2014), can be visualized (Fig. 4).

In the coastal zone, high-amplitude antiforms are exhumed both in the oceanic platform as well as in the Coastal Cordillera (Fig. 3). These broad structures, such as the Nahuelbuta Range, have uplifted the fluvial morphology and deflected laterally main rivers draining the Pacific Ocean slope (Rehak et al. 2008; Stefer et al. 2009; Geersen et al. 2011). Even though their mechanics of uplift seem to be related to stacking of accreted oceanic materials beneath the forearc zone, sudden uplift connected to the development of large earthquake rupture zones have been informed, suggesting a possible linkage to splay faults developed in co-seismic stages branched from the subduction zone (Melnick et al. 2006; Lieser et al. 2014).

6 Regional Uplift in Northern Patagonia

South of 38–39° S, a slab tear has been described from seismic tomographies after the 27/2/10 Maule earthquake (Fig. 5) (Pesicek et al. 2012). This slab tear determines a slab window south of 38° S through a W–NW direction that coincides with the development of the Loncopué trough and an attenuation of the Moho that reduces the Andean roots at these latitudes up to 33 km (Yuan et al. 2006; Rojas Vera et al. 2010, 2014; Folguera et al. 2007; Varekamp et al. 2010). The topography associated with the Agrio fold and thrust belt, south of the slab tearing, shows a strong morphological readjustment with rivers incising over their channel slopes adjusting their equilibrium to an uplifting relief (Folguera et al. 2010).

Collision of transform zones segmenting the Chile ridge has explained diachronous deformation and exhumation from the coastal to the western retroarc zones (Folguera and Ramos 2009; Dzierma et al. 2012). In particular, out of sequence growth of contractional structures can be linked to the subduction at depth of the Mocha plateau: While coastal sectors have been uplifted in the latest Pliocene–Early Quaternary, the arc and retroarc zones exhibit a younger exhumation up to the Late Pleistocene an even locally Holocene times (Fig. 6).

In the last years, geodetic and satellite gravity (GRACE) measurements associated with the Mw 8.8 Maule earthquake (27/2/2010) have shown the role of large rupture zones along the Pacific Ocean subduction zone in active uplift from the coast to the higher Andes (Tong et al. 2010; Wang et al. 2012). On one hand, co-seismic displacements after 170 years of interseismic strain accumulation (Ruegg et al. 2009) have shown that the coastal areas emerged creating a topography that explains at least in part the relief along the western coastal zone (Fig. 7) (Farías et al. 2011; Arriagada et al. 2011). On the other hand, crustal-scale extension affects the upper plate during co- and post-seismic displacements due to a strong gradient in horizontal displacements from 7 to 8 cm/y in the western retroarc zone to more than 3 m along the coastal zone (Tong et al. 2010, Wang et al. 2007). This extension has been proposed as responsible for regional uplift of the upper plate during co- and post-seismic stages (Brooks et al. 2011; Aaron et al. 2013).

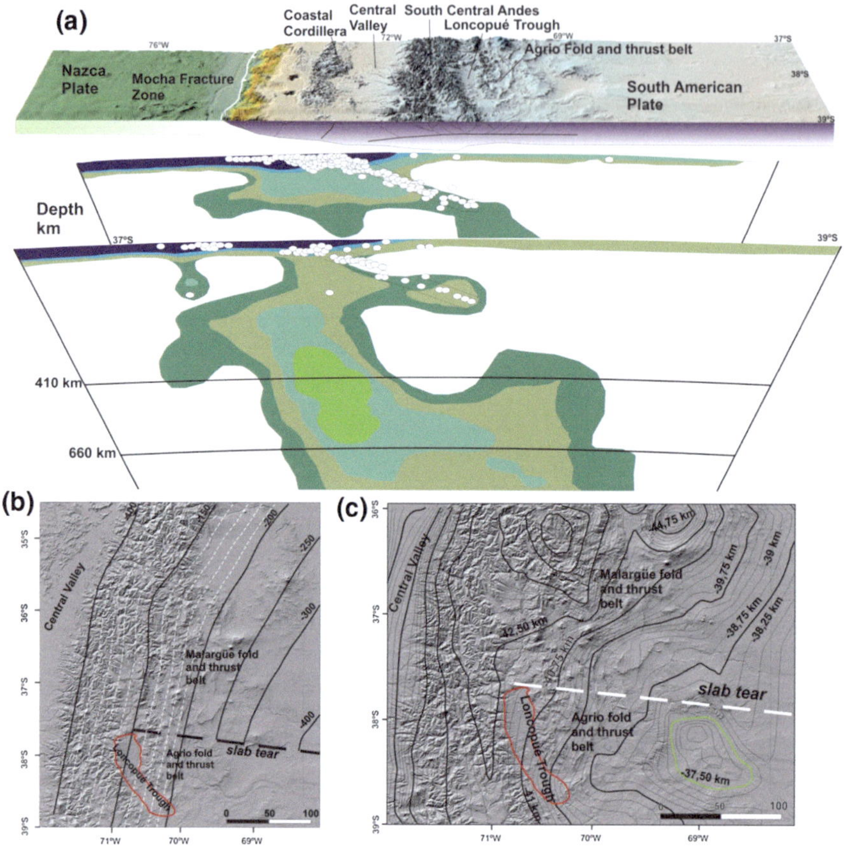

Fig. 5 **a** Vs vertical slice tomographies between 37 and 39° S at the study area (Pesicek et al. 2012) where a slab tearing has been described for the lapse 5–3 Ma (after Rojas Vera et al. 2014). Note that the lower crustal attenuation zone described by Yuan et al. (2006) beneath the Loncopué Trough coincides at surface with the area of slab tearing. **b** Subducted slab depth contours according to Pesicek et al. (2012) interrupted by the slab tearing that determines the development of an asthenospheric window south of 38° S, coincident with the northern limit of the Loncopué Trough (*red line* indicates the Loncopué trough edges); **c** Moho depth contours determined from inversion of gravity data that show the attenuation of the lower crust above the asthenospheric window south of the area of slab tear (*green line* indicates the area of maximum crustal attenuation)

7 Regional Uplift in Southern Patagonia

The Southern Patagonian region shows particular mechanisms for young to active uplift (Fig. 8). Young and buoyant oceanic crust subducted at the Pacific Ocean margin decouples a forearc sliver through the Liquiñe-Ofqui fault zone (LOFZ) (Lavenu and Cembrano 1999; Vargas et al. 2013). This fault zone runs through more than 1000 km through the arc front accommodating strike-slip to reverse

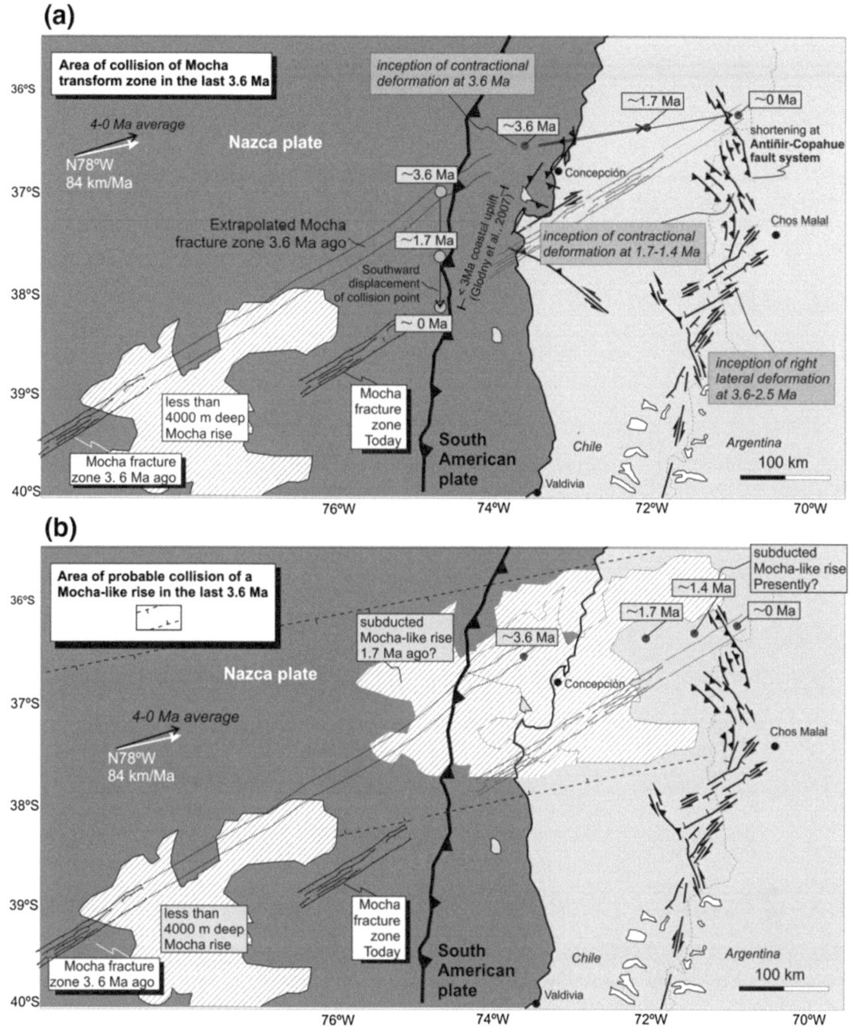

Fig. 6 Collision of the Mocha plateau associated with the Mocha transform zone segmenting the Chilean ridge and diachronous uplift of the Neuquén Andes during the late Pliocene to the Quaternary (after Folguera and Ramos 2009) **a** Southward migration of the point of inception of the Mocha transform zone into the trench in the last 3 Myr. **b** Reconstruction of the collision of the plateau Mocha in the last 3 Myr and development of associated quaternary structures between 36 and 39° S

dip-slip displacements along the North Patagonian Andes (Fig. 8). Its southern edge occurs at the point of collision of the Chile ridge against the trench from which a slab window opens beneath southern Patagonia (Fig. 8) (Russo 2010; Lagabrielle et al. 2007; Breitsprecher and Thorkelson 2009; Scalabrino et al. 2010). The latitudinal extent of this slab window coincides with anomalously high exhumation

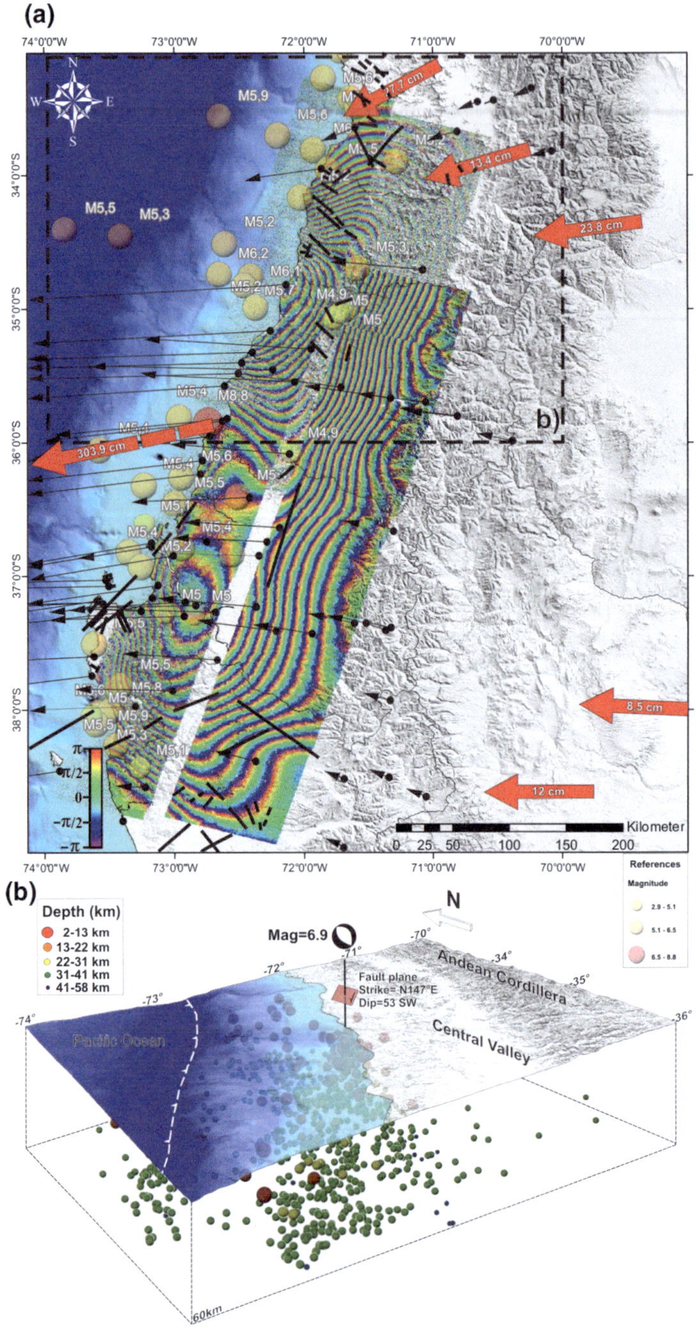
(a)
74°0'0"W
73°0'0"W
72°0'0"W
71°0'0"W
70°0'0"W
34°0'0"S
35°0'0"S
36°0'0"S
37°0'0"S
38°0'0"S
17.7 cm
13.4 cm
23.8 cm
303.9 cm
8.5 cm
12 cm
b)
Kilometer
0 25 50 100 150 200
References
Magnitude
2.9 - 5.1
5.1 - 6.5
6.5 - 8.8
(b)
Depth (km)
2-13 km
13-22 km
22-31 km
31-41 km
41-58 km
Mag=6.9
N
Fault plane
Strike= N147°E
Dip=53 SW
Andean Cordillera
Central Valley
Pacific Ocean
60km

◄ **Fig. 7** **a** Interferogram (Tong et al. 2010) and aftershocks associated with the Maule earthquake, superimposed to co-seismic displacements measured from GPS data after 170 years of interseismic deformation in the North Patagonian forearc region (Ruegg et al. 2009) (after Rojas Vera et al. 2014). Note a strong velocity gradient from 7–8 cm in the western retroarc zone to more than 3 m at the Pacific Ocean coastal zone that led to the hypothesis that co-seismic displacements associated with giant earthquakes in the subduction zone are linked to crustal stretching (Brooks et al. 2010; Farías et al. 2005; Arriagada et al. 2011). *Black lines* indicate normal faults produced during co-seismic displacements (Aaron et al. 2013); **b** aftershocks associated with the development of the extensional faulting in the Pichilemu region at the Chilean forearc and the related Maule rupture zone (see the extensional focal mechanism) (NEIC catalog, USGS 2012)

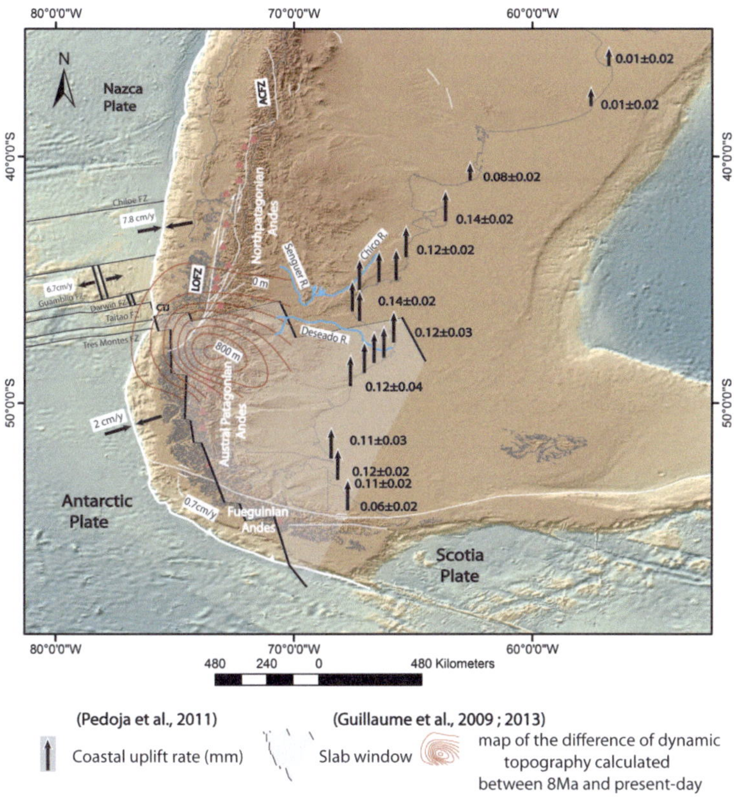

Fig. 8 Uplift mechanisms in Patagonia associated with (i) extension and isostatic rebound during the opening of an asthenospheric window (Russo 2010; Lagabrielle et al. 2007; Breitsprecher and Thorkelson 2009; Scalabrino et al. 2010), (ii) forearc detachment along the Liquiñe-Ofqui fault zone (Lavenu and Cembrano 1999; Vargas et al. 2013), both processes linked to and developed after the collision of the Chile ridge, (iii) rapid isostatic rebound after deglaciation in the high Andes coupled with low viscosity materials associated with the opening of an asthenospheric window (Ivins and James 2002, 2004; Dietrich et al. 2010), (iv) a transform plate boundary between Scotia and South American plates (Costa et al. 2006b). Uplift rates of the Atlantic Ocean after Pedoja et al. (2011) and contours of the difference in dynamic topography over the last 8 Myr and deflected river patterns from Guillaume et al. (2009, 2013)

rates through the Atlantic Ocean coast and the scarped morphology of the Patagonian cliffs suggesting a possible linkage (Codignotto 1996; Pedoja et al. 2011) (Fig. 8). In relation to this, Darwin (1846) had already observed the occurrence of shells on terraces at various elevations, which he explained by large-scale uplift over a 2000 km of coastline. A similar observation is made by Feruglio (1950) who estimated 150–200 m of uplift of the coastal terraces during the Quaternary, considering that this process should be still active along eastern Patagonia. Pedoja et al. (2011) showed that eastern Patagonia uplift is constant through time and twice the uplift of the rest of the South American Atlantic Ocean margin (Fig. 8). Thus, this enhanced uplift along the eastern Patagonian coast is interpreted from the subduction of the Chile ridge and the associated dynamic uplift (Pedoja et al. 2011). Additionally, fluvial morphological fluctuations have been recently linked to the development of an asthenospheric upwelling coming through the opened window that would be related to surface regional uplift and local topographic collapse (Fig. 8) (Murdie et al. 1993; Guillaume et al. 2009, 2013; Lagabrielle et al. 2007; Scalabrino et al. 2010).

Asthenospheric window development has also been associated with rapid isostatic rebound of the Austral Patagonian Andes after deglaciation. This rebound is measured geodetically and is explained by a low mantle viscosity determined by high heat flows associated with the opening of the slab window (Ivins and James 2002, 2004; Dietrich et al. 2010). Finally, deformation and uplift of the southern extreme of Patagonia is almost entirely linked to the activity of a transform boundary zone between Scotia and South American plates, affecting young materials and creating topography (see Costa et al. 2006b, for a revision).

8 Discussion

Mechanisms associated with regional uplift along the Andes in the last 2 Myr show to be highly contrasting. Whereas in the Central Andes contraction predominates, mainly associated with thrust activity concentrated in the eastern Andean slope, and isostatic rebound in areas of overthickened crust suffering delamination of the lower crust, to the south in the Southern Central and Patagonian Andes asthenospheric dynamics, co-seismic displacements and collision of ocean bathymetric highs become the predominant factors.

In particular, uplift across the South American plate at the Arica region, located at the central sector of the Andean subduction configuration, is strongly controlled by the neotectonic activity of the Eastern Cordillera and Subandean System bordering the Altiplano region. A nearly static trench, where the slab rollback is inhibited by a perpendicular-to-the-trench mantle flow, would provoke the westward subduction of the Brazilian craton beneath the Subandean region, inducing higher shortening rates and exhumation. Crustal thickening beneath the Altiplano region has led to delamination of lower crust and lithospheric mantle that still operates today producing isostatic readjustments at the highest Andes and consequently exhumation. This stationary trench also explains why the Atlantic Ocean passive margin is associated at these latitudes with neotectonic topography in the central Brazil that would be partially accommodating horizontal displacements imposed by the ridge push forces originated from the Atlantic mid ocean ridge.

To the south, between 27 and 33° S, flat subduction of the Nazca plate induces neotectonic activity on most of the foreland mountain systems that are forming part of the eastern Andean slope, such as the Precordillera and Sierras Pampeanas systems, whereas more limitedly over fault systems developed over the western Andean slope. Uplift is here directly controlled by activity of fault activity that breaks and uplifts the foreland area.

South of the Pampean flat subduction zone, between 34 and 38° S, uplift seems to be linked to activity of reverse faults affecting (i) the western slope of the Main Andean Cordillera evidenced by high intracrustal seismicity and (ii) the foreland area in the eastern Malargüe fold and thrust belt and San Rafael Block, following the same trend of young tectonic activity existent to the north, although mantle dynamics appear as a second-order mechanism controlling sectors where the crust yields in association with higher thermal fluxes imposed by a complex system of mantle plumes. Tectonic activity at the Tromen plateau and at the southern San Rafael block is concentrated in sites where mantle anomalies are impacting the lower crust.

South of 38° S, a strong tearing in the subducted Nazca plate is associated with neotectonic extensional systems, where the lower crust is attenuated and consequently experience isostatic readjustments. At the forearc region topography is generated in the Coastal Cordillera in association of underplated oceanic materials. Growth of the Pacific Ocean coastal zone is also influenced by co-seismic vertical displacements and by extension achieved during this and post-seismic stages imposed by the slower elastic recovering of the asthenosphere after large earthquakes in the subduction zone.

From 38 to 46° S, a forearc crustal sliver is detached from South America along the Liquiñe-Ofqui fault system presumably in relation to the oblique subduction of younger highly buoyant oceanic lithosphere attached to the Chile ridge. This fault system creates topography through a dextral transpressional mechanics of deformation through the arc and western retroarc zones.

South of 46° S, a slow subduction regime imposed by the collision of the Chile ridge in the lasts 14–12 Myr coincide with the opening of a slab window. An associated asthenospheric upwelling induces an uplifting foreland topography that is provoking the lateral migration of main rivers of Patagonia and exhuming anomalously high cliffs in the Atlantic Ocean coast. South of 50° S a transform fault boundary between the South America and Scotia plates controls younger deformations and uplifting sectors through the southern edge of South America.

9 Concluding Remarks

Regional uplift at the Central Andes is mainly governed by thrusting that accommodates horizontal displacement of the South American craton at the zones of (i) stationary trench in the Arica region, both in the eastern Andes, as well as more limitedly over the Atlantic Ocean coast, and (ii) the Pampean–Chilean flat subduction zone. Delamination of the lower lithosphere, coupled with crustal thickening associated with high shortening rates, also works at these latitudes as a mechanism of uplift.

To the south, where thrusting localized over the eastern slope of the Andes becomes inhibited, other processes appear as second-order factors that control active uplift such as distribution of mantle plumes, co- and post-seismic uplifts during and after giant earthquakes, subduction of younger oceanic lithosphere and detachment of microplates along the forearc and arc zones and finally mantle upwelling induced by slab windows linked to seismic ridge subduction and tearings in the subducted slab.

This revision exemplifies a high complexity in the patterns of distribution and different mechanisms associated with uplift and exhumation in a subduction setting. In particular, constitutes a general framework for the Southern Andes to link different kind of processes to an evolving landscape during the Quaternary.

Acknowledgmentss The authors acknowledge Carlos Costa for a revision made on an early version of the manuscript. Additionally we acknowledge Editors Dres Gasparini, Rabassa, Deschamps and Tonni, for the invitation to integrate this volume. This work was financed by PIP 11220110100506, UBACYT 20020110100019, PICT-2012-1490. This is the R-184 contribution of the Instituto de Estudios Andinos "Don Pablo Groeber", University of Buenos Aires, Conicet.

References

Aaron F, Allmendinger RW, Cembrano J, González G, Yáñez G (2013) Permanent fore-arc extension and seismic segmentation: insights from the 2010 Maule earthquake, Chile. J Geophys Res Solid Earth 118:724–739

Armijo R, Rauld R, Thiele R, Vargas G, Campos J, Lacassin R, Kausel E (2007) The West Andean Thrust, the San Ramón Fault, and the seismic hazard for Santiago, Chile. Tectonics 29 (TC2007). doi:10.1029/2008TC002427

Arriagada C, Arancibia G, Cembrano J, Martínez F, Carrizo D, Van Sint Jan M, Sáez E, González G, Rebolledo S, Sepúlveda S, Contreras-Reyes E, Jensen E, Yañez G (2011) Nature and tectonic significance of co-seismic structures associated with the Mw 8.8 Maule earthquake, central-southern Chile forearc. J Structural Geol 33:891–897

Asch G, Schurr B, Bohm M, Yuan X, Haberland C, Heit B, Kind R, Woelbern I, Bataille K, Comte D, Pardo M, Viramonte J, Rietbrok A, Giese P (2006) Seismological studies of the Central and Southern Andes. In: Oncken et al (eds) The Andes. Frontiers in earth Science, Springer, Berlin, p 567

Baker S, Gosse J, McDonald E, Evenson E, Martínez O (2009) Quaternary history of the piedmont reach of Río Diamante, Argentina. J South Amer Earth Sci 28:54–73

Bastías H, Tello G, Perucca L, Paredes J (1993) Peligro sísmico y neotectónica. 12° Congreso Geológico Argentino y 2° Congreso de Exploración de Hidrocarburos. In: Ramos VA (ed) Geología y Recursos Naturales de Mendoza. Relatorio 6(1):645–658

Bezerra F, Ferreira J, Sousa M (2006) Review of seismicity and Neogene tectonics in northeastern Brazil. Rev Asoc Geológ Argentina 61(4):525–535

Breitsprecher K, Thorkelson D (2009) Neogene kinematic history of Nazca–Antarctic–Phoenix slab windows beneath Patagonia and the Antarctic Peninsula. Tectonophysics 464:10–20

Brooks B, Bevis M, Smalley R, Kendrick E, Manceda R, Lauría E, Maturana R, Araujo M (2003) Crustal motion in the Southern Andes (26°–36°S): do the Andes behave like a microplate? Geochem Geophys Geosyst 4. doi:10.1029/2003GC000505

Brooks B, Smalley R, Bevis M, Foster J, Blanco M, Pollitz F, Folguera A, Ramos V, Barrera H, Cimbaro S, Simons M, Sladen A, Distante C (2010) Megathrust seismic events, post seismic deformation and mountain building: Andean deformation and the Maule earthquake. In: Chapman Conference, Abstracts, Concepción

Brooks B, Bevis M, Whipple K, Arrowsmith R, Foster J, Zapata T, Kendrik E, Minaya E, Echalar A, Blanco M, Euillades P, Sandoval M, Smalley R (2011) Orogenic-wedge deformation and potential for great earthquakes in the central Andean backarc. Nat Geosci. doi:10.1038/NGEO1143

Burd AI, Booker JR, Mackie R, Favetto A, Pomposiello MC (2014) Three-dimensional electrical conductivity in the mantle beneath the Payún Matru volcanic field in the Andean back-arc of Argentina near 36.5°S: evidence for decapitation of a mantle plume by resurgent upper mantle shear during slab steepening. Geophys J Int 1:1–12

Cahill T, Isacks B (1992) Seismicity and shape of the subducted Nazca plate. J Geophys Res 97 (B12):17503–17529

Calixto F, Sandvol E, Kay S, Mulcahy P, Heit B, Yuan X, Coira B, Comte D, Alvarado P (2013) Velocity structure beneath the southern Puna plateau: Evidence for delamination. Geochem Geophys Geosyst 14

Cande S, Leslie R (1986) Late Cenozoic tectonics of the Southern Chile trench. J Geophys Res 91:47–496

Cobbold P, Ruffet G, Leith H, Loeseth H, Rodrigues N, Leanza H, Zanella A (2014) Radial patterns of bitumen dykes around Quaternary volcanoes, provinces of Neuquén and southernmost Mendoza, Argentina. J South Am Earth Sci 56:454–467 (In Press)

Codignotto J (1996) Cuaternario y dinámica costera. In: Ramos VA, Turic M (eds) Geología y Recursos Naturales de la Plataforma Continental Argentina. 13° Congreso Geológico Argentino y 3° Congreso de Exploración de Hidrocarburos. Relatorio 2:17–28

Cogné N, Gallagher K, Cobbold P, Riccomini C, Gautheron C (2012) Post-breakup tectonics in southeast Brazil from thermochronological data and combined inverse-forward thermal history modeling. J Geophys Res 117 (B11413). doi:10.1029/2012JB009340

Cortés J (1999) Fallas cuaternarias oblicuas al frente montañoso en la cordillera Frontal de Mendoza (34–34°30′S). Revista Cuaternario y Ciencias Ambientales. Publicación Especial n°4

Costa C, Vita-Finzi C (1996) Late Holocene faulting in the southeast Sierras Pampeanas of Argentina. Geology 24(12):1127–1130

Costa CH, Audemard MFA, Bezerra FHR, Lavenu A, Machette MN, París G (2006a) An overview of the main quaternary deformation of South America. Rev Asoc Geológica Argentina 61:461–479

Costa C, Smalley R, Schwartz D, Stenner H, Ellis M, Ahumada E, Velasco M (2006b) Paleoseismic observations of an onshore transform boundary: the Magallanes-Fagnano fault, Tierra del Fuego, Argentina. Rev Asoc Geológica Argentina 61(4):647–657

Dávila F, Gimenez M, Nóbile J, Martínez P (2012) The evolution of the high-elevated depocenters of the northern Sierras Pampeanas (ca. 28° SL), Argentine broken foreland, South-Central Andes: the Pipanaco Basin. Basin Res 24:1–22

Darwin CR (1846) Geological observations on South America. Being the third part of the geology of the voyage of the Beagle, under the command of Capt. Fitzroy, R.N. during the years 1832 to 1836. Smith Elder and Co., London, p 279

Díaz D, Maksymowicz A, Vargas G, Vera E, Contreras-Reyes E, Rebolledo S (2014) Exploring the shallow structure of the San Ramón thrust fault in Santiago, Chile (33.5°S), using active seismic and electric methods. Solid Earth Discuss 6:339–375

Dietrich R, Ivins E, Casassa G, Lange H, Wendt H, Fritsche M (2010) Rapid crustal uplift in Patagonia due to enhanced ice loss. Earth Planet Sci Lett. doi:10.1016/j.epsl.2009.10.021

Dzierma Y, Thorwart M, Rabbel W (2012) Moho topography and subducting oceanic slab of the Chilean continental margin in the maximum slip segment of the 1960 Mw 9.5 Valdivia (Chile) earthquake from P-receiver functions. Tectonophysics 530–531:180–192

Echavarría L, Hernández R, Allmendinger R, Reynolds J (2003) Subandean thrust and fold belt of northwestern Argentina: geometry and timing of the Andean evolution. AAPG Bull 87(6): 965–985

Farías M, Charrier R, Comte D, Martinod J, Herail G (2005) Late Cenozoic deformation and uplift of the western flank of the Altiplano: evidence from the depositional, tectonic, and geomorphologic evolution and shallow seismic activity (northern Chile at 19°30′S). Tectonic 24 (TC4001). doi:10.1029/2004TC001667

Farías M, Comte D, Roecker S, Carrizo D, Pardo M (2011) Crustal extensional faulting triggered by the 2010 Chilean earthquake: the Pichilemu seismic sequence. Tectonics 30

Feruglio E (1950) Levantamiento de la costa Atlántica durante el Cuaternario. In: Descripción geológica de la Patagonia, tomo 3:180–181. Buenos Aires

Folguera A, Ramos V (2009) Collision of the Mocha fracture zone and a less than 4 Ma old wave of orogenic uplift in the Andes (36°–38°S). Lithosphere, Geological Soc Am 1(6):364–369

Folguera A, Ramos V, Hermanns R, Naranjo J (2004) Neotectonics in the foothills of the Southernmost Central Andes (37°–38°S). Evidence of the strike-slip displacement along the Antiñir-Copahue fault zone. Tectonics 23 (TC 5008), p 23

Folguera A, Zapata T, Ramos VA (2006a) Late Cenozoic extension and the evolution of the Neuquén Andes. In: Kay SM, Ramos VA (eds) Late Cretaceous to recent magmatism and tectonism of the Southern Andean margin at the latitude of the Neuquén basin (36–39°S). GSA Special paper, 407:267–285

Folguera A, Ramos V, González Díaz E, Hermanns R (2006b) Miocene to quaternary deformation of the Guañacos fold and thrust belt in the Neuquén Andes between 37° and 37°30′S. In: Kay SM, Ramos VA (eds) Late Cretaceous to recent magmatism and tectonism of the Southern Andean margin at the latitude of the Neuquen basin (36–39°S). GSA Special paper, 407: 247–266

Folguera A, Introcaso A, Giménez M, Ruiz F, Martínez P, Tunstall C, García Morabito E, Ramos VA (2007) Crustal attenuation in the Southern Andean retroarc determined from gravimetric studies (38°–39°30′S): The Lonco-Luán astenospheric anomaly. Tectonophysics 439(2007):129–147. doi:10.1016/j.tecto.2007.04.001

Folguera A, Bottesi G, Ramos V, Zapata T (2008) Crustal collapse at the retroarc zone (2–0 Ma): Tromen volcanic plateau, Southern Central Andes (36°40′–37°30′S). Tectonophysics ISAG 2005 Special Issue 459:140–160. doi:10.1016/j.tecto.2007.12.013

Folguera A, Rojas Vera E, Bottesi G, Zamora Valcarce G, Ramos VA (2010) The Loncopué trough: a Cenozoic basin produced by extension in the southern Central Andes. J Geodynamics 49:287–295

Galland O, Hallot E, Cobbold PR, Ruffet G, De Bremond d'Ars J (2007) Volcanism in a compressional Andean setting: a structural and geochronological study of Tromen volcano (Neuquén province, Argentina). Tectonics 26

Geersen J, Behrmann J, Volker D, Krastel S, Ranero C, Diaz-Naveas J, Weinrebe W (2011) Active tectonics of the South Chilean marine fore arc (35°S–40°S). Tectonics 30 (TC3006). doi:10.1029/2010TC002777

Gephart JW (1994) Topography and subduction geometry in the central Andes: clues to the mechanics of a non-collisional orogen. J Geophys Res 99:12279–12288. doi:10.1029/94JB00129

Guillaume B, Martinod J, Husson L, Roddaz M, Riquelme R (2009) Neogene uplift of central eastern Patagonia: dynamic response to active spreading ridge subduction? Tectonics 28. doi:10.1029/2008TC002324
Guillaume B, Gautheron C, Simon-Labric T, Martinod J, Roddaz M, Douville E (2013) Dynamic topography control on Patagonian relief evolution as inferred from low temperature thermochronology. Earth Planet Sci Lett 364:157–167
Guzmán C, Cristallini E, Bottesi G (2007) Contemporary stress orientations in the Andean retroarc between 34°S and 39°S from borehole breakout analysis. Tectonics 26. doi:10.1029/2006TC001958
Isacks BL (1988) Uplift of the Central Andean Plateau and bending of the Bolivian Orocline. J Geophys Res 93(B4):3211–3231. doi:10.1029/JB093iB04p03211
Ivins ER, James TS (2002) Simple model for late Holocene and present-day Patagonian glacier fluctuations and predictions of a geodetically detectable isostatic response. J Geophys Int 138:601–624
Ivins ER, James TS (2004) Bedrock response to Llanquihue Holocene and present-day glaciations in southernmost South America. Geophys Res Lett 31 (L24613). doi:10.1029/2004GL021500
Kendrick EC, Bevis M, Smalley RF, Cifuentes O, Galban F (1999) Current rates of convergence across the central Andes: estimates from continuous GPS observations. Geophys Res Lett 26:541–544
Lagabrielle Y, Suárez M, Malavieille J, Morata D, Espinoza F, Maury RC, Scalabrino B, Barbero L, Cruz RD La, Rossello E, Bellon H (2007) Pliocene extensional tectonics in the Eastern Central Patagonian Cordillera: geochronological constraints and new field evidence. Terra Nov 19:413–424. doi:10.1111/j.1365-3121.2007.00766.x
Lavenu A, Cembrano J (1999) Compressional- and transpressional-stress pattern for Pliocene and quaternary brittle deformation in fore-arc and intra-arc zones (Andes of Central and Southern Chile). J Struct Geol 21:1669–1691. doi:10.1016/S0191-8141(99)00111-X
Lieser K, Grevemeyer I, Lange D, Flueh E, Tilmann F, Contreras-Reyes E (2014) Splay fault activity revealed by aftershocks of the 2010 Mw 8.8 Maule earthquake, central Chile. Geology. doi:10.1130/G35848.1
McQuarrie N, Horton BK, Zandt G, Beck S, DeCelles PG (2005) Lithospheric evolution of the Andean fold–thrust belt, Bolivia, and the origin of the central Andean plateau. Tectonophysics 399(1–4):15–37. doi:10.1016/j.tecto.2004.12.013
Melnick D, Bookhagen B, Echtler H, Strecker M (2006) Coastal deformation and great subduction earthquakes, Isla Santa María, Chile (37°S). Geol Soc Amer Bull 118(11/12):1463–1480
Messager G, Nivière B, Martinod J, Lacan P, Xavier J-P (2010) Geomorphic evidence for Plio-quaternary compression in the Andean foothills of the southern Neuquén basin, Argentina. Tectonics 29(TC4003):1–18
Murdie R, Prior D, Styles P, Flint S, Pearce R, Agar S (1993) Seismic responses to ridge-transform subduction: Chile triple junction. Geology 21:1095–1098
Nievière B, Messager G, Carretier S, Lacan P (2013) Geomorphic expression of the Southern Central Andes forebulge (37°S, Argentina). Terra Nova 25:361–367
Oncken O, Hindle D, Kley J, Elger K, Victor P, Schemmann K (2006) Deformation of the Central Andean upper plate system—facts, fiction, and constraints for plateau models. In: Brun J-P, Oncken O, Weissert HD (eds) The Andes. Springer, Berlin, pp 1–25
Pedoja K, Regard V, Husson L, Martinod J, Guillaume B, Fucks E, Iglesias M, Weill P (2011) Uplift of quaternary shorelines in Eastern Patagonia: Darwin revisited. Geomorphol 127: 121–142
Pesicek JD, Engdahl ER, Thurber CH, DeShon HR, Lange D (2012) Mantle subducting slab structure in the region of the 2010 M8.8 Maule earthquake (30–40°S), Chile. Geophys J Int 191:317–324
Ramos VA, Kay SM (1992) Southern Patagonian plateau basalts and deformation: back-arc testimony of ridge collision. Tectonophysics 205:261–282

Ramos VA, Tomas Z, Cristallini E, Introcaso A (2004) The Andean thrust system—latitudinal variations in structural styles and orogenic shortening. In: McClay KR (ed), Thrust Tectonics and hydrocarbon systems. AAPG Memoir, 82:30–50

Ramos VA, Alonso R, Strecker M (2006) Estructura y neotectónica de las Lomas de Olmedo, zona de transición entre los sistemas Subandino y de Santa Bárbara Provincia de Salta. Rev Asoc Geológ Argentina 61(4):579–588

Rehak K, Strecker M, Echtler M (2008) Morphotectonic segmentation of an active forearc, 37–41° S, Chile. Geomorphol 94:98–116

Riccomini C, Assumpçao M (1999) Quaternary tectonics in Brazil. Episodes 22(3):221–225

Rodríguez M, Carretier S, Charrier R, Sailard M, Regard V, Herail G, Hall S, Farber D, Audin L (2013) Geochronology of pediments and marine terraces in north-central Chile and their implications for quaternary uplift in the Western Andes. Geomorphol 180–181(2013):33–46

Rojas Vera E, Folguera A, Gímenez M, Martínez P, Ruiz F, Ramos VA (2009) Evolución tectónica de la fosa de Loncopué: Estructura del depocentro cuaternario del Huecú y su relación con la sedimentación y el volcanismo. Rev Asoc Geológ Argentina 64(2):214–230

Rojas Vera EA, Folguera A, Zamora Valcarce G, Giménez M, Ruiz F, Martínez P, Bottesi G, Ramos VA (2010) Neogene to quaternary extensional reactivation of a fold and thrust belt: the Agrio belt in the Southern Central Andes and its relation to the Loncopué trough (38°–39°S). Tectonophysics 492:279–294

Rojas Vera EA, Selles D, Folguera A, Gímenez M, Ruíz F, Orts D, Zamora Valcarce G, Martínez P, Bechis F, Ramos VA (2014) The origin of the Loncopué trough in the retroarc of the Southern Central Andes from field, geophysical and geochemical data. Tectonophysics 637: 1–19

Rosenau M, Melnick D, Echtler H (2006) Kinematic constraints on intra-arc shear and strain partitioning in the Southern Andes between 38° and 42°S latitude. Tectonics 25 (TC4013). doi:10.1029/2005TC001943

Rossetti D, Souza L, Prado R, Elis V (2012) Neotectonics in the northern equatorial Brazilian margin. J South Am Earth Sci 37:175–190

Ruegg JC, Rudloff A, Vigny C, Madariaga R, de Chabalier JB, Campos J, Kausel E, Barrientos S, Dimitrov D (2009) Interseismic strain accumulation measured by GPS in the seismic gap between Constitución and Concepción in Chile. Phys Earth Planet Inter 175:78–85

Russo R (2010) Subduction of the Chile Ridge: upper mantle structure and flow. GSA Today 20 (9). doi:10.1130/GSATG61A.1

Russo RM, Silver PG (1994) Trench-parallel flow beneath the Nazca plate from seismic anisotropy. Science 263:1105–1111

Sagripanti L, Rojas Vera E, Folguera A, Gianni G, Ramos VA (2014) Out-of-sequence neotectonic reactivation of the orogenic front in the Southern Central Andes. Geomorphol (In Press)

Scalabrino B, Lagabrielle Y, Malavieille J, Dominguez S, Melnick D, Espinoza F, Suarez M, Rossello E (2010) A morphotectonic analysis of central Patagonian Cordillera: negative inversion of the Andean belt over a buried spreading center? Tectonics 29 (TC2010). doi:10.1029/2009TC002453

Schellart WP, Freeman J, Stegman DR, Moresi L, May D (2007) Evolution and diversity of subduction zones controlled by slab width. Nature 446:308–311

Schellart W, Stegman D, Farrington R, Moresi L (2011) Influence of lateral slab edge distance on plate velocity, trench velocity and subduction partitioning. J Geophys Res 116 (B10408). doi.10.1029/2011JB008535

Siame L, Bourles D, Sebrier M, Bellier O, Castano J, Araujo M, Pérez M, Raisbeck G, Yiou F (1997) Cosmogenic dating ranging from 20 to 700 ka of a series of alluvial fan surfaces affected by the El Tigre fault, Argentina. Geology 25(11):975–978

Sobolev SV, Babeyko AY (2005) What drives orogeny in the Andes? Geology 33(8):617–620

Stefer S, Moernaut J, Melnick D, Echtler H, Arz H, Lamy F, Batist M, Oncken O, Haug G (2009) Forearc uplift rates deduced from sediment cores of two coastal lakes in south-central Chile. Tectonophysics. doi:10.1016/j.tecto.2009.05.006

Søager N, Holm PM, Llambías EJ (2013) Payenia volcanic province, southern Mendoza, Argentina: OIB mantle upwelling in a backarc environment. Chem Geol 349–350:36–53

Tong X, Sandwell D, Luttrell K, Brooks B, Bevis M, Shimada M, Foster J, Smalley R, Parra H, Báez Soto JC, Blanco M, Kendrick E, Genrich J, Caccamise DJ (2010) The 2010 Maule, Chile earthquake: downdip rupture limit revealed by space geodesy. Geophys Res Lett 37

Turienzo M, Dimieri L, Frisicale C, Araujo V, Sánchez N (2012) Cenozoic structural evolution of the Argentinean Andes at 34°40′S: a close relationship between thick and thin-skinned deformation. Andean Geol 39(2):317–357

Varekamp J, Hesse A, Mandeville C (2010) Back-arc basalts from the Loncopue graben (Province of Neuquen, Argentina). J Volcanol Geotherm Res 197:313–328

Vargas G, Rebolledo S, Sepúlveda S, Lahsen A, Thiele R, Townley B, Padilla C, Rauld R, Herrera M, Lara M (2013) Submarine earthquake rupture, active faulting and volcanism along the major Liquiñe-Ofqui fault zone and implications for seismic hazard assessment in the Patagonian Andes. Andean Geol 40(1):141–171

Vergés J, Ramos VA, Meigs A, Cristallini E, Bettini F, Cortés J (2006) Crustal wedging triggering recent deformation in the Andean thrust front between 31°S and 33°S: Sierras Pampeanas-Precordillera interaction. J Geophys Res 112 (B03S15). doi:10.1029/2006JB004287

Vogt H, Vogt T, Calmes A (2010) Influence of the post-Miocene tectonic activity on the geomorphology between Andes and Pampa Deprimida in the area of Provincia de la Pampa, Argentina. Geomorphology 121:152–166

Wang K, Hu Y, Bevis M, Kendrick E, Smalley B, Barriga Vargas R, Lauría E (2007) Crustal motion in the zone of the 1960 Chile earthquake: Detangling earthquake-cycle deformation and forearc-sliver translation. Geochem Geophys Geosyst 8(10):Q10010. doi:10.1029/2007GC001721

Wang L, Shum C, Simons F, Tassara A, Erkan K, Jekeli C, Braun A, Kuo C, Lee H, Yuan D (2012) Coseismic slip of the 2010 Mw 8.8 Great Maule, Chile, earthquake quantified by the inversion of GRACE observations. Earth Planet Sci Lett 335–336:167–179

Yuan X, Asch G, Bataile K, Bohm M, Echtler H, Kind R, Onchen O, Wölbern I (2006) Deep seismic images of the Southern Andes. In: Evolution of an Andean margin: a tectonic and magmatic view from the Andes to the Neuquén basin (35–39°S) (SM Kay, VA Ramos, eds). Spec Pap Geol Soc America 407:61–72

The Marine Isotopic Stage 3 (MIS 3) in Valleys of the Undulated Pampa, Buenos Aires Province, Argentina

Adriana María Blasi, Carola Castiñeira Latorre, Gabriela Catalina Cusminsky and Ana Paula Carignano

Abstract A depositional unit called DU2 identified for the period MIS 3 (ca—30,000–60,000 year B.P.) formed by only one sedimentary facies (F3) was found in the Luján and Salto-Arrecifes rivers basins. F3 is a fluvio–lacustrine unit that overlies in erosive unconformity over eolian sediments with ages of 56,400 ± 6500 and 50,400 ± 10,200 years B.P. and is unconformably covered by another eolian vitroclastic sandy loess deposit, dated as 32,000 ± 4000 years (Infrared Stimulated Luminescence, IRSL) (Blasi et al. 2009a). It represents the recurrence of ephemeral fluvial streamlets and the development of temporary pools by subsequent damming of channels. It corresponds lithologically to sandy muddy gravel, gravelly muddy sand, gravelly mud olive to pale olive, feldspar and quartz sands, bearing extinct mollusks such as *Heleobia ameghini* and *Diplodon lujanensis*. Radiocarbon chronologies obtained on monospecific samples of *Cyprideis salebrosa hartmanni* and *Heleobia ameghini* yielded ages of 37,710 ± 840 years ^{14}C B.P. and >40,000 years ^{14}C B.P., respectively. Furthermore, the age obtained through the IRSL technique was of 44,000 ± 6500 years. Based upon the analyzed bioproxies (malacological, phytoliths and diatomological content) F3 accumulated under variable climatic conditions, ranging from temperate to colder and from subhumid to drier. According to the exhaustive stratigraphic identification, it is proposed that in N-E Buenos Aires Province, the so-called Undulated Pampa region, the

A.M. Blasi (✉)
CIC- División Mineralogía, Sedimentología & Petrología, Museo de La Plata, Universidad Nacional de La Plata, Paseo Del Bosque S/N, 1900 La Plata, Argentina
e-mail: ablasi@fcnym.unlp.edu.ar

C.C. Latorre
CONICET-División, Mineralogía, Sedimentología & Petrología, Museo de La Plata, Universidad Nacional de La Plata, Paseo Del Bosque S/N, 1900 La Plata, Argentina

G.C. Cusminsky
INIBIOMA-CONICET, Centro Regional Universitario Bariloche, Universidad Nacional Del Comahue, Quintral 1250, 8400 San Carlos de Bariloche, Argentina

A.P. Carignano
CONICET-División Paleozoología Invertebrados, Facultad de Ciencias Naturales Y Museo, Universidad Nacional de La Plata, Paseo Del Bosque S/N, 1900 La Plata, Argentina

G.M. Gasparini et al. (eds.), *Marine Isotope Stage 3 in Southern South America, 60 ka B.P.–30 ka B.P.*, Springer Earth System Sciences,
DOI 10.1007/978-3-319-40000-6_7

sediments that were accumulated during MIS3 occur only in the central portion of the studied fluvial basins. This prompted two hypotheses related to the existence of a particular drainage pattern for the Late Pleistocene, different from the present one, and subsequent tectonic controls that allowed the identification of DU2 sediments only in some of the analyzed sections.

Keywords Undulated Pampa region · Buenos Aires Province, Argentina · MIS 3 · Late Pleistocene · Infrared stimulated luminescence (IRSL) · Paleoenvironment reconstruction · Paleoclimatic conditions · Fluvial and lacustrine sedimentation · Luján river · Salto-Arrecifes rivers · Analysis of proxy records

1 Introduction

The sedimentological, paleobiological, and chronological studies of the Late Pleistocene-Holocene cropping out in the middle fluvial valleys of the Undulated Pampa region of the Buenos Aires Province (locally known as the "Pampa Ondulada Bonaerense") allowed to identify different depositional units (DU), bounded by unconformities. Depositional units and their facies, as defined for the Luján river basin, were associated by Blasi et al. (2010) based on their stratigraphic position, chronological ages, and paleoclimatic and paleoenvironmental interpretations with different marine isotopic stages (MIS 4–2). As pointed out by Sowers (2000) the marine isotope record, which is the most complete record available of Quaternary climate cycles, is the standard to which we correlate other Quaternary paleoclimatic records.

The studied sequence lies in between the continental deposits and those of the corresponding marine isotopic stage. It offers a very valuable independence concerning the use of stratigraphic nomenclature (biostratigraphy, lithostratigraphy, chronostratigraphy, sequential stratigraphy, etc.) of different stratigraphic schemes established for the study area, generally based on relative chronologies (Blasi et al. 2009b). As pointed out by these authors, the large amount of formal and even informal stratigraphic terms for naming stratigraphic units locally and the use of synonyms for regional correlations eventually generated a nomenclature chaotic situation. These authors proposed to define stratigraphic schemes through the description of DU limited by unconformities, their lithofacies and facies variations.

The results obtained in the characterization of depositional unit 2 (DU2) are herein presented and explained. It is formed by one fluvial-lacustrine facies (F3) chronologically delimited and exposed in certain restricted areas of two large fluvial basins of the Undulated Pampa (the Luján and the Salto-Arrecifes rivers basins). Its association with the Marine Isotope Stage 3 (MIS 3) is discussed in this paper.

During the interstadial MIS 3, belonging to the last glacial stage, several climatic variations occurred with alternating warmer phases, the Dansgaard-Oeschger events (DO) and colder phases, known as the Heinrich events (H), defined according to different proxies analyzed in Antarctic and Arctic polar ice cores (Bond et al. 1993;

Clark et al. 2007; Van Meerbeeck et al. 2009; Rabassa and Ponce 2013; Sinddall et al. 2008). Climate variations taking place during that period somehow have affected the region under study and their consequences were recorded in the analyzed DU.

2 Geological Setting

The study area, in northeastern Buenos Aires Province, Argentina, is located within the "Pampa Levantada" ("raised Pampa") region (Pasotti 1971). It corresponds geomorphologically to the Pampa subregion called "Undulated Pampa" (Pasotti and Castellanos 1967) that covers part of the provinces of Buenos Aires, Córdoba and Santa Fe, with an area of approximately 44,000 km^2 (Fig. 1). It is characterized by the development of undulating landforms due to the presence of low hills aligned in a SW-NE direction and eroded by creeks, streamlets, and rivers. This undulating surface was produced by fluvial erosion and minor tectonic block movements, with differential displacement, associated with large faults related to the crystalline basement (Pasotti 1971, 1973, 2000).

Among other fluvial basins draining this region, those of the herein studied Luján and Salto-Arrecifes rivers are the most relevant for Quaternary studies. Both flow into the Paraná de las Palmas, which is one of the distributaries of the lower portion of the Paraná river, demarcating the western edge of the Río de La Plata delta (Fig. 1).

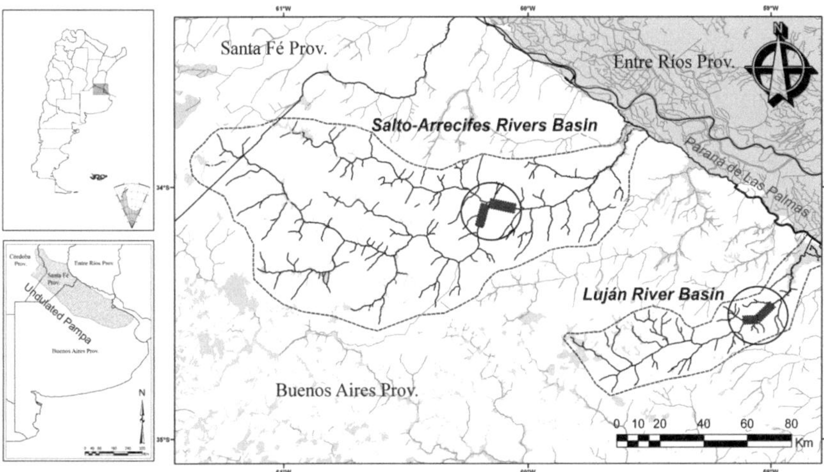

Fig. 1 Location map of the Luján and Salto-Arrecifes basins showing location of MIS 3 sections and geographic features

The Luján and Salto-Arrecifes stream basins initiate from small streamlets which in turn have their sources in shallow lacustrine water bodies. The main collectors appearing in the medium part of the basins are confined in most of the sections by very high cutbanks. Successive sediments exposed in the cutbanks show the paleoenvironmental and paleoclimatic evolution of the Late Pleistocene and Holocene of the Undulated Pampa valleys (Prieto et al. 2004; Blasi et al. 2010).

3 Materials and Methods

Information was gathered from seven localities at the Luján river basin and six other ones at the Salto-Arrecifes rivers basin (Fig. 1). The sedimentary facies exposed at the cutbanks of the studied localities were characterized by their colorimetric features (Munsell Color Chart), lithology, morphologic expression, and biological content. Samples were taken from each known facies, for sedimentological analysis as well as for their paleobiological and chronologic content (^{14}C standard, AMS radiocarbon dating and IRSL).

Sedimentological analysis was accomplished according to standard methods (Carver 1971). The analysis comprised the elimination of organic material with a solution of 30 % H_2O_2, and of cementing materials with a solution of 35 % HCl. For dispersion, 4 % $(NaPO_3)_6$ and mechanical shaking were used. Grain size analysis for sand fraction was performed by sieving at half Φ intervals and by pipetting for the silt clay fraction grain size classification (Craver 1971). The gravel, sand, silt, and clay content percentages were used for the grain size classification according to Folk (1954). Mineralogical analysis was performed in the very fine sand fraction (0.125–0.062 mm) by polarization microscopy. Clay mineralogy was performed in preparation of total and oriented samples, using a Phillips Difractometer X-rays PW3710 Cu tube.

Samples for biosiliceous particle counting and identification were treated with $(NaPO_3)_6$ for sediment disaggregation and clay removal. Then, 15 ml of 35 % HCl was added and the solution was allowed to stand for 24 h to eliminate carbonate precipitation. Samples were rinsed several times with distilled water. Next, 10 ml of 30 % H_2O_2 was added to eliminate organic matter and then the samples were boiled for 4 h and rinsed five times with distilled water. Permanent slides were mounted in Naphrax for counting and identification. A minimum of 400 biosiliceous particles was counted at ×1000 magnification in each sample with an Olympus BX 40 microscope.

Phytoliths were identified according to Bozarth (1992), del Puerto et al. (2006), del Puerto (2009), Fredlund and Tieszen (1994), Fernández et al. (2006), Gallego and Distel (2004), Twiss (1992), and Zucol (1998, 2000, 2001). Although phytoliths cannot be assigned to individual grass taxa, different ratios of phytoliths (pooid, chloridoid and panicoid) may serve as climatic indices for paleoclimatic interpretations of a given region. In this sense, where the C3-types occur, the ratio

C3 to C4 (Twiss 1992) can be used as an index of temperature following the equation: TI = (Pooid)/[(Pooid + Panicoid +Chloridoid) * 100].

Higher values suggest a cooler climate in high latitudes or altitudes (where C3 are prevalent), and lower values suggest warmer temperatures that would be found in lower latitudes or elevations. Similarly, by comparing chloridoid to the sum of chloridoid and panicoid phytoliths a humidity index: HI = (Chloridoid)/[(Chloridoid + Panicoid) * 100] may be obtained. Values close to 100 indicate an arid climate, whereas lower values indicate a humid climate. Diatom species were identified and classified according to Frenguelli (1941, 1945), Metzeltín and García-Rodríguez (2003), Metzeltín et al. (2005) and Witkowski et al. (2000). Chrysophycean cysts were identified according to Duff et al. (1995) and sponge spicules were recognized according to Ezcurra de Drago (1993).

Samples were washed under 0.5 mm mesh opening sieve to obtain and concentrate malacological species. Samples were observed under binocular glass (40×) where species were identifiedand measureded.

To the study of the ostracod fauna, samples were disaggregated with water and washed under a sieve of 63 microns (Tyler Screen System No. 230). All adults and juvenile specimens found in 1 g of dry sediment were picked up. The species were determined according to Moore and Pitrat (1961), Van Morkhoven (1963), Ramírez (1967), Bertels and Martínez (1990) and Ferrero (1996, 2009), among others.

Finally, in order to determine radiocarbon ages and ages through IRSL, samples were processed according to Blasi et al. (2010).

4 Analyzed Stratigraphic Sections

The sedimentary and paleobiological record of DU2 (Facies 3) exposed at the cutbanks of the Luján and Salto-Arrecifes rivers basins was sampled in a few places along the sections studied in both basins (Fig. 1). In these rivers, the deposits corresponding to this interval are circumscribed to the middle portion of the fluvial basins and were associated by Blasi et al. (2010) to the Marine Isotopic Stage 3. In the Luján river basin, it crops out in four clearly constrained sampling sites and in the Salto-Arrecifes rivers basin, in two sites (Table 1). Depending on the hydrologic state of the rivers, they are exposed in the middle to lower portion of the cutbanks (Fig. 2).

5 Results

The depositional unit 2 is formed by facies F3 (Blasi et al. 2010) and appears in massive lens shaped layers with variable thickness, from 1 to 1.30 m (Fig. 2). It rests in erosive to paraconcordant unconformity over the underlying depositional unit DU1/DU2 (Fig. 2). Below this unit, depositional unit 3 (DU3), (facies F4–F5)

Table 1 Studied sections in the Luján river and Salto-Arrecifes rivers, with the corresponding profiles

		Studied localities
UD2 Facies 3	Luján river	Profile Arroyo Sin Nombre (PASN) (34° 34′ 54″ S/59° 10′ 20″ W)
		Profile Benedictinos (PBN) (34° 34′ 42″ S/59° 10′ 04″ W)
		Profile Puente Oeste (PW) (34° 34′ 27.27″ S/59° 07′ 40.34″ W)
		Profile Molino Viejo (PMV) (34° 34′ 8.59″ S/59° 07′ 29.04″ W)
	Salto-Arrecifes rivers	Profile Barranca Salto (PBS) (34° 8′ 14.3″ S/60° 11′ 16.9″ W)
		Profile Arrecifes (PAR) (34° 04′ 54.4″ S/60° 05′ 29.6″ W)

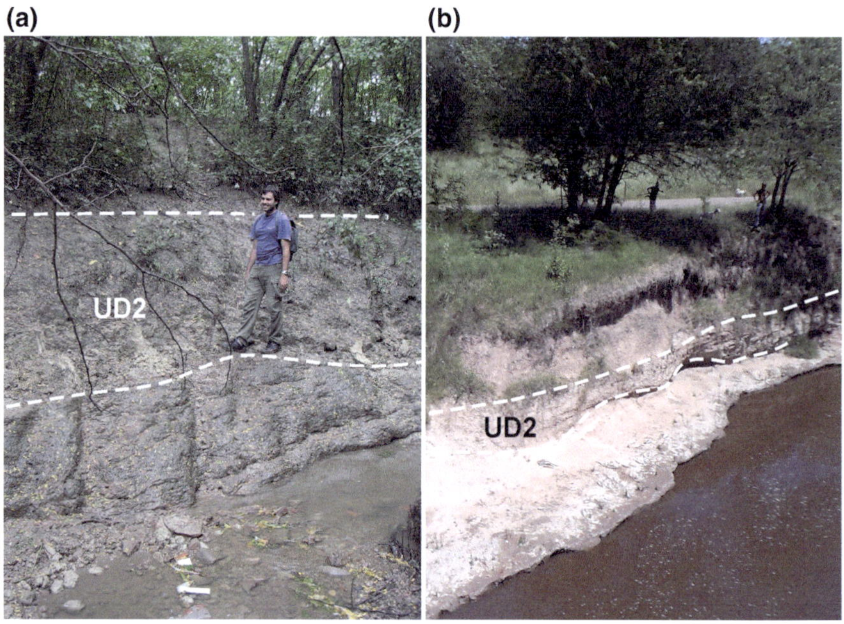

Fig. 2 Outcrops of depositional unit 2 (DU2) at the cut banks. **a** Arroyo sin Nombre at Luján river section (Blasi et al. 2010); **b** Arrecifes river section

and/or lacustrine Holocene deposits occur in stratigraphic uniformity (Blasi et al. 2010). It is lithologically composed of sandy muddy gravel, gravelly muddy sand, and olive green gravelly mud (5Y 5/6) (Figs. 3, 4).

The gravel fraction is composed of fine-grained sedimentary lithic fragment (muddy intraclast), bioclasts (such as reworked bone fragments), and rounded to subrounded calcium carbonate clasts (calcretes fragments), larger than 2 mm. The sandy fraction shows muddy intraclasts, rounded calcrete fragments, as well as, siliciclastic and bioclastic grains. Among the latter, quartz is predominant over

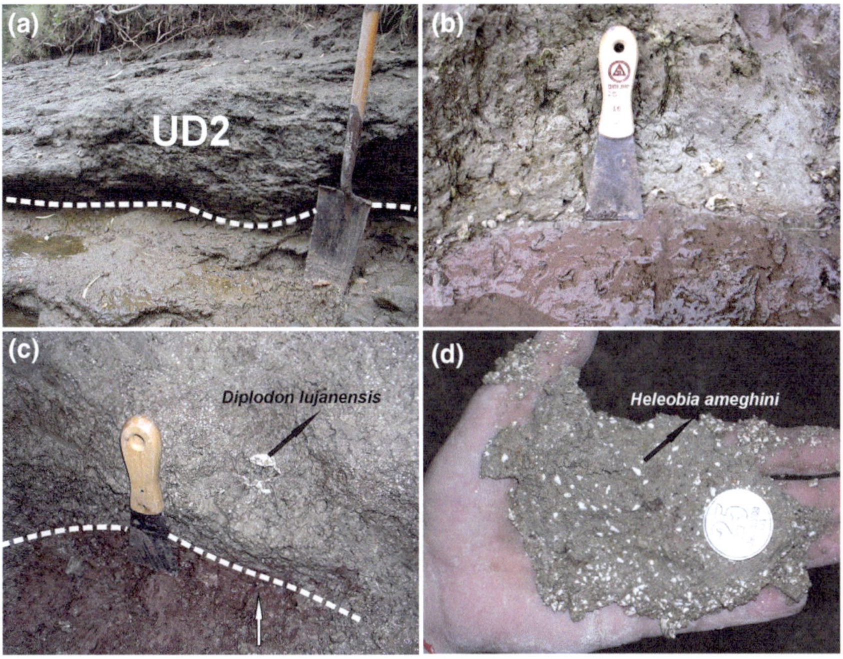

Fig. 3 Depositional unit 2 (DU2); **a** erosive contact between UD2 and UD1; **b** Net contact between *olive green* DU2 and *red* DU1, basal gravels of calcretes lithoclast; **c** UD2 *olive green*, gravelly muddy sand with *Diplodon lujanensis* shells; **d** Abundant shells of *Heleobia ameghini* in *olive green*, gravelly muddy sand

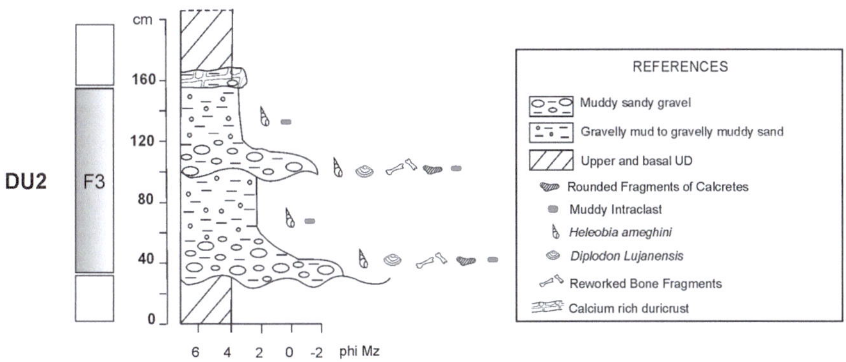

Fig. 4 Schematic model for the stratigraphic sequence of depositional unit 2—MIS 3 outcrops and lithostratigraphic column with depositional unit 2 (DU2) and Facies 3 characteristics

feldspar (mostly plagioclases) and some bioclasts such as small broken shell fragments and undetermined bone fragments. There are scarce volcanic glass shards. The clay fraction is represented by quartz, plagioclase, and clay minerals.

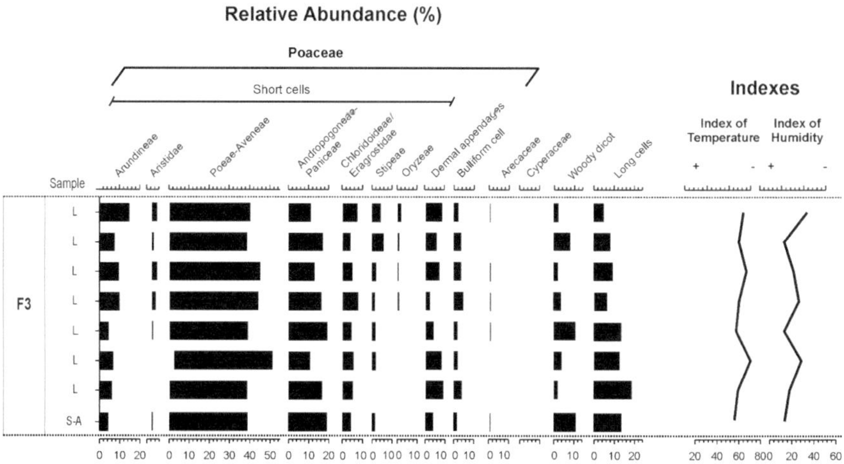

Fig. 5 Phytolytic diagram. Relative percentages of abundance of different gramineae morphotypes in samples from diverse Luján river sections and Barranca Salto section. Index of temperature and humidity, in the different samples showing fluctuating in climatic conditions (modified from Blasi et al. 2010)

The crystals which appear with a small amount are gypsum/anhydrite and calcite/dolomite. Illite is the dominant clay mineral, and kaolinite occurs in a smaller proportion.

In the biosilica content there is a broad predominance of gramineae phytoliths. According to the identified morphotypes, short winter gramineae cells, mainly Avenae and Poeae tribes prevail. They are followed by morphotypes assigned to species of the subfamily Arundinoideae, mainly from the tribe Arundinae. This family has C_4 species within the Aristidae (Aristide) tribe and C_3 in the Arundinae tribe (e.g. *Cortaderia* Stapf). Other gramineae C_3 morphotypes correspond to the Oryzeae tribe. C_4 species are mainly represented by morphotypes produced by the subfamilies Panicoideae and Chloridoideae (Fig. 5).

The diatom record shows abundance of the planktonic species *Cyclotella meneghiniana* Kützing. Other frequent oligohalobus and mesohalobous bentonic taxa are *Amphora copulate* (Kützing) Schoeman and R.E.M. Archibald, and *Navicula peregrina*, (Ehrenberg) Kützing, respectively. This association is also formed by epiphyte species such as *Cocconeis placentula* Ehrenberg with oligohalobus tychoplanktonic features, the halobius *Fragilaria capucina* Desmazières, and oligohalobus indifferent aerophiles such as *Pinnularia borealis* Ehrenberg (Table 2) (Fig. 6).

Two extinct mollusk species were identified: the gastropod *Heleobia ameghini* (Doering) and the bivalve *Diplodon lujanensis* Ihering, (Blasi et al. 2010; De Francesco and Blasi 2012). In the Luján river basin only one and abundant ostracod, *Cyprideis salebrosa hartmanni* Ramírez, was found. In the DU2 of the Salto-Arrecifes basin, *C. salebrosa hartmanni* represents the dominant fraction of

Table 2 Relative frequency and ecological characteristics of the diatom taxa

		Diatom taxa	Abundance	Habitat	Salinity
UD2	Facies 3	*Amphora copulata* (Kütz.) Schoeman & Arch	F	B	OI
		Anomoeoneis sphaerophora (Ehr.) Pfitzer	S	B	OH
		Cocconeis placentula Ehr.	F	E	OI
		Cyclotella meneghiniana Kütz	A-VA	P	OH
		Cymbella cymbiformis Ehr.	S	E	OI
		Epithemia adnata (Kütz) Bréb.	A	E	OI
		Navicula peregrina (Ehr.) Kütz	F-A	B	M
		Nitzschia amphibia Grun.	A	Aer	OI
		Pinnularia borealis Ehr.	F	Aer	OI
		Surirella minuta Bréb.	R	B	Po
		Tabularia tabulata Agardh	F	B	OH
		Gomphonema lujanensis Reichardt & Maidana	R	B	OI

References F frequent, *S* scarce, *A* abundant, *VA* very abundant, *R* rare, *B* benthic, *E* epiphytic, *P* planktonic, *Aer* aerphilous, *Ol* oligohalobus indifferent, *OH* oligohalobus halophyllus, *M* mesohalobus, *Po* polyhalobus

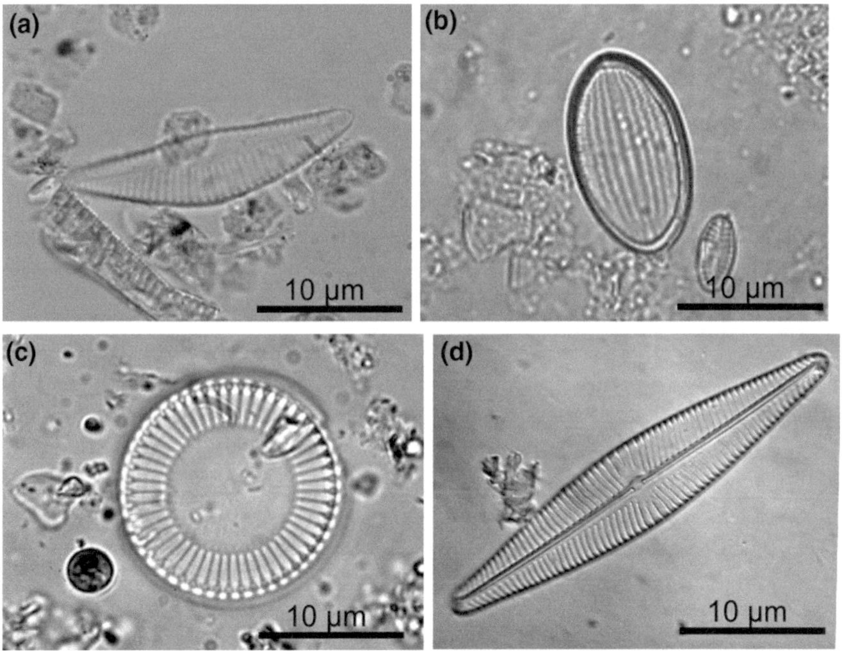

Fig. 6 The most abundant taxa of diatoms found in DU2. **a** *Amphora copulate*; **b** *Cocconeis placentula*; **c** *Cyclotella meneghiniana*; **d** *Epithemia adnata*

the ostracod association; accompanied in a lesser extent by species of *Limnocythere* Brady and *Ilyocypris* Brady and Norman. The ostracod association recovered from the DU2 is characterized by adult and juvenile specimens suggesting an autochthonous fauna (Boomer et al. 2003) (Fig. 7). Noteworthy, in the Arrecifes site (Table 1) and particularly in the DU2 deposits, bone remains that may be assigned to *Megatherium* Cuvier were found (S. Vizcaino, pers. comm.).

Radiocarbon ages were obtained from specimens of *Cyprideis salebrosa hartmanni, Heleobia ameghini,* and *Diplodon lujanensis*. In two cases, the ages were beyond the limits of the radiocarbon dating method. One age was also obtained by means of the IRSL technique (Table 3).

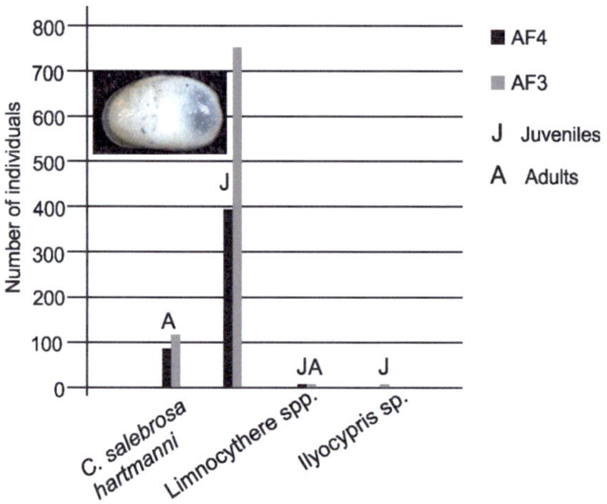

Fig. 7 Ostracod species and their abundance in DU2 for the Arrecifes river section; illustrated ostracod: *Cyprideis salebrosa hartmanni*, female left valve

Table 3 Radiocarbon and IRSL dating

Age (a: ^{14}C yr BP) (b: IRSL)	Laboratory number	Sample	Geographic coordinates	Stratigraphic assignment
37.710 ± 840 a	Beta—217826	*Cyprideis salebrosa hartmanni*	34° 34′ 08″ S 59° 07′ 29″ W	UD2 Luján river (Blasi et al. 2010)
>40.000 a	LP—1733	*Helobia ameghini*	34° 34′ 55″ S 59° 10′ 20″ W	
44.000 ± 6500 b	UNL—1928	feldespar	34° 34′ 08″ S 59° 07′ 29″ W	
>40.000 a	LP—2985	*Diplodon* sp.	34° 04′ 4.4″ S 60° 05′ 29″ W	UD2 Salto-Arrecifes rivers

Beta Beta Analytic; *LP* LATYR, La Plata; *UNL* University of Nebraska, Lincoln

It should be noted that the green color of DU2 was used to correlate outcroppings of the same color appearing in different basins of other geographic areas, independently from their stratigraphic or temporal location. It was also used as an informal stratigraphic nomenclature. Di Lello et al. (2009) determined by Mössbauer spectroscopy that the dark green color of a hermetically kept sample dried in argon atmosphere showed a couplet corresponding to about 10 % spectra of ferrous iron (Fe^{2+}). Some authors suggested that this could come from minerals of the original sediment (such as ferrous illite), neoformed minerals, and incorporation and/or adsorption of ferrous ion in crystalline structures under negative Eh conditions.

6 Paleoenvironmental Reconstruction

The DU2 facies 3 was interpreted as a mixed or hybrid deposit (Pettijohn et al. 1987) with two origins: the deposit of detritus transported by channelized floods and sheet floods of sporadic flash floods generated by strong rains under arid climatic conditions or strongly seasonal, and decantation of very fine sand and muddy sediments in lentic bodies formed afterwards.

Channelized or sheet flows could have been drained to lower areas of the land carving numerous gullies or streamlets in older deposits (the DU1/facies F2). This mechanism is considered as the greatest provider and transporter of a large amount of sediments (sediment delivery, sensu Hooke and Mant 2002) in environments characterized by water deficit. At a later stage, gullies or streamlets could have experienced damming by deposition of the bedload (tractionload), originating the development of temporary pools or small lentic bodies. With another climatic fluctuation, a new cycle with channelized flows developed and subsequent small lentic bodies may have restarted. Intraformational clast-size gravel may have formed during the first stage of intense erosion due to sheet flows and streamlets runoff. These are formed by erosion and reworked older sediments, which act as source of fine-grained sedimentary lithoclast (muddy intraclast), fragments of calcretes (locally known as “tosca”) and bone fragments. This coarse bedload is transported under high hydrodynamic flow conditions. At a later stage, sandy and/or sandy/muddy sediments may have accumulated among the high energy open gravels once a reduction of energy for load transport is produced. There may have also been some input of fine eolian vitroclastic material defined as sandy loess by Zárate and Blasi (1991) that may have been masked by large amounts of local alluvium deposits.

Temporary pools were formed by damming of streamlets, no major depositional processes occurred, and the diversities of diatoms, ostracods and mollusks are low. During wetter periods, there were probably flooding events and generation of lentic bodies with larger water surface. During the most arid periods, these meso- to eutrophic water lentic bodies may have undergone environmental stress due to the sharp fluctuation of water level, salinity, and temperature. That was probably the reason for the development of monospecific microfauna of ostracod, mollusk and diatom, such as *Cyprideis salebrosa hartmanni*, *Heleobia amheghini,* and *Diplodom lujanensis*, as well as frequent mesohalobuis algae (*Navicula peregrina*) which are tolerant of greater salinity. As mentioned before the ostracod association was dominated by *C. salebrosa hartmanni* together with species of *Limnocythere* and *Ilyocypris*. *Cyprideis* is aeuryhaline genus which lives also in fresh to brackish waters, and is very adaptable to this type of environmental conditions. Ornellas and Wurdig (1983) found this species in hyposaline environments (0–29 psu) with a temperature range from 15 to 25° C in sandy silt sediments rich in organic matter and littoral vegetation. On the other hand, Ferrero (1996, 2009) suggested that this species is typical to brackish environments, and Ramírez (1967) mentioned it in limnic environments of Buenos Aires Province. The faunal association, recovered in this study suggests similar conditions, from fresh to oligohaline environments. Also, the abundance of aerophilous diatoms suggests that the water bodies or pools may have been intermittent or temporary; whereas the strong seasonal rain under temperate to colder climatic conditions would be evidenced by the larger presence of gramineae phytoliths C3 and a high representation of chloridoid morphotypes within C4.

7 Correlation

Blasi et al. (2010) correlated DU2 with the different stratigraphic units defined by different authors in the Luján area. In this way, DU2 is related to the "Pampeano Lacustre" terrain of "Villa de Luján" (layer 3), as it has been defined by Ameghino (1880–1881: 568) and Layers 5, 6, and 7 defined in Paso de Azpeitía (Ameghino 1880–1881: 567). Furthermore, DU2 correlates with the deposits assigned to the true "Piso Pampeano Lacustre" (layers 9, 8 and 7) by Ameghino (1884: 165) and with the "Piso Lujanense" (following Ameghino 1889).

This unit is also correlated with the yellowish green, sandy and nodular, lacustrine sediment bearing mammal remains of the "Bonaerense lacustre" or "Lujanense" beds as defined by Rovereto (1914). It also correlates with the "Prebonaerense" stage of Frenguelli (1921: 66) and/or the "Lujanense" beds of Frenguelli (1928). There are also similarities with the Units1 A and B as defined by Dangavs and Blasi (1995) in the Luján river, (Blasi et al. 2010).

Recently, Toledo (2011) included an unconformity named as "Interlujanense" in the stratigraphic scheme of Ameghino (1880–1881), overlying a greenish, fining upward sequence with radiocarbon ages from 46,500 ± 4000 years ^{14}C B.P. to 32,000 ± 1400 years ^{14}C B.P., (the "Lujanense Verde Inferior" sequence). Above the unconformity is another fining upward deposit, (the "Lujanense Rojo" sequence), that starts with a basal conglomerate.

In this sense, most of the UD2 discussed in this paper may be correlated with the depositional sequence "Lujanense Verde Inferior" ("Lower Green Lujanense") or the Jáuregui Member of the Luján Formation (Toledo 2011), whereas an upper portion of the DU2 (the upper conglomerate level shown in Fig. 3) may be correlated with the level defined by Toledo (2011) as basis for the "Lujanense Rojo" ("Red Lujanense depositional sequence") or basal section of the La Eloísa Member of the Luján Formation.

It should be noted that the Luján Formation was defined by Fidalgo et al. (1973) for the Río Salado basin (Blasi et al. 2009b) and that this lithostratigraphic unit was subdivided in two members: the Guerrero and Río Salado members.

However, Toledo (2011) created another Luján Formation, to define the Late Pleistocene interval of the previously Luján Formation as defined by Fidalgo et al. (1973). Similarly, the Luján Formation is renamed by Toledo (2011) (in its Holocene or Río Salado member portion, Fidalgo et al. 1973) as a new lithostratigraphic unit called the La Plata Formation. However, this author was not in compliance with the provisions of the Argentine Stratigraphic Nomenclature Code (Comité Argentino de Estratigrafía 1992), Sect. 31.13 (page 29) on Lithostratigraphic Units, which specifies that no identical name given to a previous lithostratigraphic unit can be used for a new one.

8 Why Is UD2 (MIS3) Present Only in Short and Middle Areas of the River Basins?

The depositional unit DU2 has only been recognized in restricted and discontinued areas of sections located in the middle portion of the fluvial basins (Fig. 1). This information is quite significant for the understanding of the evolution of the area from the Late Pleistocene to the Holocene, and contributes to the paleontological research, limiting the expectations of finding fossil remains plausible of being assigned to the interval 60,000–30,000 years B.P. The DU2 could be observed only in streches about 7 km long in the main course of the Luján river basin and up to the present, this is the same for the Salto-Arrecifes rivers basin (Fig. 1).

In order to explain the reasons for the absence of outcrops of DU2 upstream and downstream of the mentioned stretches, some hypotheses may be proposed to be tested in future works. Although they are not the object of this work, so far, it is believed that these hypotheses are relevant in the context of this chapter.

Hypothesis 1: This phenomenon could be related to the existence, within the interval of accumulation of DU2 of the "parallel ravine model" posed by Pasotti (1971, 1973).

Pasotti (1972, 1973) posed for the first time the question that the "Undulated Pampa" could show the overlapping of two hydrological models formed at two different times. During the Pleistocene, the "Collinear" paleomodel of "parallel ravine model" and/or "last paleomodel" would have been active; in turn, during the Holocene, the hydrographic network developed the present ("grilled") configuration. The collinear model is represented by a series of straight paleo-ravines parallel to each other, regularly separated, which formed no hierarchical networks, SW-NE bound, with high gradient draining a eastern more bound, "broad and uniform" Pampa (Pasotti 1971, 1973, 1974, 2000). Pasotti (1972) stated that these models appear from the Saladillo river up to Arroyo del Medio (at the Santa Fe-Buenos Aires provincial border) and in the "Pampa Bonaerense", "farther south than the Arrecifes river". According to this author, the present networks cut the paleo-ravines at differing angles (straight and obtuse) which, in some cases, match those of the older network, but only in short stretches. In 1971–1973 Pasotti stated: "I think this network is the last of Pleistocene age and I named it as the last paleo-model" (Pasotti 2000: 4; free translation by the present authors). "The present hydrographic networks superpose the previous one, cutting the ravines or streamlets with differing angles that due to tectonics match only as an exception and in short stretches" (Pasotti 1971, 1973).

These ideas allowed the present authors to formulate the second hypothesis:

Hypothesis 2: The small area where DU2 can be found is due not only to the present network overlapping with the Pleistocene draining network (ravines) but also to a differential tectonic control that caused that this unit could only be observed at the base of cut banks in stretches coincident with raised tectonic blocks.

Neotectonic records of the Argentine Pampa plains were studied by Brunetto and Iriondo (2007). These authors developed several analyses including studies of the structural alignments, identification of geomorphological units and topographic features, from the interpretation of satellite images and field observations, thus being able to demarcate structural blocks. They also referred that the changes exerted by such structural elements on the direction of the superficial drainage allowed for the existence of some degree of deformation during the Late Pleistocene.

More recently, Racca (2010) in his regional scale studies on Arroyo del Medio fluvial basin (provincial boundary between the Provinces of Santa Fe and Buenos Aires), related the origin of greater-sized landforms (alignment) within the basin to neotectonic factors. Therefore, according to the this author, the study area would be affected by large movements due to very recent tectonic events (Holocene) that may have modified the existing of extremely flat topography (Late Pleistocene) and Pasotti's collinear model (1972). This highlights that the macromorphology of the area under study would have tectonic origin, minimizing the possible influence of paleoclimatic variability.

For the area of the Río de La Plata, Cavallotto (2002) concluded that through the pre-Holocene general topographic characteristics, the existence of a tectonic control shown by transversally oriented river alignments had been identified. On the other hand, the differences in radiometric values of the elevation of the Holocene maximum sea transgression registered along the Argentine coast would also reflect perhaps a neotectonic influence (Codignotto et al. 1992).

9 Conclusions

The depositional unit 2 (DU2) overlies an erosive unconformity on eolian sediment with IRSL ages of ~56,400 ± 6500 and 50,400 ± 10,200 years. These sediments are also overlain by sandy loess deposits of the same eolian genesis (UD3, in Blasi et al. 2010) dated ca. 32,500 ± 4100 years (IRSL) (Blasi et al. 2010).

Ages obtained for the lower and the upper DU2 deposits suggest an association with the Marine Isotopic Stage 3.

For this unit (DU2) a fluvial-lacustrine facies (F3) has been identified, representing ephemeral fluvial streams with gullies or streamlets formed during instantaneous rainstorm episodes and development of temporary pools of eutotrophic to distrophic characteristics formed by subsequent damming of the channels. The dominant lithology is sandy gravelly mud. These sediments are hybrid or mixed deposit according to Pettijohn et al. (1987), with predominance in the gravel size fraction of muddy intraclast, rounded fragments of calcretes ("tosca") and pebbled of bones. The sandy fraction is mainly quartz in composition. The scarce presence of volcanic glass in its petrographic characteristics should also be noted, whereas clay minerals are mainly illite.

Olive green coloring is highlighted by the presence of reduced iron ions in the illitic clay matrix (Di Lello et al. 2009). Some authors suggested that it could have originated in the alteration of detrital illite clay and/or adsorption of ferrous ion in crystalline structures under Eh negative conditions. Therefore, this attribute could repeat itself in any sector of the stratigraphic columns that may have been modified by the same processes.

Another relevant feature is the presence of mollusk species such as *Diplodon lujanensis* and *Heleobia ameghini*, that lack representation in modern fauna and which constitute an exceptional case for the Late Quaternary of Argentina (De Francesco and Blasi 2012).

Ages ascribed to the MIS3 were obtained from specimens of *Cyprideis salebrosa hartmanni*, *Heleobia ameghini,* and *Diplodon lujanensis*. In two cases, the obtained ages were beyond the accepted, reliable limits of the radiocarbon dating method. Age was also obtained through the IRSL technique. Climatic conditions inferred for the DU2 deposits were from rather temperate to colder, with subhumid–humid and subhumid—drier phases or strongly seasonal.

Likewise, the following questions were raised: Did the Luján and the Salto-Arrecifes rivers basins flow in the Late Pleistocene (during MIS3) as gullies or ephemeral stream lets subparallel to each other with high gradient and drainage courses with SW-NE direction? Have the geofractures associated with the crystalline basement and minor fractures in the sedimentary filling been active during the Holocene? And therefore, is there tectonic control over the present configuration of drainage networks and exposition of sequences associated with the MIS3? Can the superposition of both models be located in some uplifted blocks, and thus explain that they only appear in the referred stretches?

Acknowledgments The authors thank E. Apolinaire for his assistance with the edition of the maps. Special thanks to the Editorial Committee for the invitation to participate in this publication. Reviews by Cecilia Deschamps, Jorge Rabassa and Germán M. Gasparini referees improved the manuscript. This work was supported by CONICET-PIP 5086 and CONICET-PIP 0342; Ostracod studies (by G.C. and A.C.) were supported by ANPCyT-PICT 2010-0082 and 2014-1271, CONICET-PIP 819 and CONICET-PIP 0021, and UNCo B 166.

References

Ameghino F (1880–1881) La Antigüedad del Hombre en el Plata. Oficina de Gobierno Provincia de Buenos Aires, La Plata

Ameghino F (1884) Excursiones geológicas y paleontológicas en la provincia de Buenos Aires. Boletín Academia Nacional de Ciencias de Córdoba VI:161–257

Ameghino F (1889) Contribución al conocimiento de los mamíferos fósiles de la República Argentina. Academia Nacional de Ciencias (Córdoba) 6:1–102

Bertels A, Martínez D (1990) Quaternary ostracodes of continental and transicional littoral shallow marine environments. Courier Forschungsinstitut Senckenberg 123:141–159

Blasi A, Hanson P, Fucks E, Prieto A, Young A (2009a) Infrared stimulated luminescence (IRSL) dating of late Pleistocene deposits from the middle course of the Luján river. Argentina. In: Actas IV Congreso Argentino de Cuaternario y Geomorfología, XII Congresso da Associação Brasileira de Estudos do Quaternário and II Reunión sobre el Cuaternario de América del Sur, La Plata, p 228

Blasi A, Prieto A, Fucks E, Figini A (2009b) Análisis de las nomenclaturas y de los esquemas estratigráficos del Pleistoceno tardío—Holoceno en la cuenca del río Lujan, Buenos Aires, Argentina. Ameghiniana 46(2):373–390

Blasi A, Castiñeira C, del Puerto L, Prieto A, Fucks E, de Francesco C, Hanson P, García-Rodríguez F, Huarte R, Carbonari J, Young A (2010) Paleoambientes de la Cuenca Media del río Luján (Buenos Aires, Argentina) durante el último Período Glacial (EIO 4-2). Lat Am J Sedimentol Basin Anal 17(2):85–112

Bond G, Broecker W, Johnsen S, Mc Manus J, Labeyrie L, Jouzel J, Bonani G (1993) Correlations between climate records from North Atlantic sediments and Greenland ice. Nature 365:143–147

Boomer I, Horne DJ, Slipper IJ (2003) The use of ostracods in Palaeoenvironmental studies, or what can you do with an ostracod shell? Paleontol Soc Pap 9:153–180

Bozarth S (1992) Classification of opal phytoliths formed in selected dicotyledons native to the Great Plains. In: Rapp G, Mulholland S (eds) Phytolith Systematics. Plenum, Nueva York

Brunetto E, Iriondo M (2007) Neotectónica en la Pampa norte (Argentina). Revista de la Sociedad Geológica de España 20(1–2):17–29

Cavallotto JL (2002) Evolución holocena de la llanura costera del margen sur del Río de la Plata. Rev Asoc Geol Argent 57:376–388

Carver R (1971) Procedures in Sedimentary Petrology. Wiley, New York

Clark P, Hostetler SW, Pisias NG, Schmittner A, Meissner KJ (2007) Mechanisms for a $\sim$ 7-Kyr climate and sea-level oscillation during marine isotope stage 3. In: Schmitter A, Chiang J, Hemming S (eds) Ocean circulation: mechanisms and impacts. American Geophysical Union, Geophysical Monograph 173, Washington

Codignotto JO, Kokot R, Marcomini S (1992) Neotectonism and sea-level changes in the coastal zone of Argentina. J Coastal Res 8(1):125–133

Comité Argentino de Estratigrafía (CAE) (1992) Código Argentino de Estratigrafía. Asociación Geológica Argentina Serie B N°, Buenos Aires, 20:1–64

Dangavs N, Blasi A (1995) El Lujanense y Platense (sensu Ameghino) en el Río Luján, Luján, Provincia de Buenos Aires. In: Actas IV Jornadas Geológicas y Geofísicas Bonaerenses, Junín, (I):109–117

De Francesco C, Blasi A (2012) Redescripción y significado paleoambiental de *Heleobia ameghini* (Doering, 1882) (Gastropoda: Rissooidea) en el Pleistoceno tardío de la provincia de Buenos Aires, Argentina. Ameghiniana 49:17–25
del Puerto L (2009) Silicofitolitos como Indicadores Paleoambientales: Bases comparativas y reconstrucción paleoclimática desde el Pleistoceno tardío en el SE del Uruguay. Pedeciba-Facultad de Ciencias de la Universidad de la República, Montevideo
del Puerto L, García-Rodríguez F, Inda H, Bracco R, Castiñeira C, Adams JB (2006) Paleolimnological evidence of Holocene paleoclimatic changes in Lake Blanca, southern Uruguay. J Paleolimnol 36:151–163
Di Lello C, Blasi A, Mercader R, Desimoni J (2009) El color de los sedimentos fluvio-lacustres del Pleistoceno tardío en la cuenca media del río Luján, provincia de Buenos Aires, Argentina. I Reunión Argentina de Geoquímica de la Superficie, Córdoba, 1:17
Duff K, Zeeb B, Smol J (1995) Atlas of chrysophycean cysts. Developments hydrobiology 99. Kluwer Academic, Dordecht
Ezcurra de Drago I (1993) Distribución geográfica de las esponjas Argentinas (Porifera: Spongillidae, Potamolepidae y Metaniidae). In: Boltovskoy L (ed) Relaciones zoogeográficas y vías de poblamiento. Instituto de Limnología "Dr. Raúl A. Ringuelet", La Plata
Fernández M, Zucol A, Osterrieth M (2006) Phytolith assemblages and systematic associations in grassland species of the south-Eastern Pampean Plains, Argentina. Ann Bot 98:1155–1165
Ferrero L (1996) Paleoecología de ostrácodos holocenos del estuario del río Quequén Grande (Provincia de Buenos Aires). Ameghiniana 33:209–222
Ferrero L (2009) Foraminíferos y ostrácodos del Pleistoceno tardío (Mar Chiquita, provincia de Buenos Aires, Argentina). Ameghiniana 46:637–656
Fidalgo F, De Francesco F, Colado U (1973) Geología superficial en las hojas Castelli, J. M. Cobo y Monasterio (Pcia. de Bs. As.). V Congreso Geológico Argentino, Actas, Carlos Paz, IV:27–39
Fredlund GG, Tieszen LT (1994) Modern phytolith assemblages from the North American Great Plains. J Biogeogr 21:312–335
Frenguelli J (1921) Los Terrenos de la Costa Atlántica en los alrededores de Miramar (Prov. de Buenos Aires) y sus correlaciones. Boletín de la Academia Nacional de Ciencias de Córdoba XXIV:325–485
Frenguelli J (1928) Observaciones geológicas en la región costera sur de la Provincia de Buenos Aires. Anales de la Facultad de Ciencias de la Educación II:1–145
Frenguelli J (1941) Diatomeas del Río de la Plata. Rev Museo Plata 3:213–334
Frenguelli J (1945) Las diatomeas del Platense. Revista del Museo de La Plata, III:77–221
Folk RL (1954) The distinction between grain size and mineral composition in sedimentary rock nomenclature. J Geol 62(4):344–359
Gallego L, Distel R (2004) Phytolith assemblages in grasses native to Central Argentina. Ann Bot 94:865–874
Hooke JM, Mant JM (2002) Morpho-dynamics of ephemeral streams. In: Bull LJ, Kirkby MJ (eds) Dryland rivers. Wiley, Chichester
Metzeltín D, García-Rodríguez F (2003) Las Diatomeas Uruguayas. DIRAC, Montevideo
Metzeltín D, Langebertalot H, García-Rodríguez F (2005) Diatoms of Uruguay: taxonomy, biogeography, diversity. In: Lange-Bertalot H. (ed) Iconographia Diatomologica 15. A.R.G. Gantner Verlag, Koenigstein
Moore RC, Pitrat CW (1961) Treatise on invertebrate paleontology. Part Q. Arthropoda 3. Crustacea, Ostracoda. Geological Society of America and University of Kansas Press, New York
Ornellas LP, Wurdig N (1983) *Cyprideis salebrosa hartmanni* Ramirez, F., 1967, A new subspecies from Brazil and Argentina. Pesquizas 15:94–112
Pasotti P (1971) Influencia de un Paleomodelo de red hidrográfica en la llanura de Santa Fé. In: Quinto Congreso Nacional del Agua, Santa Fe
Pasotti P (1972) Sobre la existencia de un Paleomodelo de red hidrográfica en la Provincia de Santa Fe. Boletín Filial Rosario, Sociedad Argentina de Estudios Geográficos GÆA 5, Rosario

Pasotti P (1973) Sobre la presencia del último paleomodelo de red hidrográfica de edad pleistocénica en la llanura de la Provincia de Santa Fe. Publicaciones del Instituto de Fisiografía y Geología 57, Rosario

Pasotti P (1974) La Neotectónica en la llanura pampeana. Fundamentos para el mapa neotectónico. Publicaciones del Instituto de Fisiografía y Geología 58, Rosario

Pasotti P (2000) La geomorfología de la llanura pampeana en territorio santafesino. Boletín del Instituto de Fisiografía y Geología 70 (1–2):11–13, Rosario

Pasotti P, Castellanos A (1967) Rasgos morfológicos generales de la llanura pampeana. Conferencia Regional Latinoamericana de U.G.I., México, In: Boletín de la Filial Rosario de la Sociedad Argentina de Estudios Geográficos 3

Pettijohn FJ, Potter PE, Siever R (1987) Sand and sandstone. 2nd edn. Springer, New York

Prieto AR, Blasi A, De Francesco C, Fernández C (2004) Environmental history since 11,000 ^{14}C yr B.P. of the northeastern Pampas, Argentina, from alluvial sequences of the Luján river. Quatern Res 62(2):146–161

Rabassa J, Ponce JF (2013) The Heinrich and Dansgaard-Oeschger climatic events during Marine Isotopic Stage 3: Searching for appropriate times for human colonization of the Americas. Quatern Int 299:94–105

Racca JMG (2010) Geomorfología de la Cuenca del Arroyo del Medio. Boletín del Instituto de Fisiografía y Geología, Rosario, 72–75:13–42

Ramírez F (1967) Ostrácodos de Lagunas de la Provincia de Buenos Aires. Revista del Museo de La Plata X:5–54

Rovereto G (1914) Studi di Geomorfología Argentina. IV. La Pampa. Bolletino della Societá Geologica Italiana XXXIII:75–129

Sinddall M, Rohling EJ, Thompson WG, Waelbroeck C (2008) Marine isotope stage 3 sea level fluctuations: data synthesis and new outlook. Rev Geophys 46:1–29

Sowers J (2000) Correlating quaternary landforms and deposits to global climate change. In: Noller JS, Sowers JM, Lettis WR (eds) Quaternary geochronology. Methods and Applications. American Geophysical Union, Washington, DC

Toledo M (2011) El Legado Lujanense de Ameghino: revisión estratigráfica de los depósitos Pleistocenos-Holocenos del valle del río Luján en su sección tipo. Registro paleoclimático en la Pampa de los estadios OIS 4 al OIS 1. Revista de la Asociación Geológica Argentina 68 (1):121–167

Twiss P (1992) Predicted world distribution of C3 and C4 grass phytoliths. In: Rapp G, Mulholland S (eds) Phytolith systematic: emerging issues. Advances in archaeological and museum science, vol 1. Plenum Press, New York

Van Meerbeeck CJ, Renssen H, Roche DM (2009) How did marine isotope stage 3 and last maximum climates differ?—Perspectives from equilibrium simulations. Clim Past 5:33–51

Van Morkhoven FPCM (1963) Post-palaeozoic Ostracoda. Their morphology, taxonomy and economic use. Elsevier, Amsterdam

Witkowski A, Lange-Bertalot H, Metzeltin D (2000) Diatom flora of marine coasts. In: Lange-Bertalot H (ed). Iconograhia diatomologic, vol.7 Koeltz Scientific, Koeingstein

Zárate M, Blasi A (1991) Late Pleistocene and Holocene loess deposits of the Southeastern Buenos Aires Province, Argentina. GeoJournal 24:211–220

Zucol AF (1998) Microfitolitos de las Poaceae Argentinas: II. Microfitolitos foliares de algunas especies del género Panicum (Poaceae, Paniceae), en la provincia de Entre Ríos. Darwiniana 36:29–50

Zucol AF (2000) Fitolitos de poaceae de Argentina: III. Fitolitos foliares de especies del género Paspalum (Paniceae), en la Provincia de Entre Ríos. Darwiniana 38:11–32

Zucol AF (2001) Fitolitos III. Una nueva metodología descriptiva. Asociaciones fotolíticas de Piptochaetium montevidense (Stipeae, Poaceae). Boletín de la Sociedad Argentina de Botánica 36:69–85

Sea Level Changes During Marine Isotopic Stage 3 (MIS 3) in Argentina

Federico Ignacio Isla and Enrique Jorge Schnack

Abstract Marine Isotopic Stage 3 (MIS 3) was a period of rapid climatic changes and sea level fluctuations. Regarding these fluctuations some doubts were based on the limit of the radiocarbon dating method (about 50,000 years B.P.). However, the modelling of the isotopic oxygen ratios is also indicating sea level fluctuations. In this sense, only at certain depths it is possible to accept these ages at stable coasts, and taking note about the taphonomic conditions within the sequence. Shells located at depths higher than 60 m on stable and wide continental shelves as that of the Northern Patagonia have been analyzed here in that sense.

Keywords Sea level changes · MIS 3 · Tectonic effects · Patagonia

1 Introduction

During the 1970s and 1980s radiocarbon dates performed on mollusc shells from coastal deposits yielded ages between 30,000 and 40,000 years B.P. As these ages were in concordance with an amelioration of climate, a Mid-Wisconsinan marine transgression was proposed. Some radiocarbon laboratories certified the validity of such ages whereas others doubted about them as thcy wcrc too close to the maximum older limit of the ^{14}C dating method (Radtke 1988; Schnack and Pirazzoli 1990). In many cases, other dating methods (U/Th or amino acid racemization) confirmed these doubts; several molluscs were then attributed to the Sangamonian

F.I. Isla (✉)
Instituto de Geología de Costas y del Cuaternario, Instituto de Investigaciones Marinas y Costeras, Facultad de Ciencias Exactas y Naturales, Universidad Nacional de Mar del Plata, 7600 Mar del Plata, Argentina
e-mail: fisla@mdp.edu.ar

E.J. Schnack
Laboratorio de Oceanografía Costera, Facultad de Ciencias Naturales y Museo, Universidad Nacional de La Plata, 1900 La Plata, Argentina
e-mail: enrique.schnack@gmail.com

G.M. Gasparini et al. (eds.), *Marine Isotope Stage 3 in Southern South America, 60 ka B.P.–30 ka B.P.*, Springer Earth System Sciences,
DOI 10.1007/978-3-319-40000-6_8

(Tyrrhenian) highstand (Schellmann and Radtke 1997). Only at coasts with rapid uplift, a highstand of 40,000 years B.P. could have been above present sea level. Based on assumptions of uniform uplifting trends, deposits at depths of −40, −20, and −10 m below mean sea level (MSL) could be accepted (Isla 1988).

At present, and independently from tectonic trends, the modelling of sea level can be estimated from oceanic paleotemperatures derived from oxygen isotopic proportion contents from benthic or planktonic forams (Chappell and Shackleton 1986; Siddall et al. 2008, 2010). Several fluctuations were accepted with maximum highstand of −60 m. However, studies from individual cores, and others based on average records, indicate that there were at least 4 fluctuations of the order of 30 m during the Marine Isotopic Stage 3 (MIS 3) interval (Siddall et al. 2008).

In the present paper, an update of this information is analyzed considering data collected from the continental shelf of Argentina.

2 Hydroisostasy, Glacioisostasy, and Local Tectonics

To discriminate sea level variations from those changes associated to endogenic processes, it is worth to recognize the origin of these fluctuations. In this sense, the uplift assigned to the "broad-shelf effect" or hydroisostasy (Siddall et al. 2008) was estimated about 0.8 mm/year for the Argentine continental shelf (Guilderson et al. 2000).

Regarding the tectonic conditions for the Patagonian coast, there is a certain controversy between those who denied it for the last 7000 years (Schellmann and Radtke 2010), those who believed in a differential behaviour of tectonic blocks (Codignotto et al. 1992), and those who suggested that the regional uplift has taken place since the Pleistocene with a mean rate of 0.12 mm/year (Pedoja et al. 2011), but peaking to a maximum of 1.2 mm/year at Tierra del Fuego (Isla and Bujalesky 2008). The higher uplifting trends of northern Santa Cruz were produced by the volcanism associated to the subduction of the Chile Ridge (Isla et al. 2015).

3 Methods

Several radiocarbon data which form mollusc shells were considered in this review. However, some dates should be handled with care because they were close to the limit of the ordinary decayment method (Martínez et al. 2001). Conventional radiocarbon decayment and Accelerator Mass Spectrometry have the same dating limit, approximately 50,000 years B.P. (Linick et al. 1989), although some laboratories extended this limit by isotopic enrichment (Walker 2005).

Original descriptions and photographs about the cores obtained by the Vemma vessel in 1961 can be downloaded at http://www.ngdc.noaa.gov/mgg/curator/data/vema/vm17/.

Core VM 17-112 (404 cm long) was collected from 40° 32′ S and 60° 19′ W at 59 m depth, and it was revised by one of the present authors (EJS). Core VM 17-117 (418 cm long) was collected from 82 m depth at 41° 41′S and 59° 19′W. Both cores are bearing shells at their bottoms dated between 36,310 and 53,860 years BP.

Original data from shells collected in Northern Patagonia were already described (Rutter et al. 1989). Purified aminoacids from mollusc shells were extracted using acidified ethanol and pentafluoropropionic anhydride. They were afterwards separated by gas chromatography using a Chirasil-Val capillary column. D/L ratios of aspartic acid and leucine were compared for different molluscan species, either fossil or living (Rutter et al. 1989).

4 Results

4.1 Radiocarbon Data

Considering the sources of error described above, it is necessary to be sure that the shells sampled at the continental shelf were not reworked. This is the case of the radiocarbon dates performed on samples from the core Vemma 17-112, obtained at 59 m depth and described by C. Fray and T. Willis in 1962 (Fig. 1). From the first metre of the core, mollusc fragments obtained from a gravel layer yielded an age of 10, 070 ± 50 years B.P. (cal. age 10,620 years; Guilderson et al. 2000). Toward the bottom of the core, another layer with gravel and shell fragments were described as a beach with beach-rock fragments. This layer gave ages of 36,310 ± 430 and 44,920 ± 680 years B.P., therefore suggesting a Late Pleistocene marine highstand that could correspond to the MIS 3 interstadial in relation to the depth (approximately 62.88 m; Guilderson et al. 2000).

The radiocarbon dates from core VM 17-117 gave ages of 52,730 ± 2960 and 53,860 ± 2030 years B.P. for the level located 85.5 m below MSL. Large shells fragments of the genus *Donisia, Pecten, Venus* and *Tellina* were scattered within muds. Another shell fragment at a depth of 84.3 m gave a modern age of 8900 ± 50 years B.P. (original description by M. Morgenstein 1965).

Radiocarbon data performed on mollusc shells from the continental shelf of Argentina permitted to propose a Holocene sea-level rise curve (Guilderson et al. 2000; Isla 2013). Beyond the Pleistocene-Holocene transition, the data showed significant variations during the MIS 3 interstadial, corresponding to cores VM 17-112 and VM 17-117 (Fig. 2). These fluctuations could be assigned to the Dansgaard–Oeschger (D–O) variations.

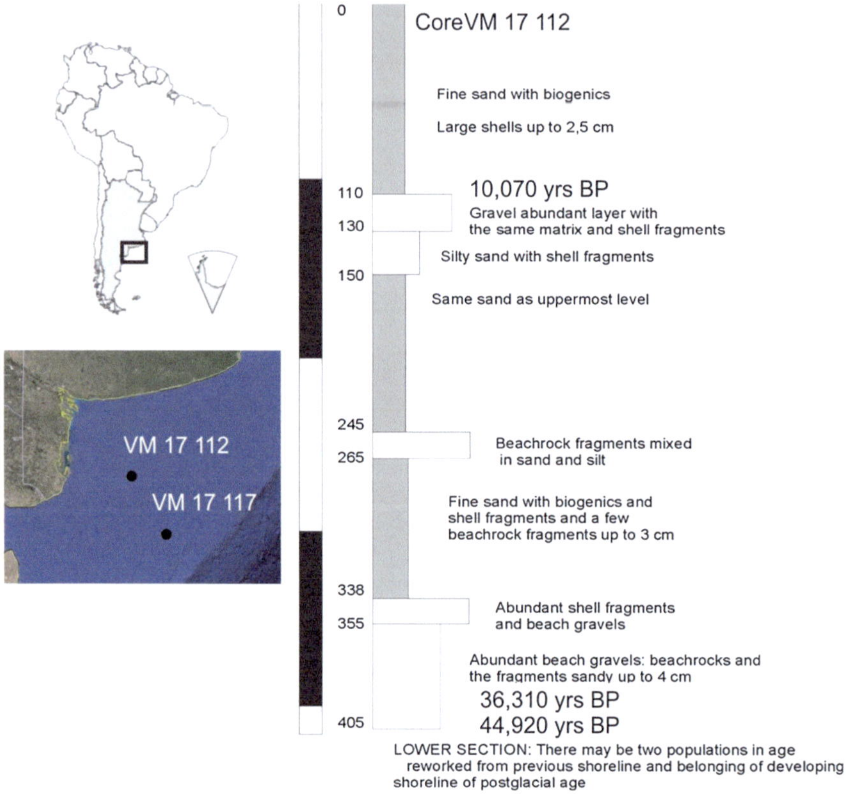

Fig. 1 Location map of both cores considered and description of core V 17-112 (59 m depth)

4.2 *Racemization Data*

Pleistocene and Holocene mollusc samples were discriminated by their isoleucine-aspartic ratios in Buenos Aires province, Argentina (Fig. 3a). Holocene, Last Interglacial and an older highstand were discriminated in Patagonia by using amino acid ratios (Rutter et al. 1989). In some cases, these ratios were mixed and Pleistocene and Holocene highstands were difficult to differentiate (Fig 3b). However, no correlation was then intended with the isotopic scale. Later work, using ERS and associated methods identified MIS 1, MIS 5e, MIS 7, MIS 9, and even MIS 11 along the Patagonian coast (Schellmann and Radtke 1997, 2000, 2003). For the continental shelf, there is a good distinction between molluscs that were at different depths. Shells collected at depths shallower than −70 m have higher ratios whereas those collected at deeper positions have lower ratios (Fig. 3c).

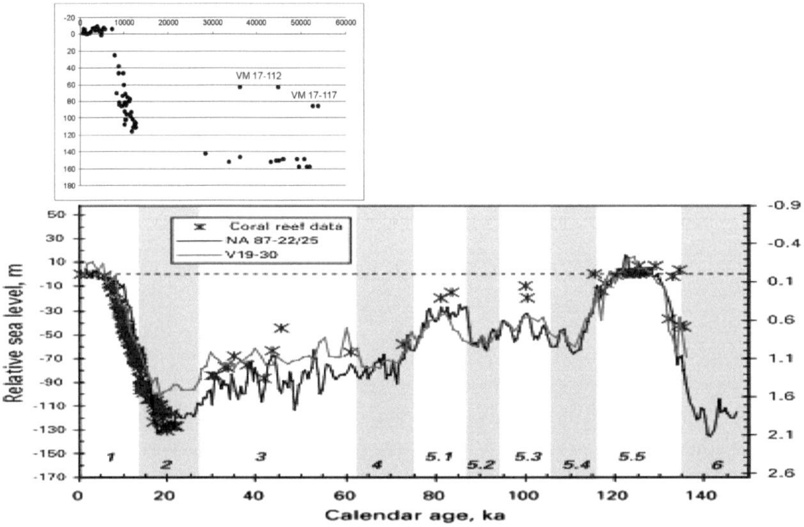

Fig. 2 **a** Radiocarbon dates from the Argentine continental shelf (modified from Guilderson et al. 2000 and Isla 2013) with dates of cores VM 17-112 and VM 17-117. **b** Sea level curve smoothed from oxygen isotopic relationships from calcareous organisms content in the cores from the North Atlantic Ocean (NA 87-22) and the Pacific Ocean (V 19-30), and U/Th dates reported from coral reefs (modified from Waelbroeck et al. 2002; Schellmann et al. 2004)

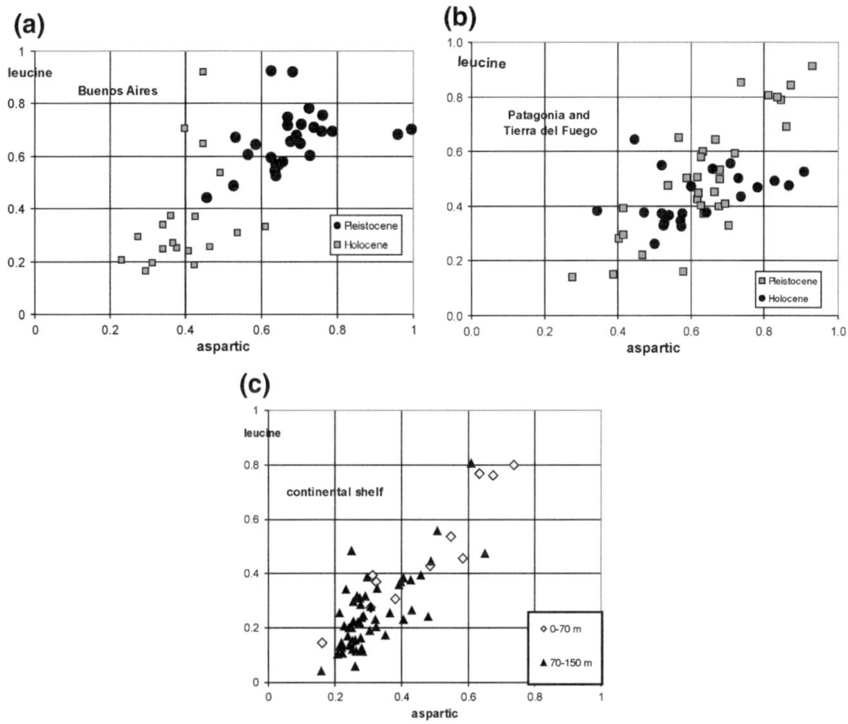

◀ **Fig. 3** Isoleucine-Aspartic relationships from mollusc samples collected from Buenos Aires province (**a**), Patagonia (**b**) and Argentine continental shelf at different depths (**c**) (Rutter, Schnack, and Zárate unpublished data 1990)

5 Discussion

Shells collected from cores extracted from the inner shelf of the state of Paraná (Brazil) yielded ages of 46,700 + 5800/−3350 (CENA-434) years B.P. (Veiga 2005). Another core from the surroundings at depths of 12–14 m gave an age of 40,600 + 2250/−1750 (CENA 433) years B.P. (Veiga 2005). At 8 m depth, Souza (2005) dated a Pleistocene substrate at 30,900 ± 900 years B.P. Similarly, from the base of cores collected from the São Sebastião Channel, São Paulo State, Brazil, radiocarbon dates were suggesting evidence corresponding to MIS 3 (Klein and Mahiques 2003; Mahiques et al. 2010). Similar ages were accepted as belonging to a Late Pleistocene transgression at the Maricá coastal plain, Río de Janeiro. In this area, and analyzing cores from percussion and mechanical borings and GPR records, two Late Pleistocene highstands were recognized: the MIS 3 highstand was assigned to three units (III, and IV a and b), whereas the MIS 5e was assigned to Unit I (Da Silva et al. 2014).

At the Chilean coast, the MIS 3 highstand was reported at the island of Santa María at maximum altitudes of 30–50 m (Melnick et al. 2006) suggesting a conservative uplifting trend of 1.5 mm/year (Jara-Muñoz and Melnick 2015).

Sea level models derived from oxygen isotope ratios measured at benthic organisms do not respond linearly. A lag effect of several thousand years, also called the "legacy effect," should be considered (Siddall et al. 2010). At the same time, the response of sea level to temperature changes is minimum during glacial and interglacial full stages, but maximum during their transitions (Siddall et al. 2010). The relationships between estimated paleotemperatures and sea level should be handled with caution for the case of wide continental shelves where the isostatic component could be of great concern. In this line of reasoning, it should be handled that during the early stages of deglaciation (Early Holocene) the relative sea level is very sensitive to isostatic adjustments.

Ice cores from Greenland showed a correlation between oxygen isotope ratios and methane content. However, there is no perfect fit between Antarctic and Greenland ice cores (Wolff et al. 2010). A more detailed definition of the Greenland interstadials was recently published considering also a record of the calcium ion concentrations as a reflection of the atmospheric dust (Rasmussen et al. 2014).

MIS 3 stratigraphy has been precisely stated in the Northern Hemisphere in regard to the frequency of dated volcanic ash fallouts (Blockley et al. 2014). Unfortunately, the frequency of volcanic eruptions at the Southern Hemisphere is poorly known (Rampino et al. 1979).

6 Conclusions

1. Regarding the highstand levels reached during the MIS 3 events, it is possible to record their deposits at the centre of the Argentine continental shelf at depths below 50 m.
2. In uplifting coasts it is possible to record the MIS 3 highstand at a higher altitude.
3. Taphonomic analyses of shell beds should be recommended in order to discriminate the deposits of the MIS 3 and MIS 5 highstands.

Acknowledgments Nat Rutter (University of Alberta, Edmonton, Canada) performed the amino acid analyses and participated in the field studies. Nat Rutter and M.A. Zárate contributed in several activities, the latter also in the preparation of the materials at University of Alberta, Edmonton, Canada. A large part of this work was financed through a grant from the National Geographic Society—4101/89, awarded to E.J. Schnack.

References

Blockley SP, Bourne AJ, Brauer A, Davies SM, Hardiman M, Harding PR, Lane CS, MacLeod A, Matthews IP, Pyne-O'Donnell SDF, Rasmussen SO, Wulf S, Zanchetta G (2014) Tephrochronology and the extended intimate (integration of ice-core, marine and terrestrial records) event stratigraphy 8e128 ka b2k. Quat Sci Rev 106:88–100

Chappell J, Shackleton NJ (1986) Oxygen isotopes and sea level. Nature 324:137–140

Codignotto JO, Kokot RR, Marcomini SC (1992) Neotectonism and sea-level changes in the coastal zone of Argentina. J Coastal Res 8(1):125–133

Da Silva ALC, Da Silva MAM, Gamboa LAP, Rodrigues AR (2014) Sedimentary architecture and depositional evolution on the Quaternary coastal plain of Maricá, Rio de Janeiro, Brazil. Braz J Geol 44(2):191–206

Guilderson TP, Burckle L, Hemming S, Peltier WR (2000) Late Pleistocene sea level variations derived from the Argentine Shelf. Geochem Geophys Geosyst 1, 2000G000098

Isla FI (1988) Where was the sea level 30–40,000 years ago? The Patagonian point of view. Quat S Am Antarct Pen 6:33–64

Isla FI (2013) The flooding of San Matías Gulf: the Northern Patagonia sea-level curve. Geomorphology 203:60–65

Isla FI, Bujalesky GG (2008) Coastal geology and morphology of Patagonia and Fueguian Archipielago. In: Rabassa JR (ed) The Late Cenozoic of Patagonia and Tierra del Fuego, Elsevier Sci. Publ., Ch. 10, 227–240

Isla F, Espinosa M, Iantanos N (2015) Evolution of the Eastern flank of the North Patagonian Ice Field: the deactivation of the Deseado River (Argentina) and the activation of the Baker River (Chile). Zeitschrift für Geomorphologie 59(1):119–131

Jara-Muñoz J, Melnick D (2015) Unraveling sea-level variations and tectonic uplift in wave-built marine terraces, Santa María Island, Chile. Quat Res 83:216–228

Klein DA, Mahiques MM (2003) Evidências de variações do nível relativo do mar durante o último ciclo Glacial, na porção norte do Canal de São Sebastião, litoral norte do Estado de São Paulo. II Congresso sobre Planejamento e Gestão das Zonas Costeiras dos Países de Expressão Portuguesa, IX Congresso da Associação Brasileira de Estudos do Quaternário e II Congresso do Quaternário dos Países de Língua Ibéricas, Florianópolis, pp 1–5

Linick TW, Damon PE, Donahue DJ, Jull AJT (1989) Accelerator mass spectrometry: the new revolution in radiation dating. Quat Int 1:1–6
Mahiques MM, Sousa SH, Furtado VV, Tessler MG, Toledo FAL, Burone L, Figueira RCL, Klein A, Martins CC, Alves DPV (2010) The southern Brazilian shelf: general characteristics, quaternary evolution and sediment distribution. Braz J Oceanogr 58:25–34
Martínez S, Ubilla M, Verde M, Perea D, Rojas A, Guérequiz R, Piñeiro G (2001) Paleo-ecology and geochronology of Uruguayan coastal marine Pleistocene deposits. Quat Res 55:246–254
Melnick D, Bookhagen B, Echtler HP, Strecker MR (2006) Coastal deformation and great subduction earthquakes, Isla Santa María, Chile (37° S). Geol Soc Am Bull 118(11–12):1463–1480
Pedoja K, Regard V, Husson L, Martinod J, Guillaume B, Fuks E, Iglesias M, Weil P (2011) Uplift of Quaternary shorelines in Eastern Patagonia: darwin revisited. Geomorphology 127:121–142
Radtke U (1988) How to avoid "useless" radiocarbon dating. Nature 333(6171):307–308
Rampino MR, Self S, Fairbridge RW (1979) Can rapid climate change cause volcanic eruptions? Science 206:826–828
Rasmussen SO, Bigler M, Blockley SP, Blunier T, Buchardt SL, Clausen HB, Cvijanovic I, Dahl-Jensen D, Johnsen SJ, Fischer H, Gkinis V, Guillevic M, Hoek WZ, Lowe JJ, Pedro JB, Popp T, Seierstad IK, Steffensen JP, Svensson AM, Vallelonga P, Vinther BM, Walker MJC, Wheatley JJ, Winstrup M (2014) A stratigraphic framework for abrupt climatic changes during the Last Glacial period based on three synchronized Greenland ice-core records: refining and extending the INTIMATE event stratigraphy. Quat Sci Rev. doi:10.1016/j.quascirev.2014.09.007
Rutter N, Schnack EJ, Fasano JL, Isla FI, Del Río L, Radtke U (1989) Correlation and dating of Quaternary littoral zones along the Patagonian coast, Argentina. Quat Sci Rev 8:213–234
Schellmann G, Radtke U (1997) Electron spin resonance (ESR) techniques applied to mollusc shells from South America (Chile, Argentina) and implications for palaeo sea-level curve. Quat Sci Rev 16:465–475
Schellmann G, Radtke U (2000) ESR dating stratigraphically well-constrained marine terraces along the Patagonian Atlantic coast (Argentina). Quat Int 68–71:261–273
Schellmann G, Radtke U (2003) Coastal terraces and Holocene sea level changes along the Patagonian Atlantic coast. J Coastal Res 19(4):983–996
Schellmann G, Radtke U (2010) Timing and magnitude of Holocene sea-level changes along the middle and south Patagonian Atlantic coast derived from beach ridge systems, littoral terraces and valley-mouth terraces. Earth Sci Rev 103:1–30
Schellmann G, Radtke U, Potter EK, Esat TM, McCulloch MT (2004) Comparison of ESR and TIMS U/Th dating of marine isotope stage (MIS) 5e, 5c, and 5a coral from Barbados—implications for palaeo sea-level changes in the Caribbean. Quat Int 120:41–50
Schnack EJ, Pirazzoli P (1990) Quaternary sea-level changes. Global Planet Change 82:65–68
Siddall M, Rohling EJ, Thompson WG, Waelbroek C (2008) Marine Isotope Stage 3 sea level fluctuations: data synthesis and new outlook. *Rev Geophy* 46, RG4003: 29 pp
Siddall M, Kaplan MR, Schaefer JM, PutnamA Kelly MA, Goehring B (2010) Changing influence of Antarctic and Greenlandic temperature records on sea-level over the last glacial cycle. Quat Sci Rev 29:410–423
Souza MC (2005) Estratigrafia e evolução das barreiras holocênicas paranaenses, sul do Brasil. Unpublished Thesis, Federal University of Paraná, Curitiba, 95 pp
Veiga FA (2005) Processos morfodinâmicos e sedimentológicos na plataforma continental rasa paranaense. Unpublished Thesis, Federal University of Paraná, Curitiba, 193 pp
Waelbroeck C, Labeyrie L, Michel E, Duplessy JC, McManus JF, Lambeck K, Balbon E, Labracherie M (2002) Sea-level and deep water temperature changes derived from benthonic foraminifera isotopic records. Quat Sci Rev 21:295–305
Walker M (2005) Quaternary dating methods. Wiley, International Library of Archaeology. J 286 pp
Wolff EW, Chappell J, Blunier T, Rasmussen SO, Svenson A (2010) Millennial-scale variability during the last glacial: The ice core record. Quat Sci Rev 29:2828–2838

Paleogeographic Evolution of the Atlantic Coast of South America During Marine Isotope Stage 3 (MIS 3)

Juan Federico Ponce and Jorge Rabassa

Abstract The preparation of a digital model showing the rising and lowering of relative sea level, by means of using the Global Mapper 10 program, allowed an approximate reconstruction of the paleogeographic evolution of the Atlantic coast of South America during the Marine Isotopic Stage 3 (MIS 3). To elaborate this digital model, the curve of global sea level variations as proposed by Lambeckand Chappell (2001), for the last glacial cycle, was taken into consideration. The model shows the development of an extensive coastal plain, which extended almost continuously from Staten Island (Isla de los Estados; southeastern end of Argentina) until the Panama isthmus. The surface of this coastal plain varied from a maximum expansion of around 1,182,000 km^2, when sea level achieved its minimum level of approximately −80 m below present sea level (b.p.s.l.) within the Marine Isotope Stage 3, in between 40,000 and 30.000 cal. years B.P., and a minimum area of approximately 954,000 km^2, when sea level was at its highest position of −60 m b. p.s.l. (in between 57,000 and 63,000 cal. years B.P.). These figures represent an overall surface variation in the order of only 20 % between both extreme paleogeographic configurations, proving that the coastal plain was a permanent, stable feature of the landscape of eastern South America, not only during the Late Pleistocene glacial stages MIS 4 and MIS 2, but even also during MIS 3. Its average width varied between 76 ± 73 km and 61 ± 71 km, showing significant latitudinal variations. Thus, South America increased its total surface by 6.6 and 5.3 % during MIS 3. This coastal plain had its maximum extent at the latitude of the eastern Argentina province of Buenos Aires (between approximately 35° and 40° South latitude), when the Río de la Plata estuary did not exist. Its minimum amplitude occurred in central Brazil and in front of the Caribbean coast of Venezuela and Colombia. The information provided by the present model sustains the importance of the Atlantic coastal plain of South America as a distinctive feature of the South

J.F. Ponce (✉) · J. Rabassa
CADIC-CONICET, Bernardo Houssay 200, 9410 Ushuaia,
Tierra del Fuego, Argentina
e-mail: jfedeponce@gmail.com; jfponce@cadic-conicet.gob.ar

J.F. Ponce · J. Rabassa
ICPA-UNTDF, Onas 450, 9410 Ushuaia, Argentina

G.M. Gasparini et al. (eds.), *Marine Isotope Stage 3 in Southern South America, 60 ka B.P.–30 ka B.P.*, Springer Earth System Sciences,
DOI 10.1007/978-3-319-40000-6_9

American landscape and proves its paleogeographic continuity even during the warmest periods of Marine Isotope Stage 3. These facts suggest that the coastal plain, most of it submerged today, has been a major element of the landscape during most of thc Quatcrnary.

Keywords Marine Isotope Stage 3 · South America · Sea level change · Interstadial · Continental shelf · Atlantic Ocean coast

1 Introduction

The South American Continental Shelf (SCS) is located immediately to the east, in the southern portion, and to the northeast and north, in the northern part of the studied area of the South American Atlantic present coasts (Fig. 1). This shelf extends approximately between latitude 12° 00′ N/54° 20′ S and longitude 34° 45′ W/77° 30′ W. Toward the deepest parts of the ocean, it is bounded by the continental talus. The SCS has an approximate surface of 2,050,000 km^2, a maximum length of 12,000 km, a minimum width close to only 1 km in the central part of Venezuela, and a maximum width of 880 km, N of the Malvinas-Falklands islands (Fig. 1). It shows a maximum depth of around −230 m below present sea level (b.p. s.l.), immediately W of these islands, but most of it is less than 100 m deep. The SCS is characterized by a quite gentle slope (less than 0.5°) and very low internal relief (Fig. 1). The SCS is one of the more extensive submarine platforms in the world, most of which has been repeatedly abandoned by the sea during glacial times and flooded again during glacial terminations. On an average, it is reasonable to state that at least around 800,000 years of each million years, the SCS was emerged and exposed to the atmosphere.

During MIS 3, sea level became permanently established below present sea level (see Rabassa and Ponce 2013 and this volume, and the references cited there). Thus, a great portion of the SCS was lastingly exposed generating an extensive coastal plain along the present South American Atlantic coast. Due to the fast and frequent changes of sea level during MIS 3, the surface extent of this coastal plain was highly variable along such period of time.

The various global sea level variation curves proposed by several authors for the last glacial cycle (Fig. 2) show significant differences between them. However, when these curves are carefully compared, it may be observed that sea level during MIS 3 would have oscillated approximately between a minimum of −80 m b.p.s.l. and a maximum of −60 m b.p.s.l. In the present work, the paleogeographic configurations of the South American Atlantic Coast, which roughly corresponded to these maximum and minimum depths for sea level as they were reached during MIS 3, are presented and discussed.

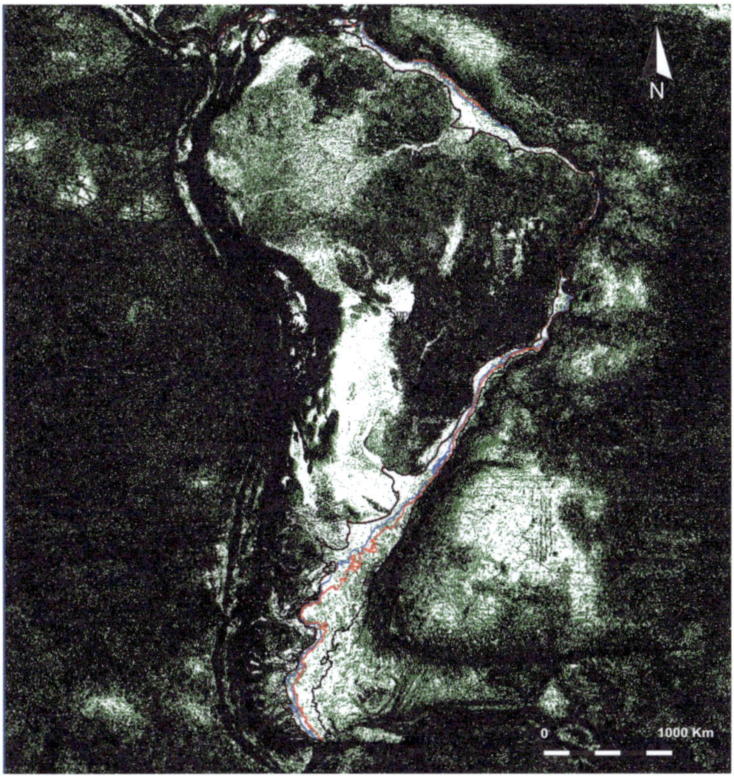

Fig. 1 Slope map of South America. *White areas* indicate slopes between 0° and 0.2°, *green areas* between 0.2° and 0.5°, and *black areas*, slopes steeper than 0.5°

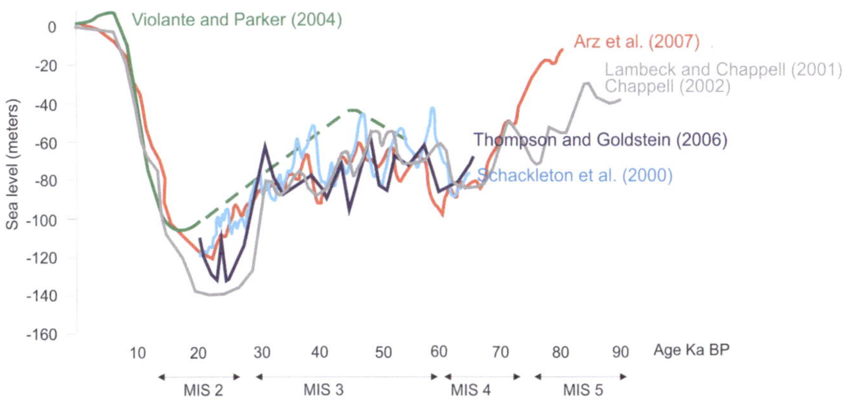

Fig. 2 Late Pleistocene sea level curves: the last glacial–interglacial cycle (modified from: Shackleton et al. 2000; Lambeck and Chappell 2001; Violante and Parker 2004; Thompson and Goldstein 2006; Arz et al. 2007)

2 Methodology

The program Global Mapper 10 was used for the preparation of a digital model of sea level changes, taking into consideration the curve of global sea level rise as proposed by Lambeck and Chappell (2001) for the last glacial cycle. The digital land elevation models of the Shuttle Radar Topography Mission (SRTM), W100S10.BATHYMETRY.SRTM and W060S10.BATHYMETRY.SRTM., with a 1 × 1 km resolution pixel were analysed with this program. Using these tools, paleogeographic maps were drawn showing the maximum and minimum position of sea level during MIS3. The indicated calibrated ages are following Lambeck and Chappell (2001) data. The different positions of the paleocoastallines for the chosen moments should be taken as minimum and tentative because these paleogeographic models are not taking into consideration neither the erosive action of marine currents through time nor tectonic processes or glacioisostatic rebound, if any. The elevation and plane measurements were completed by means of the Global Mapper 10 program, as well the slope map of South America.

The Atlantic Ocean coast of South America was divided in six sub-regions, with the aim of obtaining a better analysis of the variations and paleogeographic characteristics of it. These sub-regions are: Southern (Argentine Patagonia), Southeastern (Province of Buenos Aires, Argentina, and southern Uruguay), Central (southeastern coast of Brazil), Northeastern (eastern coast of Brazil), Northern (northern coast of Brazil and the Guyanas), and Northwestern (the Caribbean coasts of Venezuela and Colombia) (Fig. 3).

3 The Paleogeographic Model

The different curves of global sea level variations proposed by several authors for the last glacial cycle (Fig. 2) show that sea level during MIS 3 would have oscillated between an approximate minimum of −80 m b.p.s.l. and a maximum of −60 m b.p.s.l. Given these conditions, an extensive coastal plain was developed at varying distances east from the present South American Atlantic coast, when part, if not all, of the present South American submarine platform became emerged (Fig. 4). The coastal plain became extended in an almost continuous mega-landform from Isla de los Estados (Staaten Island, southern end of Argentina) up to the Panamá isthmus, with a total length of approximately 11,900 km. Its extension was highly variable along the entire MIS 3. Southwards, the ancient coastal plain presented a larger development at the latitude of the present Argentine provinces of Río Negro and Buenos Aires, and along the present coast of Uruguay. The paleogeographic model herein presented shows also a possible development of a very large number of smaller island located eastwards from the Argentine provinces of Buenos Aires, Río Negro and Chubut. Northwards, the plain reached its greater

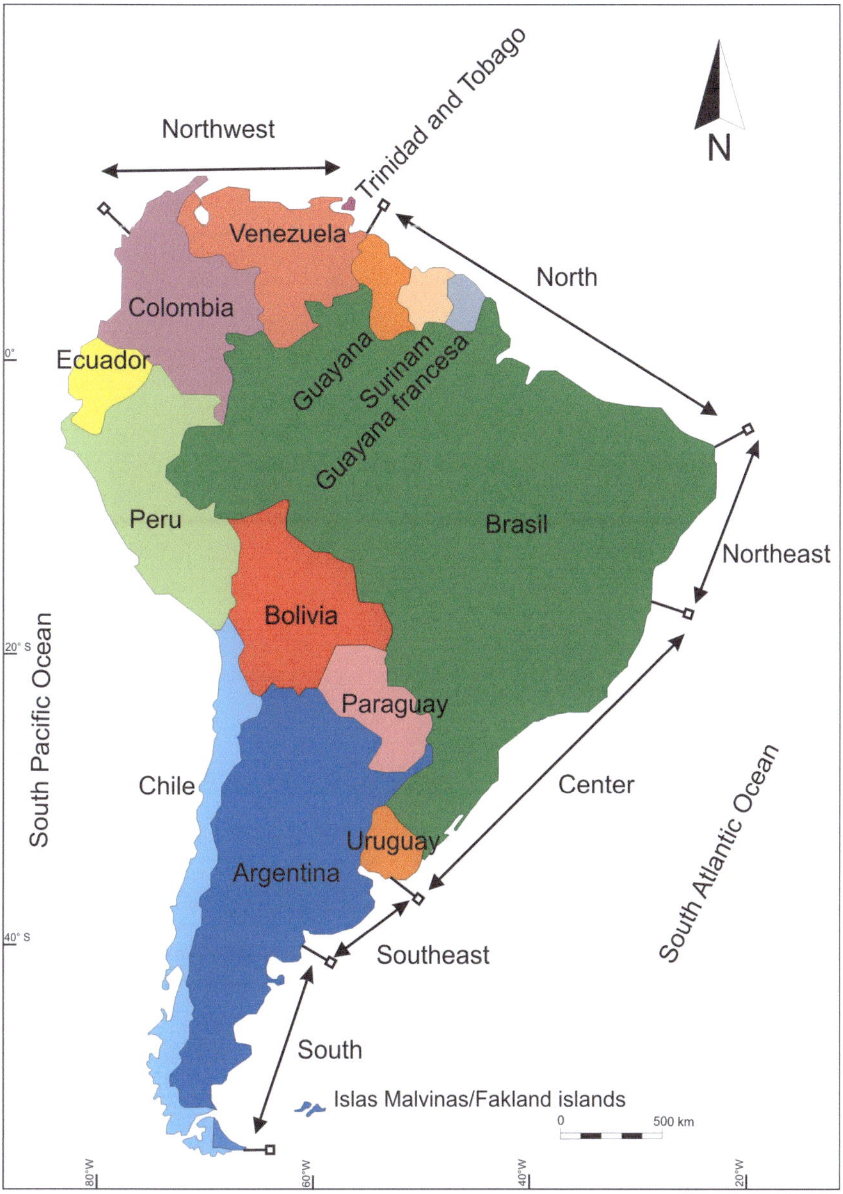

Fig. 3 Map of South America with the proposed boundaries of the different sub-regions

extent between the Brazilian states of Rio Grande do Norte and the eastern extreme of Venezuela with its maximum development in front of the present coasts of the Brazilian states of Amapá, Pará, and Maranhão .

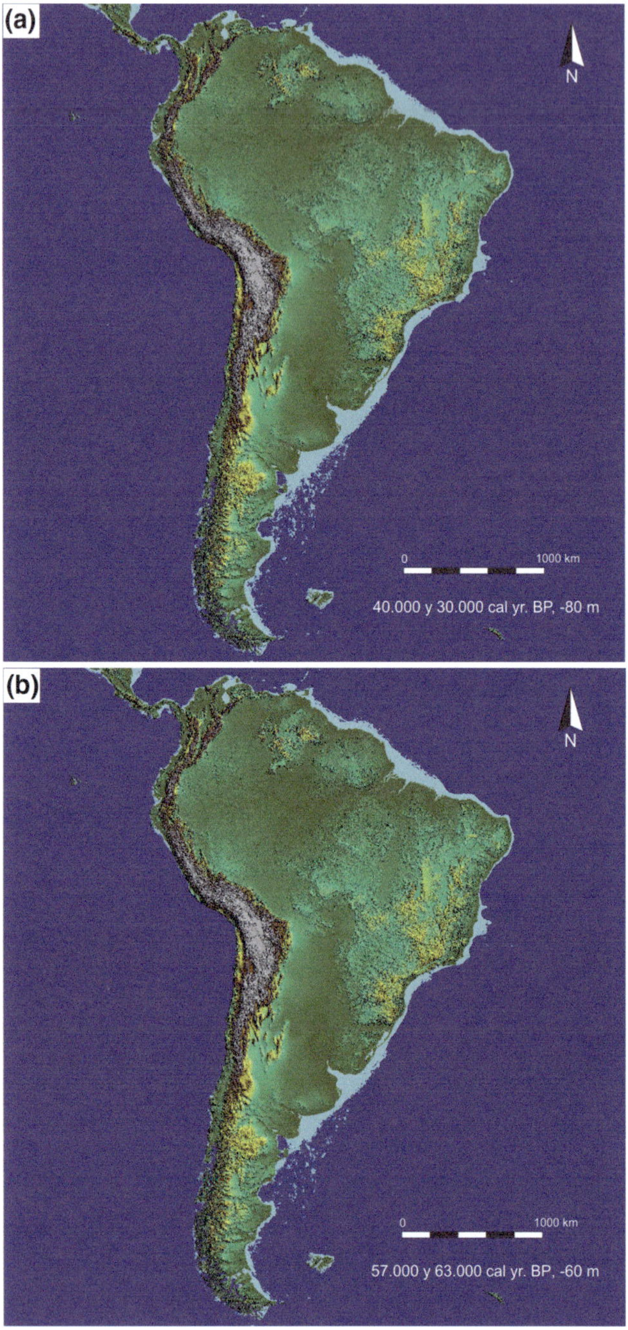

Fig. 4 Paleogeographic configuration of South America during MIS 3; **a** minimum relative sea level; **b** maximum sea level position during MIS 3

Accordingly, the paleogeographic configurations of the Atlantic Ocean coast of South America are presented, taking into consideration the maximum and minimum sea level elevations reached during MIS 3.

4 The Maximum Extension of the Coastal Plain (Between 40,000 and 30,000 Years B.P.)

The ancient coastal plain achieved its maximum development during MIS 3, between 40,000 and 30,000 years B.P., when sea level reached its lowest position during this period (−80 m b.p.s.l.) (Fig. 4a). During this lapse the plain achieved an approximate surface of 1,182,900 km^2, equivalent to the 6.6 % of the present surface of South America. The average width of the plain was 76 $\pm$ 73 km, with a maximum which varied from 300 to 400 km in the southern sector of South America, along the present Atlantic coast of the Province of Buenos Aires, Argentina, and southern Uruguay. In front of the Buenos Aires coast (in the southeastern zone), the average width was 240 $\pm$ 182 km. The large deviation from the means is due to high irregularity of the South American submarine platform in terms of extent and slope. According to this paleogeographic configuration, the mouth of the Rio de la Plata was located at 440 km toward the East of its present position. Closer to the Uruguayan coast and southern Brazil (central zone), the average width of the coastal plain was of approximately 78 $\pm$ 46 km. In Patagonia (southern zone) the average width reached approximately 57 $\pm$ 34 km. The paleogeographic model shows also the development of a large number of small islands in front of the present Patagonian coast. Southwards, the Isla Grande of Tierra del Fuego was still united to the rest of the continent. Under these conditions, the surface of the Islas Malvinas/Falkland Islands would have been of around 31,500 km^2, doubling their present surface (Fig. 5).

In the central portion of Brazil (central zone), the coastal plain would have had its minimum size. In this sector, the average width of the plain would have been of around 25 $\pm$ 9 km (Fig. 6).

Toward the North, along the northeastern zone, the mean width of the plain was close to 113 $\pm$ 54 km. In this sector, the maximum width of the plain reached 315 km in front of the coast of the state of Pará, near the present mouth of Rio Amazonas. Following this paleogeographic configuration, the MIS 3 Amazonas mouth was located at around 290 km northeast of its present position. In the northwestern sector of South America, from the central part of Venezuela towards the boundary of Colombia and Panamá, the coastal plain had a weak development with a mean width of 35 $\pm$ 48 km. During these times, the Gulf of Venezuela and the Lago Maracaibo would have not existed as such, according to our model. Several islands presently located along the coasts of Venezuela, such as the Tortugas Islands and Isla Margarita, as well as the archipelago of Trinidad-Tobago, would have been united to the rest of the continent (Fig. 7).

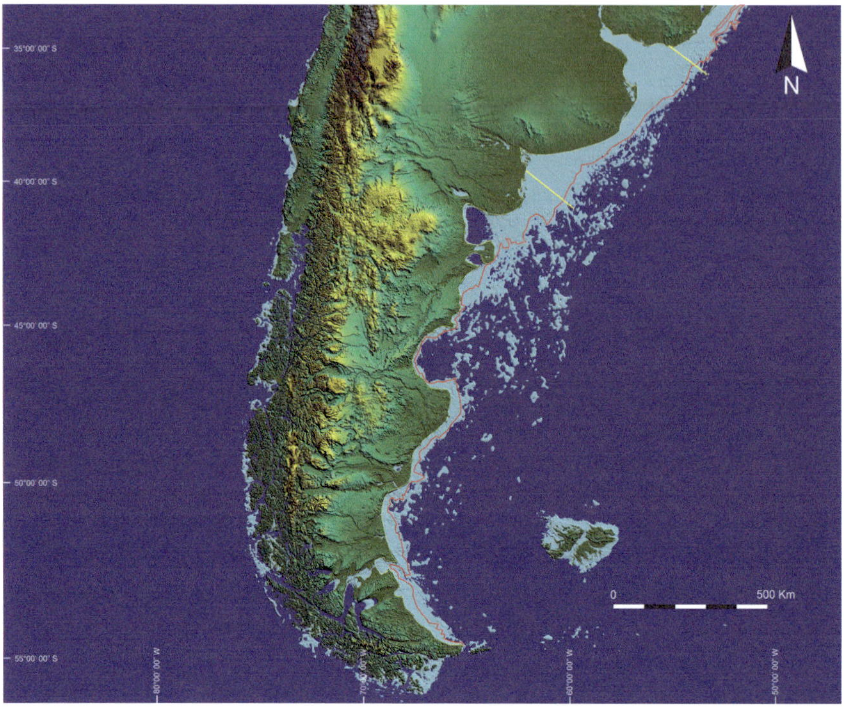

Fig. 5 A detailed view of the paleographic configuration of southern South America during the maximum development of the coastal plain during MIS 3. The continuous *red line* indicates the outer boundary of the plain during its minimum expansion in MIS 3

After MIS 3, a similar paleogeographic configuration would have taken place during the Late Glacial times, around 14,100 cal. years B.P.

5 The Minimum Extension of the Coastal Plain (Between 57,000 and 63,000 Cal. Years B.P.)

The shorter extent of the coastal plain took place between 57,000 and 63,000 cal. years B.P., during the maximum level reached by sea level during MIS 3 (Fig. 4b). Following this paleogeographic configuration, the South American coastal plain had a total surface of around 954,000 km^2, just a 5.5 % of the present surface of South America and around 20 % smaller than during its maximum extent in MIS 3. The mean width of the plain during these times was 61 ± 71 km. In the southern sector of South America, the plain had a quite large extension only in front of the Buenos Aires and Uruguay coasts, mainly near the present location of the city of Bahía Blanca and the region of the present Rio de la Plata estuary, where it had a

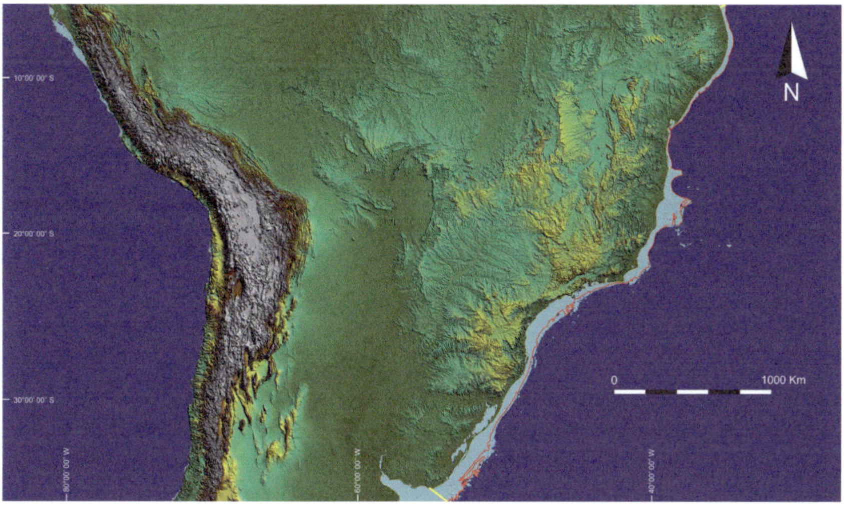

Fig. 6 A detailed view if the paleogeographic configuration of central South America during the maximum development of the coastal plain during MIS 3. The continuous *red line* indicates the outer boundary of the plain during its minimum expansion in MIS 3

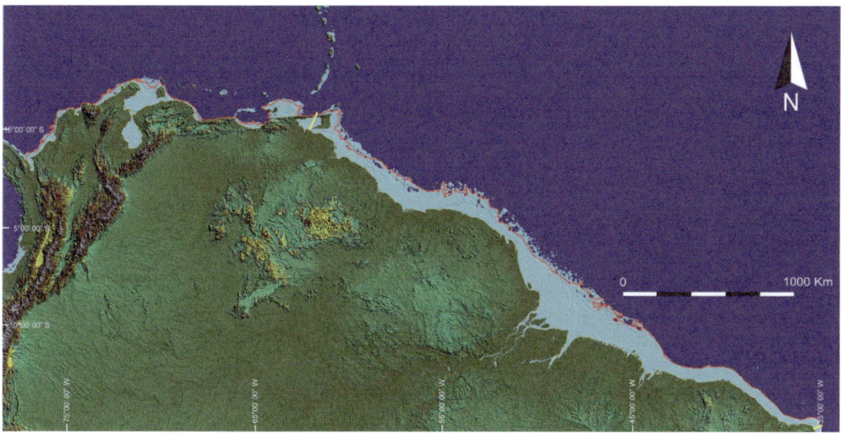

Fig. 7 A detail of the paleogeographic configuration of northern South America during the maximum development of the coastal plain in MIS 3. The continuous *red line* indicates the outer boundary of the plain during its minimum expansion in MIS 3

maximum width of 210 and 290 km, respectively, and a surface of roughly 33 % less than during its maximum expansion. Concerning this paleogeographic configuration, the MIS 3 mouth of the Rio de la Plata was located at 400 km eastwards from its present location. In Patagonia (southern zone), the mean width of the plain was reduced to less than half its previous configuration (21 $\pm$ 16 km). In this sector of South America, the coastal plain presented the largest difference in its surface during MIS 3, becoming reduced in up to 47 %. The Isla Grande de Tierra del Fuego was still united to the rest of the continent. This paleogeographic configuration showed a smaller development of islands in front of the Argentine coasts, mainly in front of Buenos Aires province (Fig. 5). In front of the coasts of Uruguay and southern Brazil (central zone, Fig. 6), the extent of the coastal plain was 68 ± 43 km, only 19 % less that its equivalent during its maximum expansion.

In the northeastern zone, the extent of the plain had no significant differences during MIS 3, since in these times its mean width was 23 ± 9 km, that is, only 2 km less than during the lowermost sea level position.

Toward the north, in the northern coast of Brazil, Guyana, Suriname, French Guiana and easternmost Venezuela (northern zone), the average width of the plain was closer to 105 ± 54 km and its surface was only 12 % less than in the configuration of the maximum development. In this sector, the maximum width was 270 km nearby the mouth of the Rio Amazonas. Finally, in the northwestern sector, during that time interval, the width of the coastal plain was 28 ± 37 km, only a 15 % less than its equivalent during the minimum MIS 3 relative sea level (Fig. 7).

After MIS 3, this paleogeographic configuration was reached again toward the end of the Late Glacial times, around 12,500 cal. years B.P.

6 Final Remarks

The paleogeographic model herein presented shows the development of an extensive coastal plain east of the present Atlantic coast of South America, during MIS 3. During this entire period, the coastal plain was larger than today. Its surface was highly variable along this whole period due to the fast variations of relative sea level recorded during such period. The maximum extension of this plain was reached between 40,000 and 30,000 years B.P., with a total surface close to 1,182,900 km^2. The minimum expansion was achieved between 57,000 and 63,000 cal. years B.P., reaching around 954,000 km^2. These values represent 62 % and 50 % of the total surface exposed during the Last Glacial Maximum. The plain showed its maximum development at the latitude of the present Argentine provinces of Buenos Aires and Río Negro, and along the present coast of Uruguay. Both in these sectors and Patagonia, the largest changes in the position of the coastline were recorded. Toward the north, along the northeastern coast of Brazil, the variations in the position of the coastline were almost negligibly.

The generation of the paleogeographic evolution model provides new information for future paleoenvironmental, paleoclimatic, and paleobiogeographic reconstructions and it will be of great help to understand the biota migrations which took place in South America during MIS 3.

References

Arz HW, Lamy F, Ganopolski A, Nowaczyk N, Pätzold J (2007) Dominant Northern hemisphere climate control over millennial-scale glacial sea-level variability. Quat Sci Rev 26(3–4): 312–321. doi:10.1016/j.quascirev.2006.07.016

Lambeck K, Chappell J (2001) Sea level change through the lastglacial cycle. Science 292: 679–686

Rabassa J, Ponce JF (2013) The Heinrich and Dansgaard-Oeschger climatic events during marine isotopic stage 3: searching for appropriate times for human colonization of the Americas. Quatern Int 299:94–105

Rabassa, J, Ponce JF, (this volume) The Heinrich and Dansgaard-Oeschger climatic events during Marine Isotopic Stage 3

Shackleton NJ, Hall MA, Vincent E (2000) Phase relationships between millennial-scale events 64,000–24,000 years ago. Paleoceanography 15:565–569. doi:10.1029/2000PA000513

Thompson WG, Goldstein SL (2006) A radiometric calibration of the SPECMAP timescale. Quat Sci Rev 25(23–24):3207–3215. doi:10.1016/j.quascirev.2006.02.007

Violante RA, Parker G (2004) The post-glacial maximum transgression in the de la Plata river and adjacent inner, Argentina. Quatern Int 114:167–181

The Continental Record of Marine Isotope Stage 3 (MIS 3; ~60–25 ka) in Central Argentina: Evidence from Fluvial and Aeolian Sequences

Marcelo Zárate, Adriana Mehl and Alfonsina Tripaldi

Abstract Marine Isotope Stage 3 (MIS 3) is characterized by high climatic variability resulting from numerous centennial to millennial scale events. The environmental and climatic reconstruction of this interval is restricted by the sparsity of high-resolution (centennial-scale) terrestrial records in most of South America. This contribution is an attempt to reconstruct the general environmental and climatic conditions of southern South America during MIS 3 by means of continental records located in central Argentina; this is an extensive and heterogeneous region made up of diverse geomorphological settings under different climatic conditions. Therefore, the main features of several aeolian and fluvial records situated in different geomorphological settings across the region are overviewed. The results indicate the predominance of regional aggradation during MIS 3 with differences in the accumulation rates and dominance of either aeolian or fluvial deposits depending on the geomorphological setting. The aggradation process was interrupted by stability intervals evidenced by paleosols in the San Rafael plain, the San Luis paleo-dunefield, the eastern Sierras Pampeanas piedmont and the eastern Pampean plain. The paleosols might represent lapses of decreasing aeolian input and perhaps more humid conditions. In addition, paleobiological indicators from alluvial sequences suggest higher temperatures and water availability between 35 and 31 ka in the Andean piedmont, while dry subhumid or strongly seasonal conditions with alternating subhumid-humid phases were inferred in the eastern Pampean plain during MIS 3. These intervals tend to cluster during the second part of MIS 3, and might reflect the environmental responses to some of the climatic

M. Zárate (✉) · A. Mehl
INCITAP (CONICET-UNLPam), Avenida Uruguay 151,
6300 Santa Rosa, La Pampa, Argentina
e-mail: mzarate@exactas.unlpam.edu.ar

A. Tripaldi
IGEBA (CONICET)—Department of Geology,
University of Buenos Aires, Ciudad Universitaria, C1428EHA Buenos Aires, Argentina

G.M. Gasparini et al. (eds.), *Marine Isotope Stage 3 in Southern South America, 60 ka B.P.–30 ka B.P.*, Springer Earth System Sciences,
DOI 10.1007/978-3-319-40000-6_10

oscillations that occurred during MIS 3. Detailed analysis and a more adjusted chronology are needed to correlate the aeolian and fluvial episodes along with the stability intervals at regional and continental scales.

Keywords Late Pleistocene · Andean piedmont · Pampean plain · Paleoclimate · Argentina

1 Introduction

The Quaternary geological history is characterized by cyclical climatic changes of different frequency and intensity consisting in the alternation of cold periods (glacial intervals) and temperate-warm spans (interglacial intervals). These cyclical changes generated periodic reorganizations of the landscape and the environmental system. Among others, the last glacial cycle (~ last 125 ka), one of the best-studied periods in paleoclimatology (Veres et al. 2013), is a relevant analog documented by exceptional records of high resolution (*e.g.,* ice cores, deep sea sediment cores), a useful tool to foresee possible future environmental and climatic scenarios at different time scales.

The last glacial cycle has been chronostratigraphically subdivided into several marine isotope stages (MIS) according to the variations of the $\delta^{18}O$ on deep sea sediments (Pisias et al. 1984; Martinson et al. 1987). MIS 3 is the 60–25 ka interval characterized by sea levels between 60 and 90 m below the present (Siddall et al. 2008; Isla and Schnack, this volume) and general milder climates than those dominant during the preceding MIS 4 and the succeeding MIS 2 (the last Glacial Maximum). MIS 3 is well documented by both terrestrial and marine records that suggest a complex interstadial interval, in turn composed of numerous shorter climatic oscillations (stadials and interstadials of less hierarchy). In this respect, the Greenland ice cores contain the record of abrupt climatic changes during the Late Pleistocene, involving rapid warming events (8–15° C). They correspond to 25 centennial-scale climatic oscillations (interstadials and stadials), also known as Dansgaard-Oeschger events (D-O), 15 of which occurred during the 60–25 ka interval (Bradley 2015, and references therein). In the Northern Hemisphere, these changes are clearly reflected in terrestrial settings by pollen assemblages that suggest the alternation of cold and warm phases of variable duration (*e.g.,* Gómez Orellana et al. 2007, and references therein). In the southern hemisphere, the Antarctic ice cores also register a millennial scale climatic variability (Wolff et al. 2010). The comparison of the climatic conditions at both hemispheres, however, suggests a complex global pattern, when Antarctica appears to warm up, Greenland becomes cold; these opposite responses are thought to be consistent with a mechanism involving ocean heat transport (Wolff et al. 2010). In most of South

America, as well as in many areas of Asia and Northern Africa, the environmental and climatic reconstruction of MIS 3 is restricted by the sparsity of high resolution (centennial-scale) terrestrial records (Voelker 2002).

The present paper is an attempt to infer the general conditions during MIS 3 east of the Andes Cordillera (central Argentina) by means of the evidences provided by continental records. Although natural archives of high resolution are still very scarce, the aeolian deposits, broadly distributed across the region, along with the Late Pleistocene fluvial records offer a general insight into the major responses of the environmental system between 60 and 25 ka. Aeolian and fluvial sequences of central Argentina have been the subject of numerous contributions in the last 15 years, with the focus on the characteristics and significance of the deposits, and the regional evolution of the landscape; a major goal of the studies was to obtain an adjusted chronology of the records, previously calibrated by scarce numerical ages mainly coming from the late Glacial-Holocene interval. Therefore, this paper overviews the main features of several aeolian and fluvial records situated in different geomorphological settings of central Argentina between ~30 and 40° S with the purpose of inferring the paleoenvironmental and paleoclimatic conditions. The significance of the records at regional and continental scales is discussed, and finally several possible lines of research are examined and proposed to conduct future research in the region.

2 Environmental and Geomorphological Setting

The region under analysis, situated in central Argentina, is a vast and heterogeneous territory made up of diverse geomorphological settings; it extends from the Andean piedmont to the eastern Pampean plain (~1000 km W-E), and from the Sierras Pampeanas piedmont of Córdoba (~30° S) to the central area (~35° S) of the Pampean plain (~400 km N-S) (Fig. 1, Fig. 3). At present, a climatic gradient is evident across the region characterized by more humid conditions in the east-northeast that pass into more arid conditions toward the west-southwest.

The eastern Andean piedmont is a complex geomorphological setting that consists of several landforms characterized by their geological and structural features that determine the occurrence of heterogeneous environmental conditions in short distances (Fig. 2). The San Rafael tectonic block (a fragmented foreland block) is here included as part of the Andean piedmont in order to simplify and facilitate the discussion. The piedmont is drained by major rivers (Atuel, Diamante, Tunuyán, Mendoza) with their headwaters in the high Andes Cordillera, that are tributaries of the Desaguadero–Salado fluvial system (Fig. 2). The present climate is arid to semiarid (~200–500 mm of annual precipitation) with a shrub vegetation

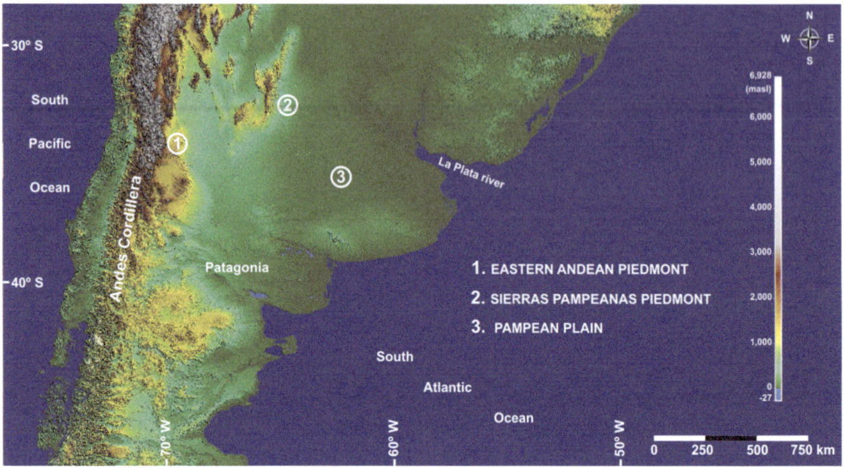

Fig. 1 General location of central Argentina and the three main geomorphological settings corresponding to the eastern Andean piedmont, the Sierras Pampeanas piedmont and the Pampean plain

cover completely modified by intense agriculture in the surrounding of the major rivers (agricultural oasis).

The eastern Sierras Pampeanas piedmont (Córdoba and San Luis provinces) is a gently sloping landscape that grades into the Pampean plain eastwards and southwards (Fig. 3).

The piedmont is dissected by fluvial systems that drain the Sierras Pampeanas, and formed major alluvial fans in the Late Quaternary (Carignano 1999). Currently, the area has a mean annual precipitation between 700 and 800 mm, mainly concentrated during the summer, and a mean annual temperature of 18° C. At present, most of the piedmont is deeply modified by intense agriculture and affected by severe fluvial erosion with formation of deep and long gullies.

The Pampean plain is a large, setting of low relief, that gently slopes eastward at a regional scale. The boundary with neighboring geomorphological domains is transitional, grading into the piedmont of the Sierras Pampeanas northwestward and the Patagonian environment southward; the Desaguadero-Salado fluvial system is considered the geomorphological boundary with the Andean piedmont (Zárate 2009). The present climate varies from humid and temperate conditions in the eastern Pampas (900–1100 mm) to semiarid-arid conditions (250–300 mm) in the western Pampas at the border with the Andean piedmont. This climatic condition is reflected by the vegetation cover, at present almost completely altered by agriculture, which grades from grasslands to shrubs in the west-southwest. The Pampean plain is subdivided into several units on the basis of geomorphological and climatic

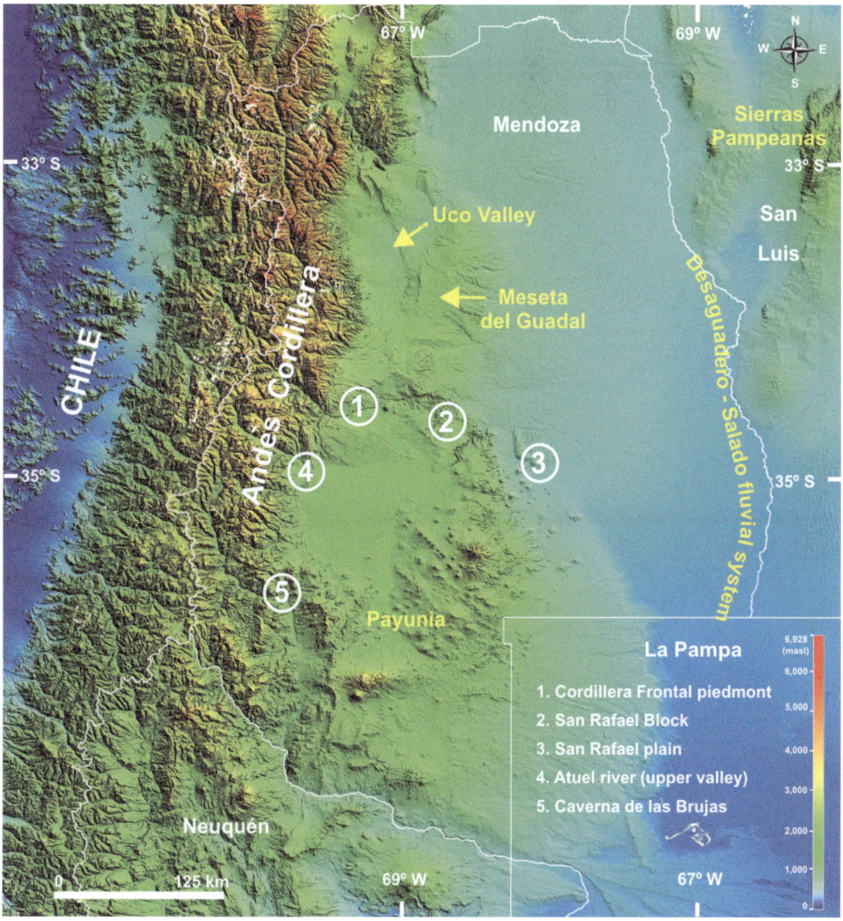

Fig. 2 Andes Cordillera and piedmont of Mendoza province with location of the study areas and sites

conditions. Most of the central and western Pampean plain is dominated by aeolian landforms consisting of extensive sand mantles, and dune fields grading into a fringe of a loess mantle toward the northern, eastern, and the southeastern sectors. No significant fluvial systems are present with the exception of the Río Quinto that drains the southern part of Sierras Pampeanas (Fig. 1). The northern Pampean plain is drained by fluvial systems with their headwaters in Sierras Pampeanas; eastwards, several minor streams tributaries of the Paraná-Rio de la Plata are present. Further south, the plain is drained by the Río Salado system.

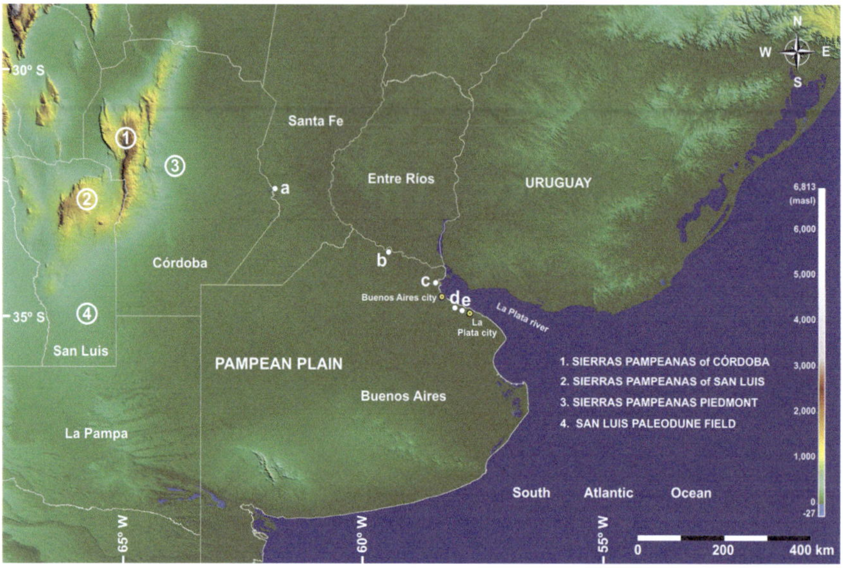

Fig. 3 The Sierras Pampeanas piedmont and Pampean plain with location of the sites mentioned in the text: **a** Tortugas, **b** Baradero, **c** Luján valley, **d** Hudson, **e** Gorina

3 Study Sites

3.1 The Andean Piedmont

Two areas (the Valle de Uco and the San Rafael plain) have been the subject of multidisciplinary studies on the sedimentology, geomorphology, paleobiology, and geochronology (^{14}C dates, OSL dates, cosmogenic isotope dates) of the deposits and landforms.

The Valle de Uco, situated at the piedmont of the Frontal Cordillera, is a Quaternary sedimentary basin at around 750 m above present sea level (a.s.l.) with a flat and gently steeping topography incised by the Río Tunuyán and its tributaries (Fig. 2). The studied successions crop out at the banks of several streams; the most continuous and thickest exposures are located along the margins of Arroyo (i.e., creek) La Estacada (~33° 29′ S, 69° 01′ W). The exposed alluvial sections encompassing the time interval of MIS 1 to MIS 3, are dominantly composed of homogeneous light brownish gray to light brown (10YR 6/2 and 6/3) massive sands; the deposits were mostly accumulated by sheet fluid overflows that affected overbank areas and probably temporary inactive channels of sandy-like braided streams in distal fan environments, punctuated by volcanic episodes (tephras), some minor aeolian activity and limnic levels. The numerical ages obtained suggest that aggradation was already underway at least since ~50 ka B.P. (Mehl and Zárate 2012). At La Bomba site (33° 28′ S, 69° 03′ W), the lower section of the alluvial succession comprises a 3 m thick deposit of horizontally stratified fine sand-sandy silt deposits, alternating with several limnic levels; it contains abundant remains of mollusk shells of four different species and dated by AMS between 35 and 31 ka B. P. (De Francesco et al. 2007). The dominance of semiaquatic and hygrophilous land snail species indicates the development of very shallow water bodies, probably a damp habitat that suggests a relatively humid and mild interval (De Francesco et al. 2007). Upwards, aggradation continued with accumulation of channel deposits consisting of matrix-supported conglomerates with both horizontally and cross-bedded structures followed by channel-fill deposit surrounded by fine-grained sheet-like deposits likely from a floodplain area. Massive fine sands and silty sands were aggraded later during the last part of MIS 3 and MIS 2.

The San Rafael plain is an aggradational environment composed of fluvial and aeolian deposits. Two sections exposed by channel incision along the riverbanks of Arroyo Agua del Chancho (34° 27.19′ S, 68° 23.47′ W) were examined for sedimentological analysis using OSL dating for their chronological calibration (Tripaldi et al. 2011). The numerical dates indicate that the sediments were deposited during the last ~58 ka. Five periods of sedimentation (1, 2, 3, 4, 5) were differentiated on the basis of the relative dominance of fluvial and aeolian processes. Periods 1, 2, 3 occurred during MIS 3 while periods 4 and 5 are broadly equivalent to MIS 2 and 1, respectively. The OSL chronology indicates that fluvial aggradation was dominant in ephemeral streams, mainly controlled by unconfined sheet flows between *ca.* 58 and 39 ka (period 1) and from *ca.* 36 to 24 ka (period 3) which was characterized by more active aeolian processes than during period 1 as revealed by the occurrence of more mixed fluvial–aeolian facies. An inferred interval of stability and soil formation (period 2), with root casts and oxidation features that indicate moister conditions, separated these fluvial deposits, and lasted from ~39–36 ka. Period 4, between 24 and 13 ka (MIS 2), was dominated by aeolian processes.

The MIS 3 scenario of the Andean piedmont is completed by the geomorphological evolution of several major rivers calibrated by cosmogenic isotope dating. In the mountain catchment-fan system of the Río Las Tunas (~33° 20′ S), at a proximal location of the Cordillera Frontal piedmont, Pepin (2010) reported the formation of two entrenched terraces between 20 and 20–15 ka (MIS 2) which were interpreted of climatic origin. The incision events affected the glacio-fluvial conglomerate deposits of the Las Tunas Formation resulting from a long lasting aggradation period starting ~600 ka B.P. (Pepin 2010). Thus, continuous aggradation is dominant during most of the last glacial cycle, including MIS 3.

West of the San Rafael plain, along the reach of the Río Diamante (~34° 41′ S) that dissects the piedmont of Cordillera Frontal, Baker et al. (2009) identified five fill and strath terrace systems. According to the chronology obtained, the formation of the three youngest terraces occurred during MIS 4 (Qt3), and MIS 2 (Qt4 and Qt5). Aggradation occurred during MIS 4 followed by an incision at the end of this stage. Aggradation reinitiated later, sometimes during MIS 3 and continued until MIS 2, ending with the formation of a strath terrace (Qt4 strath) at 22 ± 7 ka followed by the deposition of Qt4 fill terrace 13 ± 3 ka: Baker et al. (2009) pointed out that likely, the most recent episode of alluvial aggradation was associated with the MIS 2 glacial advance.

Further south, the upper Río Atuel valley (~35° 05′ S, El Sosneado) that drains a sector of Cordillera Principal, exhibits synorogenic Quaternary deposits arranged into three alluvial cones (Q_{6C}, $Q_{5C,}$ and Q_{0C}) and four terrace levels (Q_{4T}–Q_{1T}) (Messager 2010). Q_{5C} and Q_{4T}, considered of Late Pleistocene age were correlated with Qt3 of the Río Diamante (Baker et al. 2009); thus, these landforms are interpreted to be formed during MIS 4, Q_{3T} and Q_{2T} terraces were assigned to MIS 2 and correlated with the Río Diamante terraces Qt4 and Qt5 of Baker et al. (2009), respectively. According to the geomorphological reconstruction of the upper Atuel valley by Messager (2010), MIS 3 was an interval of dominant fluvial aggradation.

3.2 *Eastern Piedmont of Sierras Pampeanas*

The loess mantle piedmont has been the subject of several contributions dealing with the sedimentological characteristics and the geochronology (OSL, TL) of the deposits at three sites (Lozada, Corralito, and Monte Ralo). Lozada (31° 39′ S; 64° 08′ W, 390 m a.s.l), situated on an alluvial paleofan, is a 9.3-m thick section exposed by a road along the wall of an artificially excavated channel. The succession chronologically calibrated by OSL dates, consists of a surface soil and a basal truncated paleosol formed during MIS 5, separated by over 4 m of light color (yellowish brown, 10YR 5/4) sediments, texturally heterogeneous including silty clay loam, silt loam and fine sandy loam (Kemp et al. 2006). The basal MIS 5 paleosol was truncated sometime between <78.2 ± 4.5 and >90.1 ± 5.8 ka, and buried by fluvial sediments during the time span of MIS 4 and MIS 3. Prior to 50,1 ± 2.5 ka (MIS 3), loess accumulation started and continued until the mid-Holocene; the micromorphological occurrence of sorted layers and embedded aggregates evidence phases of reworking and possible incorporation of other fluvial components during MIS 3 (Kemp et al. 2006).

The Monte Ralo (31° 55′ S; 64° W; 450 m a.s.l.) and Corralito (32° 00′ S and 64° 15′ W, 470 m a.s.l.) loess sections are located further south, which comprise exposures along gullies walls, 9 and 12 m thick, respectively. Frechen et al. (2009) carried out a geochronological analysis by means of IRSL dates in order to establish a more reliable chronological framework for the sedimentary record during the last glacial/interglacial cycle. At both sections, the successions are composed of four periods of increased loess accumulation and three intercalated palaeosols or pedocomplexes. From the two sections, the Corralito succession shows the most developed paleosols. The lowermost is a pedocomplex believed to be formed during MIS 5. It is covered by loess deposits with evidence of fluvial reworking most likely during MIS 4–3. This period of sedimentation was followed by the formation of another pedocomplex, correlated with MIS 3 interstadials, in turn buried by the accumulation of loess during Pleniglacial and/or Lateglacial loess (Frechen et al. 2009). The Monte Ralo section includes two paleosols, the lower probably formed during an early Glacial or Middle Pleniglacial interstadial and the upper (weak paleosol) during the early Middle Pleniglacial (MIS 3) covered by loess deposited most likely during MIS 2. Thus, sometime during MIS 3 (Middle Pleniglacial according to the authors) at least two periods of soil formation are recorded (Frechen et al. 2009).

3.3 *The Pampean Plain*

The Late Quaternary history of the Pampean plain is reconstructed on the basis of aeolian (dunes and loess) deposits and a fluvial succession located in the eastern Pampean plain. Four loess sections at different localities of the eastern Pampean plain, and dune sections in the western Pampean plain have been studied with emphasis on their chronology that was calibrated by luminescense dates.

The San Luis paleo-dune field (Tripaldi and Forman 2007), situated in the western area of the Pampean plain (Fig. 1), is included in the Western Pampean Sand Dune field (WPSD) of central Argentina (Zárate and Tripaldi 2012), a mesic environment at 500–550 m. a.s.l. with a mean annual precipitation of 700 mm and a mean annual temperature of 17° C. The dominant geomorphological features are aeolian geoforms consisting of sand mantles and stabilized dunes, mainly with blowouts and small parabolic dunes (Tripaldi and Forman 2007). Intervals of MIS 3 are exposed in two sections of the dune area.

The Blowout section (33° 49.178′ S, 65° 43.343′ W) consists of ~10 m of fine sand where four sedimentary units were differentiated (from bottom to top B1, B2, B3, B4). The lowermost unit B1 hosts a 70 cm thick paleosol (cambic horizon) with rhizoliths, developed on very well-sorted fine sand with unexposed lower contact. This paleosol was truncated by the deposition of unit B2 composed of a very well-sorted fine sand, with faint horizontal laminations, followed by the accumulation of ~7.5 m of medium-to-fine sands (units B3, B4, B5). Quartz sand from the base of paleosol PB2 yielded an OSL datc of 32,700 ± 2150, while two ages (27,650 ± 2080 years and 25,090 ± 1630 years) were obtained from the overlying unit B2 (Tripaldi and Forman2007). Very young ages (95 ± 10 years to 15 ± 2 years) were obtained for the uppermost units (B3, B4, B5). These results indicate the dominance of aeolian activity with an interval of stability (soils formation) at the later part of MIS 3, followed by aeolian reactivation.

The Nueva Escocia section(33° 49.197′ S; 65° 43.322′ W), ~20 km to the south of the Blowout section, shows a succession of three aeolian units (NE1, NE2, NE3) of very well-sorted medium-to-fine sand that alternates with four paleosols (cambic horizons). OSL ages range from 27,330 ± 2080 years to 24,700 ± 1500 years suggesting the occurrence of variable moisture conditions and/or variable sources of aeolian sand during the interval that covers the transitional time period between MIS 3 and MIS 2 (Tripaldi and Forman 2007).

Eastward of the San Luis paleo-dune field, in the domain of the Central Pampean dunefield (CPD) (Zárate and Tripaldi 2012), Latrubesse and Ramonell (2010) reported aeolian accumulation in linear dunes in the western Buenos Aires province during MIS 3 bracketed between ~30 and 42.7 ka.

In the eastern Pampas, loess–paleosol successions have been the focus for stratigraphic, paleontological and sedimentological analysis, some of them that were chronologically calibrated by numerical dates are here summarized. The Tortugas section (32°45′ S, 61°50′ W; 100 m a.s.l.), is located close to a tributary stream of the Río Paraná (Fig. 3). The 10 m thick section originally described and interpreted by Kröhling (1999) is made up of relatively uniform silt loam deposits, primarily differentiated on the basis of color and compaction (Kemp et al. 2004). The last glacial cycle is documented by the accumulation of the Tezanos Pintos Formation subdivided into two members separated by an erosive unconformity. The sedimentation of the lower member is bracketed between 145.57 ± 9.4 ka and 68.7 ± 3.7–63.2 ± 3.4 ka followed by the formation of an argillic paleosol, later eroded by the end of MIS 4 (at least ~68.7 ± 3.7 ka B.P. Kemp et al. 2004). The accumulation of the upper member of the unit started at the very end of MIS 4 and continued during most of MIS 3, and the early part of MIS 2 (23,371 ± 4 ka and 63.273 ± 4 ka, although Kemp et al. (2004) pointed out that the deposition may have likely continued until the Holocene according to TL dates previously reported (Kröhling 1999). Both members of the Tezano Pintos Formation are the result of aeolian accumulation with evidence of water sorting and sediment reworking as well as continued bioturbation during sediment accumulation (Kemp et al. 2004).

Three other loess sections (Baradero, Hudson, Gorina) are located in north-eastern Buenos Aires province, southeast of Tortugas. The Baradero site (33° 47′ S, 59° 39′ W; 70 m a.s.l.) is a 17-m thick vertical cliff along the right margin of the Río Paraná. The lower part of the section is represented by swampy deposits modified by soil formation spanning the equivalent of at least part of MIS 5 (Kemp et al. 2006). The soil forming interval was followed by a transitional period between ~80 ka and 25 ka (MIS 4, MIS 3) marked by a change in the depositional regime characterized by increasing aeolian input. Toward the upper part of this interval, two significant breaks in sedimentation occurred documented by the development of paleosols after 55.2 ± 2.9 ka, and sometime after 27.8 ± 1.5 ka. The succeeding interval of MIS 2 was dominated by loess accumulation (Kemp et al. 2006).

The Hudson section (~34° 49′ S; 58° 06′ W) situated at ~9 m a.s.l. by the Río de la Plata coastal plain was a 4 m thick exposure on a quarry wall no longer available for observations that consisted of four lithostratigraphic units (Zárate et al. 2009). The lower part of the section is composed of clayey silts accumulated in a paludal environment. During MIS 5e the deposits were overlain by lenses of shelly marine sediments of a tidal environment, later modified by pedogenesis that gave way to the formation of a paleosol (Zárate et al. 2009). A phase of erosion followed, truncating the upper part of the paleosol. After this interval of instability, massive clayey silts accumulated in a paludal setting mostly developed during MIS 3

(<54 ka to >23 ka). Prior to 23 ka the uppermost loess mantle started to accumulate and continued until the early Holocene followed by the development of the present soil profile (Zárate et al. 2009).

The Gorina section (34° 54′ S, 58° 02′ W), 11 km southeast from the Hudson section (Fig. 3), is a 12 m thick exposure located at 17 m a.s.l. on a quarry wall, and composed of three lithostratigraphic units with a thick pedocomplex (PU2-PU3) (Kemp et al. 2006). The uppermost part of the pedocomplex is interpreted as being developed at some stage between 194 ka and 56 ka after which fine loess accumulated. According to the ages obtained most of the MIS 3 interval is absent due to an erosional episode that occurred prior to 29 ka. This was followed by the dominant accumulation of loess since the late MIS 3, extending through MIS 2 until the early Holocene ($\sim$9 ka) when the present soil profile developed (Zárate et al. 2009).

In the northeastern Pampean plain of Buenos Aires province, the MIS 3 interval is also recorded in fluvial successions. Blasi et al. (2010) focused on the stratigraphy, sedimentology and paleobiology of the sedimentary filling of the Río Luján valley ($\sim$34° 30′ S; 59° W), which was chronologically calibrated by IRSL (infrared stimulated luminescence), along with ^{14}C dates (AMS and conventional dates). The sedimentary record spans the interval between <70 ka and 11 ka B.P. Blasi et al. (2010) identified three depositional units limited by surfaces of discontinuity, overlain by Holocene deposits. The MIS 3 interval is mostly documented by depositional unit 2 and the lowermost part of depositional unit 3. Unit 2, dated between $\sim$50 ka B.P. and prior to 32.5 ka B.P., is composed of ephemeral fluvial to ephemeral lacustrine deposits accumulated in lowlands interconnected by ephemeral streams, and the formation of temporary lakes under cold and dry conditions (Blasi et al. 2010). The accumulation of depositional unit 3 started sometimes before 32 ka with accumulation of sediment in permanent lakes or ponds with variable inputs of aeolian sand and dust, under temperate and subhumid climatic conditions.

4 General Environmental Conditions in Central Argentina

A general reconstruction of conditions during MIS 3 resulting from a few available sites located along the Andes in southern South America was summarized by Voelker (2002). These conditions suggest a grassland expansion and an equatorward shift of the westerly wind belt during stadials (colder and more arid conditions) and forest expansion and poleward shift of the westerly wind belt during interstadials (warmer and more humid). Recently, the variation of δ^{18}O and δC^{13} found at a stalagmite from Caverna de las Brujas (35.8° S, 69.82° W) located $\sim$200 km south of the San Rafael plain at 1800 m a.s.l. in the Andes of southern Mendoza, provided a high resolution climatic record of the last 48 ka, suggesting that warmer conditions documented by the Greenland NGRIP core broadly match drier conditions at the cave (Akers et al. 2014). In this sense, difficulties arise when trying to transfer the northern hemisphere ice core stratigraphy to the records of the southern hemisphere, primarily because the Antarctic cores show much more moderate counterparts in a different phase (not synchronic) to the Greenland signal (Wolff et al. 2010). According to the evidence and interpretations previously mentioned, it seems that, at a centennial to millenial scale resolution, climatic and environmental cold conditions in Antarctica correlate with drier environmental spells in the eastern Andes of central Argentina, while warmer conditions prevail in the northern hemisphere. Are these conditions reflected by the aeolian and fluvial records examined?

At a regional scale, the 60–25 ka interval was characterized by the predominance of aggradation across central Argentina with differences in the accumulation rates and dominance of either aeolian or fluvial deposits depending on the geomorphological and geographical settings. In this respect, fine sandy deposits were accumulated in the Andean piedmont by ephemeral river systems with a secondary presence of aeolian deposits. In the western Pampean plain (southern San Luis), aeolian processes seem to have been the dominant mechanism of aggradation throughout the Late Quaternary; in the eastern Sierras Pampeanas piedmont of Córdoba province, MIS 3 is documented by the accumulation of finer aeolian deposits (loess) with evidence of reworking by fluvial processes, followed by predominant aeolian sedimentation since MIS 2 until the mid-Holocene. In the eastern Pampean plain several sections (Tortugas, Baradero, Hudson, and Gorina) suggest a period of nearly continuous aeolian (loess) aggradation although with different accumulation rates; the sediments were deposited in either paludal or plains settings, showing evidence of reworking by aqueous transport agents. The aggradation process was interrupted by stability intervals evidenced by paleosols in the San Rafael plain, the San Luis paleo-dunefield, the eastern Sierras Pampeanas plain and the eastern Pampean plain. These paleosols might represent lapses of decreasing aeolian input and perhaps more humid conditions.

The comparison of the records across central Argentina suggests lower accumulation rates of much finer material in the eastern Pampean plain, in agreement with the distal location of the sites in relation to the source area of sediments located at the Andean piedmont (Tripaldi and Zárate 2014, and references therein).

In terms of paleoclimatic conditions, the alluvial succession at La Estacada (Cordillera Frontal piedmont of Mendoza) includes an interval, bracketed between 35 and 31 ^{14}C ka, of apparently higher temperatures and water availability recorded by mollusk assemblages (De Francesco et al. 2007). In the eastern Pampean plain, Blasi et al. (2010) pointed out the occurrence of pronounced environmental variability during the Late Pleistocene; episodic alluvial events occurred under dry subhumid or strongly seasonal conditions with alternating subhumid-humid phases during MIS 3 while dominant aeolian deposition under subhumid dry climatic conditions prevailed during MIS 2 (Blasi et al. 2010).

General wetter and more humid conditions during MIS 3 are inferred by glacial recessions in the Patagonian Andes (Rabassa 2008, and references therein). Further north, the chronology of multiple landslide deposits and related lake sediments located in the semiarid Andean Cordillera of NW Argentina also suggest wetter and more variable conditions between 40 and 25 ka, coincident with periods of stronger activity of El Niño Southern Oscillation (ENSO) and Tropical Atlantic Sea Surface Temperature Variability (Trauth et al. 2003). The dune activity reported in the Gran Chaco of Bolivia between 36 and 33 ka was believed to be related to deflation of alluvial deposits (increased sediment availability) from a megafan environment generated under intense monsoonal effect and enhanced rainfall during MIS 3 and early MIS 2 (Latrubesse et al. 2012). In addition, other marine and terrestrial records from tropical and subtropical South America suggest more humid conditions during MIS 3 (Marwan et al. 2003).

5 Final Remarks

The potential influence of the typical climatic variability of the Northern Hemisphere in continental areas of southern South America during MIS 3 is still barely known because of the lack of high resolution records and the poor calibration control, the major hindrances to elucidate the local or regional significance of the environmental responses found at the geomorphological setting analyzed. Are the events recorded in aeolian and fluvial records of central Argentina the result of the global climatic variability or simply an environmental response to local conditions? Preliminary results from the speleothem of Caverna de las Brujas (eastern Andes) are promising since they provide the first high resolution record for the westernmost area of central Argentina.

Other potential source of information is provided by the detailed analysis and chronological calibration of the environmental stability intervals represented by paleosols. Might they reflect the environmental responses to some of the climatic oscillations that occurred during MIS 3 but, which one? Why, if there are several, are there only a couple of paleosols recorded? In this regard, considering the bracketing age intervals, the stability spells seem to cluster during the second half of MIS 3 (<40 ka), and closer to 35–30 ka. Could these episodes represent the environmental response during the transition period to the more arid and colder conditions that prevailed during MIS 2? Could the paleosols be the result of the climatic conditions mentioned by Marwan et al. (2003) that occurred ~30 ka? Future studies focused on detailed analysis and an adjusted chronology of the successions may contribute to answer these questions.

Acknowledgments The authors wish to express their gratitude to Jorge Rabassa and Eduardo P. Tonni for the invitation to one of authors (MZ) to participate at the La Plata MIS 3 symposium. We also want to thank Germán M. Gasparini for his help and patience during the preparation of this paper. Financial support was provided by EXA 234-UNLPam, PIP CONICET-2011 and UBA (UBACyT 20620100100009).

References

Akers P, Brook G, Liang F, Wang X, Auler A, Railsback B, Cheng H, Edwards RL, Benedetto C (2014) A 48,000 year record of climate change in west central Argentina from stalagmites in Caverna de las Brujas. Climate Change: The karst record, 7th International Conference, Melbourne, 29 Sept–2 Oct, 2014, Abstract with Programme, p 20–21

Baker SE, Gosse JC, McDonald EV, Evenson EB, Martínez O (2009) Quaternary history of the piedmont reach of Río Diamante, Argentina. J South Amer Earth Sci 28:54–73

Blasi A, Castiñeira Latorre C, Del Puerto L, Prieto AR, Fucks E, De Francesco C. Hanson P, García Rodríguez F, Huarte R, Carbonari J, Young A (2010) Paleoambientes de la cuenca media del río Luján (Buenos Aires, Argentina) durante el último período glacial (EIO 4–2) Lat Am J Sedimentol Basin Anal 17(2):85–111

Bradley R (2015) Paleoclimatology, reconstructing climates of the quaternary, 3rd edn. Elsevier, Amsterdam 675 p

Carignano CA (1999) Late Pleistocene to recent climatic change in Córdoba Province, Argentina: geomorphological evidence. Quat Int 57–58:117–134

De Francesco CG, Zárate MA, Miquel SE (2007) Late Pleistocene mollusc assemblages inferred from paleoenvironments from the Andean piedmont of Mendoza, Argentina. Palaeogeogr Palaeoclimatol Palaeoecol 257:461–469

Frechen M, Seifert B, Sanabria JA, Argüello GL (2009) Chronology of late Pleistocene Pampa loess from the Córdoba area in Argentina. J Quat Sci 24:761–772

Gómez Orellana L, Ramil-Rego P, Muñoz Sobrino C (2007) The Würm in NW Iberia, a pollen record from area Longa (Galicia). Quat Res 67:438–452

Isla F, Schnack E (this volume) Sea level changes during the marine isotopic stage 3 in Argentina

Kemp RA, Toms PS, King M, Kröhling DM (2004) The pedosedimentary evolution and chronology of Tortugas, a late quaternary type-site of the northern Pampa, Argentina. Quat Int 114(1):101–112

Kemp R, Zárate MA, Toms PS, King M, Sanabria J, Arguello G (2006) Late Quaternary paleosols, stratigraphy and landscape evolution in the Northern Pampas, Argentina. Quat Res 66:119–132

Kröhling D (1999) Upper quaternary of the lower Carcarañá basin, North Pampa, Argentina. Quat Int 57(58):135–148

Latrubesse EM, Ramonell CG (2010) Landforms and chronology in the Pampean sand sea, Argentina. Abstracts Volumen, 18 International Sedimentological Congress, Mendoza, Argentina, p 529

Latrubesse EM, Stevaux JC, Cremon EH, May J-H, Tatumi SH, Hurtado MA, Bezada M, Argollo JB (2012) Late megafans, fans and fluvio-aeolian interactions in the Bolivian Chaco, Tropical South America. Palaeogeogr Palaeoclimatol Palaeoecol 356–357:75–88

Martinson DG, Pisias NG, Hays JD, Imbrie JD, Moore TC, Shackleton NJ (1987) Age Dating and the orbital theory of the ice ages: development of a high-resolution 0 to 300,000-year chronostratigraphy. Quat Res 27:1–29

Marwan N, Trauth MH, Vuille MM, Kurths M (2003) Comparing modern and Pleistocene ENSO-like influences in NW Argentina using nonlinear time series analysis methods. Clim Dyn 21(3–4):317–326

Mehl AE, Zárate MA (2012) Late Quaternary alluvial records and environmental conditions in the eastern of Mendoza (33–34° S) Argentina. J South Amer Earth Sci 37:41–59

Messager G (2010) Signatures geomorphologiques de l'activite tectonique Plio-Quaternaire dans le sud des Andes Centrales, Argentine. Unpublished PhD thesis, Université de Pau et des Pays de L'Adour

Pepin E (2010) Interactions géomorphologiques et sédimentaires entre bassin versant et piedmont alluvial. Modélisation numérique et exemples naturels dans les Andes. Unpublished PhD thesis, Université de Toulouse, 282

Pisias NG, Martinson DG, Moore TC Jr, Shackleton NJ, Prell W, Hays J, Boden G (1984) High resolution stratigraphic correlation of benthic oxygen isotopic records spanning the last 300,000 years. Marine Geol 56:119–136

Rabassa J (2008) Late Cenozoic glaciations in Patagonia and Tierra del Fuego. In: Rabassa J (ed), The late Cenozoic of Patagonia and Tierra del Fuego, pp 157–198

Siddall M, Rohling EJ, Thompson WG, Waelbroeck W (2008) Marine isotope stage 3 sea level fluctuations: data synthesis and new outlook Rev Geophys 46:RG4003

Tripaldi A, Forman SL (2007) Geomorphology and chronology of Late Quaternary dune fields of western Argentina. Palaeogeogr Palaeoclimatol Palaeoecol 251:300–320

Tripaldi A, Zárate MA, Brook G, Li Giang G (2011) Late Quaternary paleoenvironments and paleoclimatic conditions in the distal Andean piedmont, southern Mendoza, Argentina. Quat Res 76:253–263

Tripaldi A, Zárate MA (2014) A review of late quaternary dune systems for tropical and subtropical South American east of the Andes. Quat Int. doi:10.1016/j.quaint.2014.06.069

Trauth MH, Bookhagen B, Marwan N, Strecker MR (2003) Multiple landslide clusters record quaternary climate changes in the northwestern Argentine Andes. Palaeogeogr Palaeoclimatol Palaeoecol 194:109–121

Veres D, Bazin L, Landais A, Toyé Mahamadou Kele B, Lemieux-Dudon B, Parrenin F, Martinerie P, Blayo E, Blunier T, Capron E, Chappellaz J, Rasmussen SO, Severi M, Svensson A, Vinther B, Wolff EW (2013) The Antarctic ice core chronology (AICC2012): an optimized multi-parameter and multi-site dating approach for the last 120 thousand years. Clim Past 9:1733–1748

Voelker AHL (2002) Global distribution of centennial-scale records for marine isotope stage (MIS) 3: a database. Quat Sci Rev 21:1185–1212

Wolff EW, Chappellaz J, Blunier T, Rasmussen SO, Svensson A (2010) Millennial-scale variability during the last glacial: the ice core record. Quat Sci Rev 29:2828–2838

Zárate MA (2009) El paisaje pampeano a través del tiempo. In: Berón M, Luna L, Bonomo M, Montalvo C, Aranda C, Carrera Aizpitarte M (eds.), V Congreso de Arqueología de la Región Pampeana. Mamul Mapu. Pasado y Presente desde la Arqueología Pampeana. Editorial Libros del Espinillo, Buenos Aires, p 19–32

Zárate MA, Tripaldi A (2012) The aeolian system of central Argentina. J Aeolian Res 3(4): 401–417

Zárate M, Kemp R, Toms P (2009) Late Quaternary landscape reconstruction and geochronology in the northern Pampas of Buenos Aires province, Argentina. J South Am Earth Sci 27:88–99

Marine Isotope Stage 3 (MIS 3) and Continental Beds from Northern Uruguay (Sopas Formation): Paleontology, Chronology, and Climate

Martín Ubilla, Andrea Corona, Andrés Rinderknecht, Daniel Perea and Mariano Verde

Abstract The Sopas Formation is a late Pleistocene continental unit that includes trace fossils, woods, fresh-water mollusks, and vertebrates with mammals being the predominant taxa. Likely, relationships with the Last Interglacial Stage or with the Last Interstadial were proposed. The paleontological content of the Sopas Formation is updated, and the climatic and environmental signals provided by the fossil content are evaluated. Radiocarbon AMS dates ranging from 33,560 ± 700 year B.P. (cal 36,089 − 39,426 year) to 39,900 ± 1,100 (cal 42,025 − 45,389 year) and TL/OSL ages from 27,400 ± 3,300 to 71,400 ± 11,000 year (being the 45–28 ka time interval better represented), support a relationship with Marine Isotopic Stage 3 (MIS 3) in most outcrops. In the fossil assemblage are taxa that indicate open habitats, savannahs, and woodlands including gallery forests and perennial rivers; living representatives of taxa related to benign climatic conditions (mostly tropical to temperate climates), some taxa that suggest arid to semiarid environments, migrants, and seasonality indicators. A replacement versus mixed faunal models is discussed in the light of available evidence.

M. Ubilla (✉) · A. Corona · A. Rinderknecht · D. Perea · M. Verde
Facultad de Ciencias, Universidad de la República (UDELAR), Montevideo, Uruguay
e-mail: martinubilla@gmail.com

A. Corona
e-mail: acorona@fcien.edu.uy

A. Rinderknecht
e-mail: apaleorinder@yahoo.com

D. Perea
e-mail: pereadnl@gmail.com

M. Verde
e-mail: icnologia@gmail.com

A. Rinderknecht
Museo Nacional de Historia Natural, Montevideo, Uruguay

G.M. Gasparini et al. (eds.), *Marine Isotope Stage 3 in Southern South America, 60 ka B.P.–30 ka B.P.*, Springer Earth System Sciences,
DOI 10.1007/978-3-319-40000-6_11

Keywords Marine isotope stage 3 (MIS 3) · Uruguay · Late Pleistocene · Sopas Formation

Abbreviations

AA	NSF-Arizona AMS Laboratory, USA
GX	Geochron Laboratories, USA
OSL	Optically stimulated luminescence
LP	Latyr, Laboratorio de Tritio y Radiocarbono, La Plata
LVD	Laboratório dataçao (LOE e TL), Sao Paulo
UIC	Luminescence Dating Research Laboratory, Department of Earth and Environmental Sciences, University of Illinois, Chicago
URU	Laboratorio de ^{14}C, Facultad de Química, Montevideo
NISP	Total number of identified specimens

1 Introduction

The Sopas Formation is a late Pleistocene continental unit that includes trace fossils, woods, fresh-water mollusks, and vertebrates with mammals being the predominant taxon (Ubilla 2004; Ubilla et al. 2004). According to biostratigraphy and numerical dating, this unit has been correlated with the Lujanian Stage/Age of the Buenos Aires province in Argentina (Late Pleistocene/Early Holocene, sensu Cione and Tonni 1999; Ubilla et al. 2004). Considering the numerical ages, the climatic and the environmental information provided by the fossil content, likely relationships with the Last Interglacial Stage or with the Last Interstadial were proposed (Ubilla and Perea 1999; Ubilla et al. 2004, 2009; Iriondo and Kröhling 2008). It seems that the faunal assemblage is older than those belonging to the Guerrero Member of the Luján Formation (Buenos Aires province, Argentina), which is in general related to the Last Glacial Maximum (Tonni et al. 1999). The Sopas Formation could be correlated with the "Secuencia deposicional Luján Verde Inferior" and in part with the "Secuencia deposicional Luján Rojo (Toledo 2011). The Touro Passo Formation in southwestern Brazil was correlated with the Sopas Formation (Bombín 1975) and yields some ages and taxa shared with this unit (Ribeiro and Scherer 2009; Kerber et al. 2011).

Currently, we have learned more about of the Interstadial Marine Isotopic Stage 3 (MIS 3) of the Last Glacial Stage (Van Meerbeeck et al. 2009; Tonni et al. 2010, 2011; Buiron et al. 2012; Rabassa and Ponce 2013; Long and Stoy 2013). This encompasses a period between ca. 60–25 ka, which is characterized by millennial climatic changes. These changes include sudden warming phases (the Dansgaard-Oeschger events) in addition to colder phases (the Heinrich events) in the northern hemisphere (Van Meerbeeck et al. 2009, 2011) and to a lesser extent in the southern hemisphere (Buiron et al. 2012; Paisani et al. 2014). The impact of these climatic processes in southern continental biota and how they are reflected in the fossil record is far from being understood.

The aims of this paper are: (a) to update the paleontological content of the northern late Pleistocene beds of Uruguay (the Sopas Formation), (b) to perform an appraisal of the climatic and environmental signals provided by the fossil content, and (c) to discuss their relationships with the MIS 3.

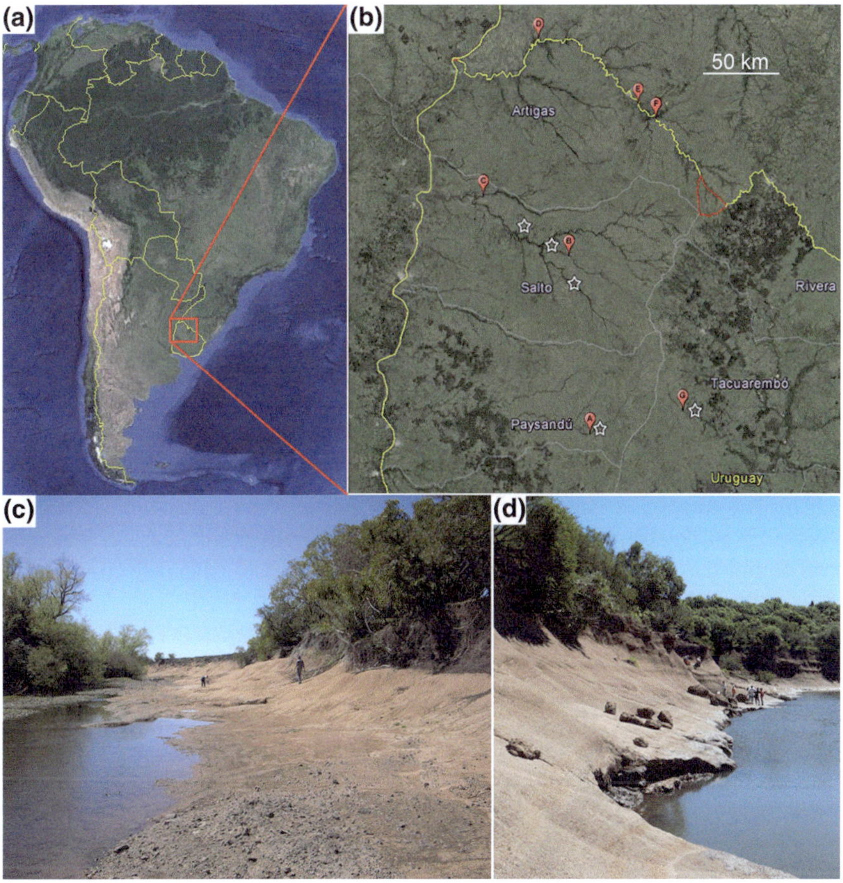

Fig. 1 **a**, **b** Geographic location of selected studied outcrops of the Sopas Formation in northern Uruguay (**b**: *a* Río Queguay, *b* Arroyo Sopas, *c* Río Arapey Chico, Paso del Buey Negro, *d–f* Río Cuareim outcrops, *g* Arroyo Malo). Stars indicate outcrops with *Castrichnus*. **c–d** outcrops of the Sopas Formation (Arroyo Sopas and Río Cuareim respectively)

2 Geographic and Geological Setting

The Sopas Formation crops out in northern Uruguay (Artigas, Salto, Tacuarembó, Paysandú, and Río Negro departments) alongside creeks and rivers and has a patchy pattern of distribution (Ubilla et al. 2004) (Fig. 1a, b). Lithological features of the analyzed outcrops have been considered in detail in Ubilla et al. (2004) and Goso and Ubilla (2004). In general, the Sopas Formation is composed of medium to coarse sandy and conglomerate basal levels belonging to fluvial facies overlain by brownish mudstones and siltstones related to floodplains deposits, and the evolution of paleosols including occasionally important levels of bioturbation. Antón (1975) described the coarse and sandy levels as the Mataojo Formation and the mudstones as the Sopas Formation. Panario and Gutiérrez (1999) and Panario et al. (2014) referred this unit to the Dolores-Sopas Formation.

The outcrops considered here are located in the following localities: Artigas: Paso del León, Mina 1, Estiba, Piedra Pintada (Río Cuareim), and Arroyo Yucutujá; Salto: Paso del Buey Negro (Río Arapey Chico), Río Arapey Grande, Arroyo Sopas, Arroyo Arerunguá, Cañada Sarandí, Ofelia Pliegas; Tacuarembó: Paso Colman, Lavié I y II (Arroyo Malo); Paysandú: Río Queguay; Río Negro: Arroyo Tres Árboles. The most studied localities are indicated in Fig. 1.

3 Materials and Methods

Samples for ^{14}C and OSL ages were taken following protocols indicated by laboratories. Calibrated ages provided in this paper were calculated using Calib 2013 including the SHCal-13 option (Stuiver et al. 2103; Hogg et al. 2013). In addition, calibrated ^{14}C and OSL ages included in Tables 1 and 2 have GPS localization data, but this information is not provided here in order to avoid depredation of fossiliferous sites. A database of the fossil content of the Sopas Formation was generated considering the specimens housed in institutional and particular collections (1053 bone remains). The contribution of each family to the **NISP** (total number of identified specimens) was quantified for mammals (Fig. 2).

4 Numerical Ages

Ubilla and Perea (1999); Ubilla et al. (2004) and Martínez and Ubilla (2004) provided some numerical ages for the Sopas Formation based on ^{14}C and TL/OSL methods. Other ages from several fossiliferous localities were obtained in the last years. All these data are compiled and analyzed here in order to build a chronologically congruent pattern (Tables 1 and 2; Appendices 1–2).

4.1 ^{14}C Ages

Recently, many finite ages based on the radiocarbon method were obtained together with minimum ages (Table 1). The samples used were wood, fresh-water mollusk shells, teeth, and bones.

From the Arroyo Malo locality many ages based on *Diplodon* shells are interpreted as minimum ages (>45 ka B.P.). However, overlying the *Diplodon* bed, we recently obtained for the first time, five finite and stratigraphically ordered ages ranging from 33,560 ± 700 year B.P. (cal 36,089−39,426 year) to 39,900 ± 1,100 (cal 42,025 −45,389 year) based on *Cyanocyclas* sp. (=*Neocorbicula*) and *Pomacea* sp. shells. These ages should be considered together with the OSL of 58–32 ka ages from the same outcrops (see below and Table 2). Two ages based on the bone mineral fraction (16,460 ± 185 year B.P.; 18,650 ± 160 year B.P.) and spatially associated with ^{14}C shell and OSL ages are totally contradictory and interpreted by laboratories as minimum ages.

In the Río Cuareim outcrops were obtained based on woods more minimum ages (>45 ka B.P.). But, there are two ages of 12,100 ± 140 year B.P. (cal 13,550–14,373 year) and 13,869 ± 54 year B.P. (cal 16,473−16,983 year) based on wood samples that came from younger facies of the Sopas Formation with very scarce fossiliferous content.

From the Cañada Sarandí (Salto) locality were obtained based on woods' minimum ages. The age of 5,599 ± 58 year B.P. based on tooth enamel of *Hippidion* cf. *H. principale* from the Paso del Buey Negro (Salto) is rejected at first glance because it is too recent and cannot be explained in the stratigraphic context (see below). In addition, an age of 12,502 ± 55 year B.P. (cal 14,234−15,001 year) from Arroyo Tres Arboles (Río Negro) was recently obtained (shell of *Cyanocyclas* sp.) associated with some extinct mammals (such as deer *Antifer* and glyptodonts).

Archeological studies performed in northern Uruguay provided several ^{14}C ages ranging from approximately 11–8.5 ka (Suárez 2011; Suárez and Santos 2010; Suárez and López 2003; López 2013; Castiñeira et al. 2010, and references therein). *Equus* sp. and *Glyptodon* sp. were reported in association with anthropogenic lithic materials in a 9,585–9,525 year B.P. level (Suárez 2011). A calibration of the 11,200 ± 500 year B.P. age (MEC 1989) provided a 2 sigma cal B.P. 11,600–14,176 year, a roughly similar age with regard to the aforementioned result for Río Cuareim in northern Uruguay (cal 13,550–14,373 year). Nevertheless, authors did not refer the sedimentary context to the Sopas Formation, except for Castiñeira et al. (2010) who related part of the analyzed sequences to this unit.

4.2 OSL/TL Ages

The sampling performed in order to produce OSL/TL ages (Table 2) was focused mainly on fossiliferous outcrops and particularly in localities with radiocarbon

Table 1 **^{14}C** conventional and **AMS*** ages from the Sopas Formation

ID lab.	Taxon sample	Locality	^{14}C age B.P. (cal B.P., 2 sigma)	Source
AA101332*	*Hippidion* (FCDPV-2450) Enamel	Buey Negro, Arapey chico, Salto	**5,599 ± 58**	**This paper**
LP-594	Wood indet	Estiba Rio Cuareim, Artigas	**12,100 ± 140 (13,550–14,373)**	Ubilla et al. (2004)
AA104912*	*Cyanocyclas* sp. shell	Arroyo 3 Árboles, Río Negro	**12,502 ± 55 (14,234–15,001)**	**This paper**
AA99843*	Wood indet	Minal, Rio Cuareim, Artigas	**13,869 ± 54 (16,473–16,983)**	**This paper**
GX-19272	Deer indet apatite	Arroyo Malo, Tacuarembó	**16,460 ± 185**	Ubilla (2001)
URU-0035	*Glyptodon* sp. mineral	Arroyo Malo, Tacuarembó	**18,650 ± 160**	Ubilla (2001)
AA104915*	*Pomacea* sp. shell	Arroyo Malo, Tacuarembó	**33,560 ± 700 (36,089–39,426)**	**This paper**
AA101329*	*Pomacea* sp. shell	Arroyo Malo, Tacuarembó	**35,530 ± 680 (38,659–41,421)**	**This paper**
AA104914*	*Pomacea* sp. shell	Arroyo Malo, Tacuarembó	**37,070 ± 810 (39,940–42,665)**	**This paper**
AA104913*	*Pomacea* sp. shell	Arroyo Malo, Tacuarembó	**38,300 ± 940 (40,865–43,932)**	**This paper**
AA104911*	*Cyanocyclas* sp. shell	Arroyo Malo, Tacuarembó	**39,900 ± 1,100 (42,025–45,389)**	**This paper**
AA101328*	*Diplodon* 1 shell	Arroyo Malo, Tacuarembó	**>45,200**	**This paper**
URU-0032	*D. peraeformis* shell	Arroyo Malo, Tacuarembó	**>45,000**	Ubilla and Perea (1999)
URU-0031	*D. peraeformis* shell	Arroyo Malo, Tacuarembó	**>45,000**	Ubilla and Perea (1999)
URU-0053	*Prosopis nigra* wood	Cañada Sarandí, Salto	**>45,000**	Ubilla and Perea (1999)
LP-490	*Prosopis* sp. wood	Piedra Pintada, Río Cuareim	**>43,000**	Ubilla and Perea (1999)
URU-0036	*Prosopis* sp. wood	Piedra Pintada, Río Cuareim	**>45,000**	Ubilla and Perea (1999)

(continued)

Table 1 (continued)

ID lab.	Taxon sample	Locality	^{14}C age B.P. (cal B.P., 2 sigma)	Source
AA101327*	Deer indet. (FCDPV-2768) enamel	Arroyo Malo, Tacuarembó	–	**This paper**
AA101331*	Ground-sloth (FCDPV-2571) ossicles	Arapey Chico,Salto	–	**This paper**
AA101330*	Deer indet. (FCDPV-2769) enamel	Arroyo Malo, Tacuarembó	–	**This paper**

^{13}C information available is provided

* to highlight which samples were dated using AMS method. All the samples without the asterisk were dated using conventional radiocarbon method

information. There are several ages that seem to be stratigraphically congruent. Nevertheless, in the available set some stratigraphic inversions were detected.

In the Arroyo Malo outcrops, ages were obtained ranging from 58,300 ± 7,400 to 32,850 ± 1,990 year. The first age mentioned here was obtained from a sample associated with the >45 ka B.P. radiocarbon ages from the *Diplodon* bed, in addition to the 34,405 ± 2,240 year sample from an overlying bed. The two OSL ages of 32 ka are based on samples taken from an overlying bed to the *Pomacea* sp. ages (cal 36,089−39,426 to 40,865−43,932 year). There are two ages of 200 and 314 ka that differ from the general pattern observed, belonging to the profiles with 58 and 32 ka, respectively, but with stratigraphic inversion. These ages are rejected taken into account the aforementioned radiocarbon and OSL information.

The base of the profile of the Arroyo Sopas, yielded an age of 43,500 ± 3,600 year and an age of 30,600 ± 5,400 year based on a sample taken from paleocave infilling sediment (Prosul 2009–2011). There are two contradictory results (14,485 ± 1,240 and 36,900 ± 6,500 year) from Paso del Buey Negro. The samples were collected from the same level, so it is necessary to increase the number of samples in future studies.

Many ages were obtained from different localities of the Río Cuareim. In Paso del León locality, the ages obtained (30,300 ± 3,700 and 71,400 ± 11,000 year) are stratigraphically inverted. At Mina 1, fossiliferous levels yielded an age of 36,100 ± 6,200 year (Prosul 2009–2011). An age of 27,400 ± 3,300 year was obtained in an isolated outcrop, of non-fossiliferous coarse beds. The age of 96,000 ± 11,000 year is based on a sample taken from an outcrop without fossil content, which seems to be an older stratigraphic bed of the Sopas Formation. The

Table 2 OSL ages from the Sopas Formation

ID lab.	Sample	Sample location	OSL Age (year)[a]	Source
UIC-3455	Medium sandy	Buey Negro, Río Arapey chico, Salto	**14,485 ± 1,240**	**This paper**
LVD-1449	Medium to coarse sandy	Río Cuareim, Artigas	**27,400 ± 3,300**	**This paper**
LVD-2657	Silty sandy	Paso del León, Río Cuareim, Artigas	**30,300 ± 3,700**	Prosul (2009–2011)
LVD-2660	Sandy-silt crotovina	Arroyo Sopas, Salto	**30,600 ± 5,400**	Prosul (2009–2011)
UIC-3458	Medium sandy	Arroyo Malo, Tacuarembó, Lavie II	**32,850 ± 1,990**	**This paper**
UIC-3451	Medium to coarse sandy	Arroyo Malo, Tacuarembó. P. Colman	**32,995 ± 1,930**	**This paper**
UIC-3332	Medium sandy	Arroyo Malo, Tacuarembó, Lavie II	**34,405 ± 2,240**	**This paper**
LVD-2655	Silty sandy	Mina 1, Río Cuareim, Artigas	**36,100 ± 6,200**	Prosul (2009–2011)
LVD-2661	Médium sandy	Buey Negro, Río Arapey chico, Salto	**36,900 ± 6,500**	Prosul (2009–2011)
LVD-647	Sandy-silt	Arroyo Sopas, Salto	**43,500 ± 3,600**	Ubilla (2004), Ubilla et al. (2004)
LVD-646	Sandy-silt	Arroyo Malo, Tacuarembó, Lavie II	**58,300 ± 7,400**	Ubilla (2004), Ubilla et al. (2004)
LVD-2658	Silty sandy	Paso del León, Río Cuareim, Artigas	**71,400 ± 11,000**	Prosul (2009–2011)
LVD-1241	Medium sandy	Río Cuareim, Artigas	**96,000 ± 11,000**	**This paper**
LVD-859	Sandy-silt	Piedra Pintada, Río Cuareim, Artigas	**180,000 ± 20,000**	Martínez and Ubilla (2004)
LVD-857	Silt	Arroyo Malo, Tacuarembó. Lavie II	**200,000 ± 25,000**	Martínez and Ubilla (2004)
LVD-2659	Silty sandy	Piedra Pintada, Río Cuareim, Artigas	**248,000 ± 26,000**	Prosul (2009–2011)
LVD-1242	Sandy-silt	Arroyo Malo, Tacuarembó. P.Colman	**314,000 ± 39,300**	**This paper**
LVD-858	Fine sandy	Piedra Pintada, Río Cuareim, Artigas	**360,000 ± 40,000**	Martínez and Ubilla (2004)

[a]For UIC ages, all errors are at one sigma and ages are calculated from AD 2010

fossiliferous locality Piedra Pintada deserves particular consideration, with ages of 360, 248 and 180 ka, that are stratigraphically coherent, which are not only clearly older than MIS 3 but also than the Last Interglacial. They are related to >45 ka B. P. radiocarbon ages (wood) and totally depart from the global pattern obtained. It is very difficult to explain these results, which should be taken with caution awaiting further analysis. It is important to highlight that the mammalian content does not differ from the other outcrops of the unit.

5 Paleontological Content

In this section, the taxonomic information of the Sopas Formation is summarized and updated (Tables 3 and 4), associated with the environmental and climatic signals provided by the various taxa identified.

5.1 Trace Fossils

This type of fossil is represented in the Arroyo Sopas locality by some burrow-like structures found associated with skeletal remains of the extinct caviine *Microcavia criolloensis* (Ubilla et al. 1999), which could be the trace-producer (Ubilla 2008). Other structures interpreted as large paleocaves were also found in the same strata. A few coprolites have been reported (Piedra Pintada, Río Cuareim) and related to medium to large predators based on their shape and caviine bones and teeth inclusions (Verde and Ubilla 2002). The hypercarnivorous canids, such as *Protocyon* or *Dusicyon avus* (Prevosti et al. 2009), could also be considered as possible producers. Other authors (Chimento and Rey 2008) claim a canid origin of these materials.

Table 3 Updated list of non-mammals vertebrates for the Sopas Formation of northern Uruguay. Based on Ubilla et al. (2004), Tambussi et al. (2005, 2009) and this paper

Teleostei
Paracanthopterygii/Acanthopterygii indet.
Testudines
Family **Testudinidae**
Chelonoides sp. Fitzgerald, 1835
Squamata
Family **Teiidae**
Tupinambis cf. *T. teguixin* (Linnaeus 1766)
Aves
Family **Rheidae**
Rhea sp. Brisson 1760
Family **Anatidae**
Chloephaga picta (Gmelin 1789)
Family **Cariamidae**
Cariama cristata (Linnaeus 1766)
Family **Psitaciidae**
Cyanoliseus patagonus (Vieillot 1817)
Family **Furnariidae**
cf. *Pseudoseisuropsis* sp. Noriega (1991)

Table 4 Updated list of mammals for the Sopas Formation of northern Uruguay

Order Didelphimorphia	Family **Caviidae**
Family Didelphidae	[a]*Cavia* sp. Pallas, 1766
[ac]cf. *Didelphis* sp. Linneus, 1758	[ac]*Galea* sp. Meyen, 1831
	[bc]*Microcavia criolloensis* Ubilla et al., (1999)
Order Xenarthra	[a]*Dolichotis* sp. Desmarest, 1820
Family **Dasypodidae**	[a]*Hydrochoerus hydrochaeris* (Linnaeus, 1766)
[ac]*Dasypus* aff. *D. novemcinctus* Linnaeus, 1758	*Neochoerus* cf. *N. aesopi* (Leidy, 1853)
Propraopus sp. Ameghino, 1881	
Family **Pampatheriidae**	**Order Litopterna**
Pampatherium typum Gervais and Ameghino, 1880	Family **Macraucheniidae**
Pampatherium humboldti (Lund, 1839)	*Macrauchenia patachonica* Owen, 1838
Family **Glyptodontidae**	Family **Proterotheriidae**
Glyptodon clavipes Owen, 1839	[c]*Neolicaphrium recens* Frenguelli, 1921
[c]cf. *Hoplophorus* Lund, 1839	*N.* cf. *N. recens*
Neuryurus rudis (Gervais, 1878)	
Panochthus tuberculatus (Owen, 1845)	**Order Notoungulata**
Family **Megatheriidae**	Family **Toxodontidae**
Megatherium americanum Cuvier, 1796	*Toxodon* cf. *T. platensis* Owen, 1837
Family **Nothrotheriidae**	
[c]*Nothrotherium* cf. *N. maquinense* (Lund, 1838)	**Order Proboscidea**
Family **Mylodontidae**	Family **Gomphotheriidae** gen. et sp. indet.
Glossotherium robustum (Owen, 1842)	
Lestodon armatus Gervais, 1855	**Order Perissodactyla**
[c]*Catonyx cuvieri* (Lund, 1839)	Family **Tapiridae**
[c]*Catonyx* sp. Ameghino. 1891	[a]*Tapirus terrestris* (Linnaeus, 1758)
Subfamily Scelidotheriinae gen. et sp. indet.	[c]*Tapirus* sp.
	Family **Equidae**
Order Carnivora	[b]*Equus (Amerhippus) neogeus* Lund, 1840
Family **Canidae**	*Hippidion principale* (Lund, 1845)
[ac]*Lycalopex gymnocercus* (Fischer, 1814)	
[c]*Dusicyon avus* (Burmeister, 1866)	**Order Artiodactyla**

(continued)

Table 4 (continued)

[c]*Protocyon troglodytes* Lund, 1838	Family **Tayassuidae**
Family **Felidae**	[ac]*Tayassu pecari* (Link, 1795)
[ac]*Felis concolor* Linnaeus, 1771	[ac]*Catagonus* wagneri (Rusconi, 1930)
[a]*Panthera* cf. *P.onca* (Linnaeus, 1758)	[bc]*Catagonus stenocephalus* (Lund in Reinhardt, 1880)
Smilodon populator Lund, 1842	Family **Cervidae**
Family **Mustelidae**	*Antifer ultra* (Ameghino, 1888)
[ac]*Lontra longicaudis* (Olfers, 1818)	[a]*Ozotoceros* aff. *O. bezoarticus* (Linnaeus, 1758)
Family **Ursidae**	*Morenelaphus brachyceros* (Gervais and Ameghino, 1880)
Arctotherium aff. *A. bonariense* (Gervais, 1852)	*Morenelaphus lujanensis* (Ameghino, 1888)
Order Rodentia	[c]*Paraceros fragilis* (Ameghino, 1888) [c]*Mazama* sp. Rafinesque, 1817
Family **Cricetidae**	Family **Camelidae**
[ac]*Reithrodon* sp. Waterhouse, 1837	*Hemiauchenia paradoxa* Gervais and Ameghino, 1880
[ac]cf. *Wilfredomys oenax* (Thomas, 1928)	[a]*Lama guanicoe* (Müller, 1776)
[a]*Lundomys molitor* (Winge, 1887)	*Palaeolama major* (Liais, 1872)
Family **Erethizontidae**	[a]*Vicugna vicugna* (Molina, 1782)
[b]*Coendou magnus* (Lund, 1839)	
Coendou cf. *C. magnus*	
Family **Echimyidae**	
Myocastor coypus (Molina, 1782)	
Family **Chinchillidae**	
Lagostomus sp. Brookes, 1828	

Based on Ubilla et al. (2004, 2009, 2011and references therein), Prevosti et al. (2009), Perea (2008), Gasparini et al. (2009, 2013), Corona (2012), Scherer (2009), and this paper
[a]Extant at generic or specific level
[b]Extinct species of extant genus
[c]Found in a single locality

The most frequent type of trace fossil is represented by *Castrichnus incolumis*, as described by Verde et al. (2007) and interpreted as earthworm aestivation chambers produced in soils (Fig. 3a). They were found in various localities (the Arroyo Sopas and Arroyo Arerunguá, Ofelia Pliegas, Arroyo Malo, and Río Queguay) (Fig. 1b). According to Verde et al. (2007), these trace fossils suggest a seasonal climate. This inference is based on the fact that some living earthworms construct identical chambers during the summer to avoid desiccation (Verde et al. 2007). Recently, Genise et al. (2013) described an identical chamber from a living earthworm in Misiones, Argentina, a subtropical rainforest area of South America. These authors stated that *Castrichnus* could be produced not only during a seasonal climate, but also when marked droughts occur. The chambers were produced during an atypical drought period in a region that lacks seasonal climate. In this sense, *C. incolumis* could indicate drought conditions even if a seasonal climate is lacking (Genise et al. 2013). Because these types of traces found in various localities in northern Uruguay require special preservation contexts, they suggest that a similar climate and environments could be involved among these different outcrops.

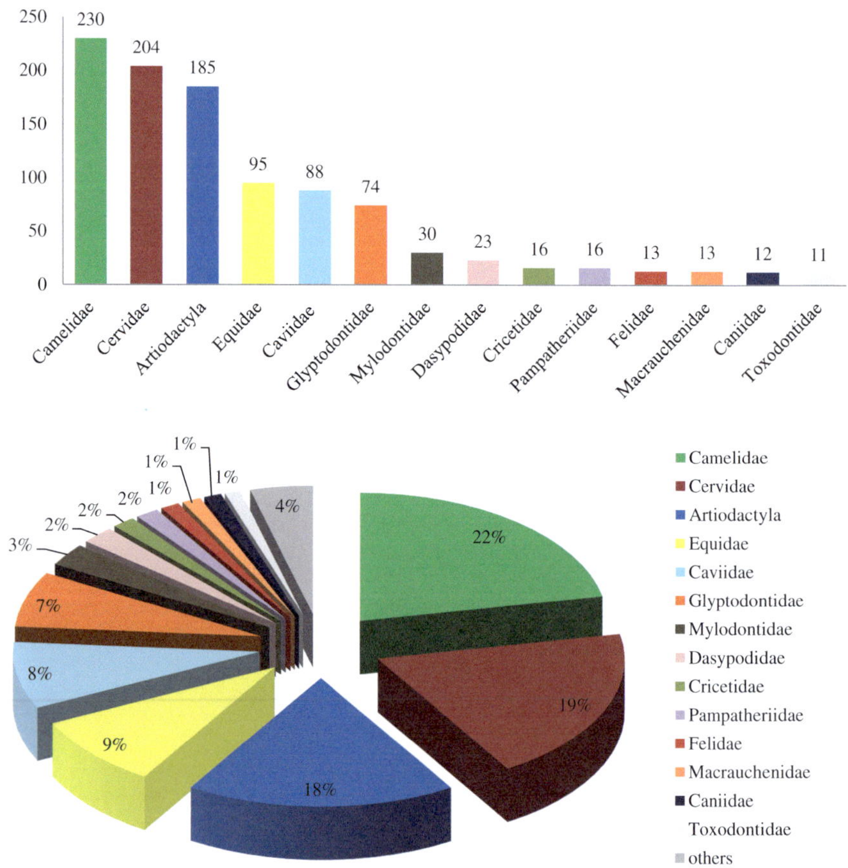

Fig. 2 *Above* Total number of identified specimens (NISP) per Family of mammals (equal or more than 1 %). *Below* Percentage of contribution per family to the total number

5.2 Wood

Very few wood remains have been found in the Sopas Formation until now (Ubilla et al. 2004; Martínez and Ubilla 2004). Though very limited information can be obtained from the available remains, some were determined as *Prosopis* (Inda and del Puerto 2002). This tree is widespread today in tropical to subtropical areas of South America, having also adapted to live in arid to semiarid soils, and proving that it is drought-resistant.

5.3 Mollusks

Fresh-water bivalves and gastropods and also a few terrestrial snails were reported for some localities of the Sopas Formation (Martínez and Rojas 2004). Bivalves are frequently found with articulated valves and gastropods are usually complete. These mollusks can be found in several localities (Arroyo Tres Árboles, Arroyo Malo, Arroyo Arerunguá, Arroyo Sopas, Arroyo Yucutujá among others). They indicate the presence of fluvial and lacustrine contexts.

5.4 Vertebrates

The vertebrates recorded in the Sopas Formation are represented by a few Teleostei indet, reptiles, and avian taxa (Table 3) and numerous mammals (Table 4). Reptiles include the large extinct terrestrial tortoise *Chelonoides* and various non-determined small turtles along with the teiid *Tupinambis* cf. *T. teguixin*. The avian taxa, even if only a few, provide interesting environmental and climatic evidences. *Rhea* and *Cariama* are indicators of open, semi-open, and wooded areas (*Cariama* needs trees to nest) (Ubilla et al. 2004; Tambussi et al. 2005). Fresh-water environments are inferred from the presence of *Chloephaga picta* (Tambussi et al. 2005) which are also indicated by the aforementioned association of mollusks in some localities. It is a southern South American species that migrates during winter to northern latitudes, up to the southern border of Uruguay. According to this record in north-central Uruguay (Tacuarembó), this anatid occupied more northern locations during the Late Pleistocene. It also indicates seasonality.

A new avian record for the Arroyo Sopas is herein reported. It is a furnariid likely belonging to the genus *Pseudoseisuropsis* that was previously referred to late Pleistocene sediments in southern Uruguay (*P. cuelloi;* Claramunt and Rinderknecht 2005). Terrestrial habits related to open and semiarid environments were inferred for *P. nehuen* (Early to Middle Pleistocene of Argentina) (Noriega 1991) and most likely similar conditions for *P. cuelloi* (Claramunt and Rinderknecht 2005).

Mammals are the dominant group, including 25 families in nine orders encompassing more than 50 species (Table 4). Many extinct taxa and also extinct species of living genera are recorded. There are some taxa not represented in the current communities of Uruguay but living today in other areas of South America showing local extinction and shifting ranges. Since the publication of Ubilla and Perea (1999), Ubilla (2004) and Ubilla et al. (2004), the diversity at the generic and specific level for the Sopas Formation has been significantly augmented by new findings (Table 4). Counting all the available specimens, the most abundant families are Camelidae, Cervidae, Equidae, Caviidae, and Glyptodontidae, reaching over 60 % of the NISP (Fig. 2). Actually, more than 50 % of the bones include the first two families and selenodont artiodactyls (18 %) whose characteristics do not allow a more specific taxonomic assignment. It has to be taken into account that the number of glyptodont bones is largely increased because of the large amount of osteoderms of their carapace.

The following taxa are firstly reported: an opossum, likely *Didelphis* (Arroyo Sopas and Paso del Buey Negro) which is part of an ongoing study; the pampatheriid *Pampatherium typum* (Arroyo Sopas) and the presence of the glyptodont *Hoplophorus,* though this finding must be confirmed (Paso del Buey Negro). This latter fossil is important because until today, it was considered restricted to the intertropical region of South America (Minas Gerais, Brazil; Porpino et al. 2010) (Fig. 3c). Surprisingly, the glyptodont *Doedicurus* has not been found yet in this unit. The ground-sloth *Catonyx* was recently found in the Sopas Formation (Corona 2012). New remains under study (upper dentition and mandible) allow confirmation that this is *C. cuvieri*, a species recorded in southeastern Brazil and southern Uruguay (Corona et al. 2013) (Fig. 3d). It is very likely that *C. cuvieri* inhabited forested areas. The first dental material of the short-faced bear *Arctotherium* aff. *A. bonariense* was discovered in the Paso del Buey Negro locality (L. Soibelzon, pers. comm. 2014) (Fig. 3b). Notably, various postcranial bones of a small to medium size bear were also found. The predator guild is also represented not only by medium to large felids but also by large and hypercanivorous canids, such as *Dusicyon* and *Protocyon* (Prevosti et al. 2009). Caviidae are well represented in the Sopas Formation, and the presence of *Dolichotis* (almost complete skull) in the Río Arapey Grande is herein confirmed (Fig. 3e).

Recently, Ubilla et al. (2011) described the first cranial remains of the extinct proterotheriid *Neolicaphrium* cf. *N. recens*. This taxon is now recorded in the Arroyo Malo, Arroyo Sopas and the Mina 1 localities. The record of peccaries was notably increased by the description of two new taxa for this unit: *Catagonus wagneri* and *C. stenocephalus,* which have important climatic and environmental significance (Gasparini et al. 2011, 2013); the presence of *Tayassu pecari* has been confirmed (Gasparini et al. 2009). This implies the presence of three species of peccaries in the same unit, which is certainly unusual in the fossil record of these mammals in South America.

The small deer *Mazama* was found in the Paso del Buey Negro locality (Fig. 3f). Today, this small deer is predominantly an inhabitant of closed forested environments. Recent reviews of camelids (Lorenzo 2009; Scherer 2009) modified the

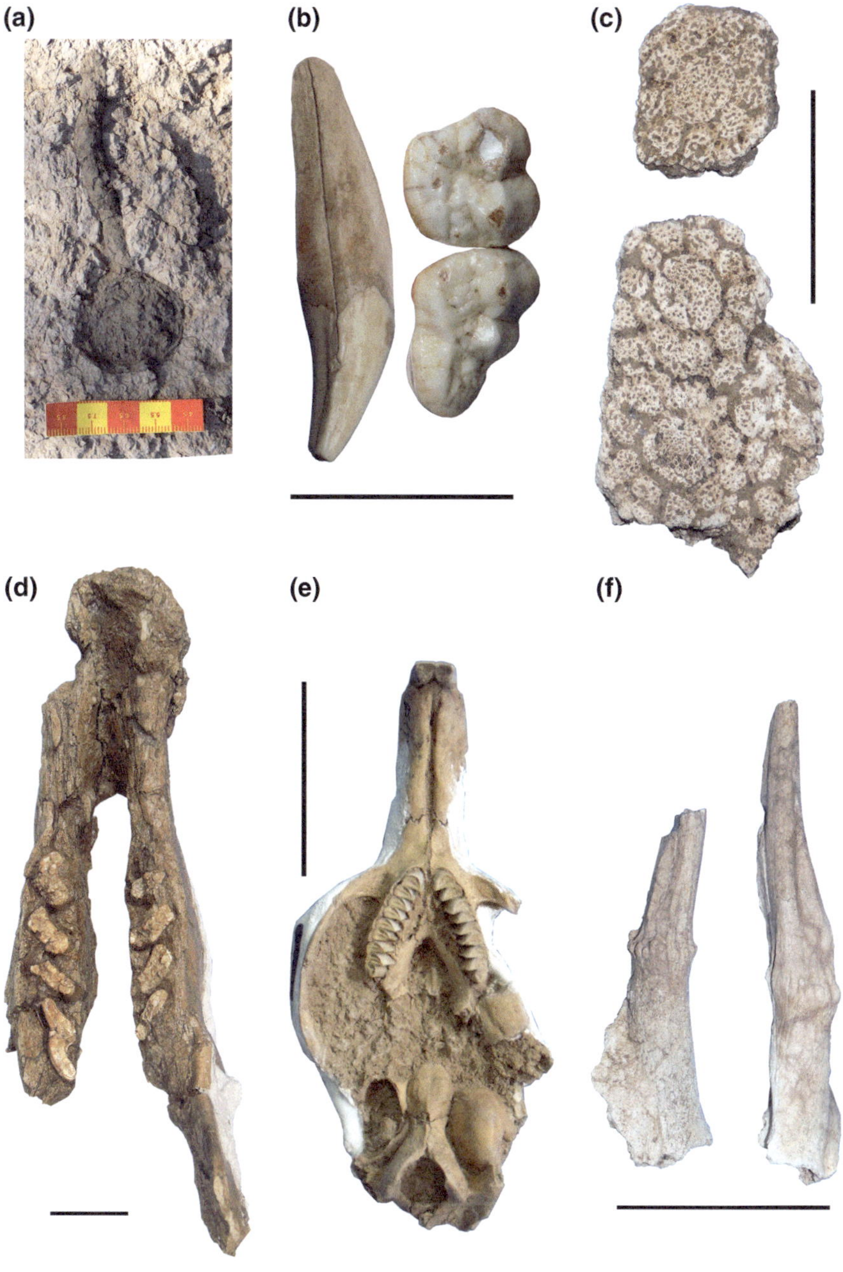

Fig. 3 **a** *Castrichnus incolumis*, **b** *Arctotherium* aff. *A. bonariense,* **c** cf. *Hoplohophorus* sp., **d** *Catonyx cuvieri*, **e** *Dolichotis* sp., **f** *Mazama* sp. **b**–**e** scale: 5 cm

taxonomic records of this group in the Sopas Formation, which are represented by *Lama guanicoe*, *Vicugna vicugna*, and *Palaeolama major*, in addition to *Hemiauchenia paradoxa*.

The "Paso del Buey Negro" at Río Arapey Chico (Salto department) (Fig. 1) is a new locality under study that deserves particular consideration. Sedimentary features are dominated by medium sandy basal levels that can laterally change to coarser beds, and mudstone strata in the uppermost portion of the outcrops. In these outcrops, very well-preserved bones of horses (*Hippidion* cf. *H. principale* and *Equus neogaeus*), including skulls, mandibles, and partially articulated postcranial bones have been found; isolated teeth of *Tapirus* sp. and remains of capybaras (*Hydrochoerus* sp.) belonging to juvenile and adults are frequently found. A few sigmodontine mandibles, the ground-sloth *Glossotherium,* grace postcranial remains of a nothrotherid sloth, peccaries similar to *Tayassu pecari,* small camelids very similar to the vicugna, antlers of *Antifer* and *Mazama* and the aforementioned teeth of *Arctotherium* cf. *A. bonariense* have been identified. This mammalian assemblage includes taxa that indicate fluvial environments along with forested to semi-forested areas; some taxa are related to tropical to temperate contexts. However, the numerical ages yield contradictory information (see Tables 1 and 2) that should be revised.

The first dental enamel $\delta^{13}C$ isotope data were provided for some ungulates (*Hippidion* cf. *H. principale*, *Equus neogaeus*, undetermined deer and a large camelid), forming part of an ongoing study (Morosi and Ubilla 2014). Predominantly browser to mixed feeding habits have been inferred, likely related to open to semi-open environments, and noticeably, no values of C_4 grassers were obtained (Morosi and Ubilla 2014).

6 Discussion

Evidence favoring correlation with the MIS 3 is discussed here, taking into account the numerical ages, the information provided by the trace fossils and the body fossil content.

The climatic/chronological pattern of the MIS 3 is well substantiated by northern hemisphere, and the various stadials (colder intervals) and interstadials (warmer intervals) are well-characterized (Van Meerbeeck et al. 2009, 2011; Rabassa and Ponce 2013; Long and Stoy 2013, among others). There are efforts to identify these events and to understand the influence of the MIS 3 in the southern hemisphere (EPICA 2006; Jouzel et al. 2007; Hodgson et al. 2009; Tonni et al. 2010; Buiron et al. 2012; Paisani et al. 2014; Gottschalk et al. 2014, among others) and inter-hemispheric connections based on Greenland–Antarctic ice-core studies (EPICA 2006; Jouzel et al. 2007). A bipolar thermal seesaw was proposed, and northern hemispheric colder events (Greenland cores) might be related to the southern hemispheric warmer events (Antarctic cores) and vice versa (EPICA 2006; Orombelli et al. 2010; Hessler et al. 2011). However, Jouzel et al. (2007), using a

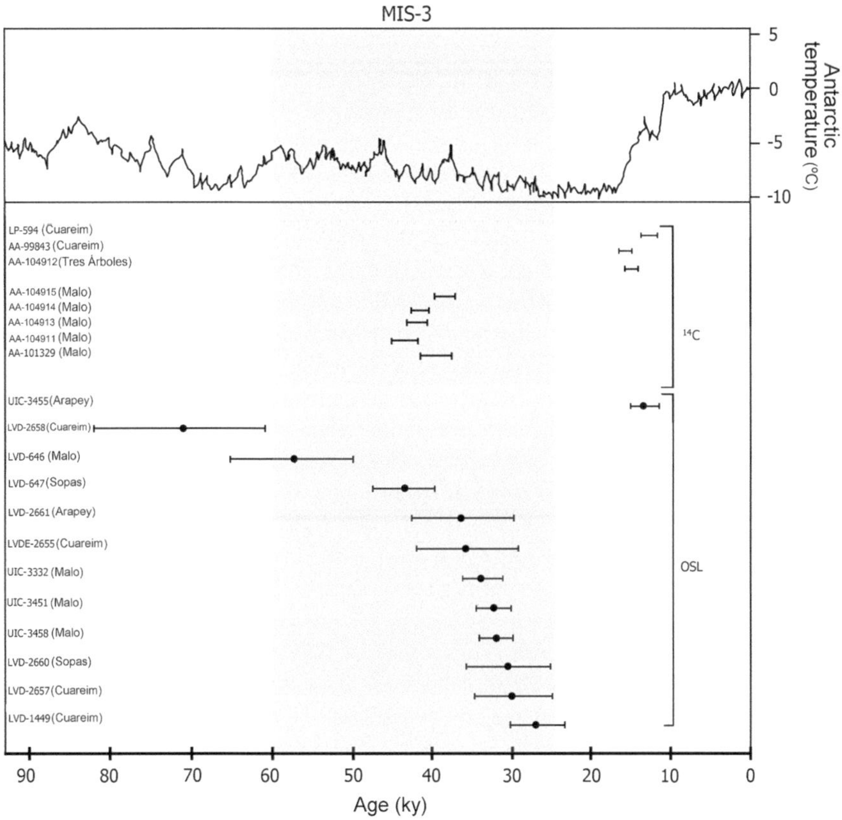

Fig. 4 *Above* The Antarctic climatic variation in the last 90 ka (redrawn from Jouzel et al. 2007). *Below* cal ^{14}C and OSL ages from the Sopas Formation and their relationships with the MIS 3 and latest Pleistocene. Horizontal bars in OSL ages indicate range

more precise calibration, claim that there is a correspondence between the warmer events in the northern hemisphere and major warmer events in Antarctica.

Aside from the three ages based on wood and shell aging with the latest Pleistocene, most radiocarbon data indicate older ages for the fossiliferous beds of the Sopas Formation. In particular, the set of ages ranging from 33,560 ± 700 year B.P. (cal 36,089−39,426 year) to 39,900 ± 1,100 (cal 42,025−45,389 year) from the Arroyo Malo locality is indicative of MIS 3 (Fig. 4). The set of TL/OSL samples taken from various outcrops of the Sopas Formation (Arroyo Malo, Arroyo Sopas, and Río Cuareim localities) also supports relationships with the MIS 3 (27,400 ± 3,300 to 71,400 ± 11,000 year) (Fig. 4). Most ages fall in the 50–25 ka time interval and it is more frequently represented the 45–28 ka time interval that includes various events in the northern and southern hemispheres (EPICA 2006; Jouzel et al. 2007; Van Meerbeeck et al. 2009; Buiron et al. 2012) (Fig. 4). The oldest samples, particularly those that are stratigraphically inverted, that depart from this chronological pattern should be considered with caution or rejected.

As it can be expected about a fluvial context, a time-averaging pattern that affects the preservation context should be assumed (Beherensmeyer et al. 2000; Ubilla et al. 2004). This can be the case for some outcrops of the Sopas Formation such as the Arroyo Malo and Paso del Buey Negro localities. In addition, geographically separated outcrops can yield different ages. In the Arroyo Sopas outcrops, when the numerical ages and the preservation pattern (bones and trace fossils associated) are taken into account, it seems that a shorter time lapse is involved. The peculiar preservation of *Castrichnus*, recorded in several outcrops, can also suggest a similar climate and environment involved in the bearing strata (Fig. 1).

Living representatives of taxa that suggest benign climatic condition (mostly tropical to temperate climates) are recorded in the mammalian assemblage. Tapirs, coendus, coypus, capybaras, river-otters, some peccaries, ocelots, and marsh-rats today inhabit tropical to temperate areas in South America. Some of these taxa also indicate fresh-water bodies along with semi-open to open context. *Rhea* is an inhabitant of open environments and *Cariama* indicates the presence of trees according to its nesting behavior; *Chlohephaga* associated with mollusks indicate fresh-water bodies. Among the extinct taxa, some of them such as some glyptodonts, both horse clades, *Macrauchenia* and *Neolicaphrium,* among others, are indicators of open to semi-open environments. The presence of this mammal assemblage at this latitude of South America in the Late Pleistocene was attributed to the influence of the Last Interglacial or the Last Interstadial, which is correlated to the MIS 3 (Ubilla et al. 2004).

On the other hand, there are some taxa that suggest arid to semiarid environments, such as some caviids (*Dolichotis, Microcavia*), some peccaries (such as *C. wagneri),* and representatives of camelids. To make things more difficult, in some outcrops (e.g., Arroyo Malo locality) we recorded representatives of tropical to temperate conditions (tapirs, marsh-rat, coendu, and river-otter) together with winter migrants such as *Chloephaga*. As previously mentioned, seasonality is substantiated by *Chloephaga* and the worm aestivation chambers. The worm traces could also indicate periods of drought.

This complex climatic and environmental pattern revealed by the fossil content of the Sopas Formation could have been driven by millennial climatic changes that were prevalent during the MIS 3. Various available ^{14}C and OSL ages that span the interval of the MIS 3 favor this hypothesis (Fig. 4).

A faunal replacement model should take into account that the alternating conditions that characterized the MIS 3 cannot be read in the profiles due to a time-averaging effect that produced an “amalgamated” fossil pattern. It must be noted that the MIS 3 formed part of the Last Glacial Cycle that was certainly colder than the Last Interglacial and the current times. The presence in the mammalian assemblage of tropical to subtropical representatives is not predicted by the postulated climatic characteristics of the MIS 3, leaving this issue open to discussion.

An alternative view to the replacement faunal model could take into account that the occurrence of tropical to temperate mammals in the Sopas Formation was facilitated when the warmer Last Interglacial (MIS 5) conditions were prevalent at this latitude. The subsequent interval of time implies colder climate that could have paved the way for the widespread of cold- and arid-adapted mammals, resulting in mixed climatic faunas. Noticeably, Alvarez-Lao (2014) described a mixed cold and temperate fauna (Iberian Peninsula) in the interval 36–30 ka as a consequence of the MIS 3 influence. The survival of tropical to temperate mammals during the MIS 3 at this latitude could have been caused by the presence of permanent streams, riparian forest, and semi-forested areas and the relatively warmer conditions established during the D-O events. Southern Brazil (Araucaria Plateau) seems to have had grasslands and savannahs in the 45–34 ka and forest in the 33–28 ka interval of time (Paisani et al. 2014). In addition, these tropical to temperate mammals are represented in the Sopas Formation by a scarce number of specimens. This pattern could be interpreted as a low abundance in the postulated mixed climatic fauna due to their relict condition. Currently, there are heterogeneous biomes in South America, such as the "Cerrado" (in Brazil), characterized by a mosaic of environments (open habitats, savannahs, and dense woodlands including gallery forests together with permanent streams) under seasonal climate, where forest species coexist with non-forest species (Carmignotto et al. 2012).

Acknowledgments We are indebted to E.P. Tonni and J. Rabassa for inviting us to attend the MIS 3 Symposium at La Plata (2013); G.M. Gasparini helped us in many editorial issues and improvement of the manuscript; L. Soibelzon gave us suggestions about Tremarctinae remains. L. Suárez (Arqueología, FHyCE, Uruguay) provided the useful literature; M. Ghizzoni (Salto) allowed us the access to information about the collection and facilitated the study of comparative material. A. Rojas and S. Martínez assisted us with the taxonomic determination of mollusk samples. E. Morosi helped us in fieldwork. Grants FCE-2009-2398, CSIC C-211-348 (M.Ubilla), and FCE-2011-6752 (D. Perea) supported this research.

Appendix 1: OSL Details According LVD Lab

ID lab	Anual dose uGy/year	Acumulated dose LOE (Gy)	Th (ppm)	U (ppm)	K
LVD-2660	1.366 ± 173	41.84	7.794 ± 0.281	2.255 ± 0.588	–
LVD-2661	1.101 ± 139	40.61	5.504 ± 0.198	1.891 ± 0.482	–
LVD-2655	2.496 ± 307	90.03	6.040 ± 0.217	2.746 ± 0.489	1.107 ± 0.161
LVD-2657	993 ± 70	30.03	3.939 ± 0.142	1.783 ± 0.210	0.036 ± 0.005
LVD-2658	1.483 ± 154	105.89	7.109 ± 0.256	2.492 ± 0.462	0.105 ± 0.015
LVD-2659	928 ± 49	230.38	3.717 ± 0.134	1.741 ± 0.150	–
LVD-1241	603 ± 9	57.9	–	–	–
LVD-1242	774 ± 19	243.7	–	–	–
LVD-1449	1.085 ± 23	29.74	–	–	–
LVD-647	1.237 ± 38.9	53.8	6.36 ± 0.19	1.73 ± 0.04	0.128 ± 0.013
LVD-646	1.073 ± 29.7	62.8	4.88 ± 0.04	1.76 ± 0.04	0.069 ± 0.007

Appendix 2: Optically Stimulated Luminescence Details on Quartz Grains of the Sopas Formation Samples According to Luminescence Dating Research Laboratory, University of Illinois (UIC)

ID lab.	Equivalent dose (Grays)[a]	U (ppm)[b]	Th (ppm)[b]	K_2O (%)[b]	H_2O (%)[c]	Cosmic dose (mGrays/year)[d]	Total dose (mGrays/year)
UIC3332	39.72 ± 2.62	0.9 ± 0.1	2.1 ± 0.1	0.99 ± 0.01	10 ± 3	0.16 ± 0.02	1.15 ± 0.08
UIC3458	45.58 ± 2.78	1.0 ± 0.1	5.0 ± 0.1	0.82 ± 0.01	10 ± 3	0.16 ± 0.02	1.39 ± 0.07
UIC3451	48.66 ± 2.44	2.0 ± 0.1	4.0 ± 0.1	0.82 ± 0.01	10 ± 3	0.14 ± 0.01	1.47 ± 0.07
UIC3455	14.31 ± 0.78	1.7 ± 0.1	3.1 ± 0.1	0.59 ± 0.01	30 ± 5	0.14 ± 0.01	1.20 ± 0.06

[a]Equivalent dose determined by the multiple aliquot regenerative dose method under blue (470 nm) excitation. Blue emissions are measured with 3-mm-thick Schott BG-39 and one, 3-mm-thick Corning 7–59 glass filters that blocks >90 % luminescence emitted below 390 nm and above 490 nm in front of the photomultiplier tube. The coarse-grained (150–250 μm or 425–500 μm) quartz fraction is analyzed

[b]U, Th and K_2O determined by ICP-Ms at Activation Laboratory Ltd., Ontario

[c]Average water content estimated from particle size characteristics assuming periodic wetting in vadose zone

[d]Cosmic dose rate component based on latitude, longitude, elevation, and burial depth of samples

References

Alvarez-Lao D (2014) The Jou Puerta Cave (Asturias, NW Spain): a MIS 3 large mammal assemblage with mixture of cold and temperate elements. Palaeogeogr Palaeoclimatol Palaeoecol 393:1–19

Antón D (1975) Evolución geomorfológica del norte del Uruguay. Dirección de Suelos y Fertilizantes, Ministerio de Agricultura y Pesca, Montevideo

Beherensmeyer A, Kidwell S, Gastaldo R (2000) Taphonomy and paleobiology. Paleobiol 26:103–147

Bombín M (1975) Afinidade paleoecológica, cronológica e estratigráfica do componente de megamamíferos na biota do quaternario terminal da Provincia de Buenos Aires, Uruguai e Rio Grande do Sul (Brasil). Comunicacoes Museum de Ciencias, Pontificia Universidade Católica do Rio Grande do Sul 9:1–28

Buiron D, Stenni B, Chappellaz J, Landais A, Baumgartner M, Bonazza M, Capron E, Frezzotti M, Kageyama M, Lemieux-Dudon B, Masson-Delmotte V, Parrenin F, Schilt A, Selmo E, Severi M, Swingedouw D, Udisti R (2012) Regional imprints of millennial variability during the MIS 3 period around Antarctica. Quat Sci Rev 48:99–112

Carmignotto A, de Vivo M, Langguth A (2012) Mammals of the Cerrado and Caatinga. Distribution patterns of the Tropical open biomes of central South America. In: Patterson B, Costa L (eds) Bones, clones and biomes. The history and geography of recent Neotropical mammals. The University Chicago Press

Castiñeira C, Zárate MA, Blasi A, Fernicola J, del Puerto L, Inda H, Bracco R, García F (2010) Aportes para una actualización de la correlación entre la Formación Sopas del Norte de Uruguay—Formación Luján de la Provincia de Buenos Aires: implicancias arqueológicas. In: Coco G, Feuillet M (eds) Arqueología de Cazadores—Recolectores en la Cuenca del Plata. Centro de Estudios Hispanoamericanos. Ed. Santa Fé

Chimento N, Rey L (2008) Hallazgo de una feca fósil en el Pleistoceno superior—Holoceno inferior del partido de General Guido, provincia de Buenos Aires. Rev Mus Argentino Cienc Nat 10:239–254

Cione A, Tonni EP (1999) Biostratigraphy and chronological scale of uppermost cenozoic in pampean area, Argentina. Quat S Am Antarctic Pen 12:23–51

Claramunt S, Rinderknecht A (2005) A new fossil furnariid from the Pleistocene of Uruguay, with remarks on nasal type, cranial kinetics, and relationships of the extinct genus *Pseudoseisuropsis*. The Condor 107:114–127

Corona A (2012) Los Scelidotheriinae (Xenarthra: Mylodontidae) de Uruguay: sistemática, distribución estratigráfica y cronología. Unpublished dissertation, PEDECIBA Universidad de la República

Corona A, Perea D, McDonald HG (2013) *Catonyx cuvieri* (Xenarthra, Mylodontidae, Scelidotheriinae) from the Late Pleistocene of Uruguay, with comments regarding the systematics of the Subfamily. J Vertebr Paleontol 33(5):1214–1225

EPICA Community Members (2006) One-to-one coupling of glacial climate variability in Greenland and Antarctica. Nature 444:195–198

Gasparini G, Ubilla M, Tonni EP (2009) Tres especies de tayassuidos (*Catagonus wagneri*, *C. stenocephalus* y *Tayassu pecari*) en el Pleistoceno tardío del norte de Uruguay (Fm. Sopas). Reunión Anual Comunicaciones Asoc Paleont Argentina, p 47

Gasparini G, Soibelzon E, Tonni EP, Ubilla M (2011) The "living fossil" Peccary *Catagonus wagneri* (Tayassuidae) and its climatic significance during the Pleistocene and Holocene. Curr Res Pleistocene 28:157–159

Gasparini G, Ubilla M, Tonni EP (2013) The Chacoan peccary, *Catagonus wagneri* (Mammalia, Tayassuidae) in the late Pleistocene (northern Uruguay, South America): Palaeoecological and palaeobiogeographic considerations. Hist Biol 25:679–690

Genise J, Cantil L, Dinghi PA, Sánchez M, Sarzetti L (2013) The aestivation chamber of the giant earthworm *Glossoscolex bergi* (Glossoscolecidae) in the subtropical rainforest of Misiones (Argentina). Ichnos 20:116–119

Goso C, Ubilla M (2004) Los depósitos continentales cuaternarios en el norte de Uruguay: estratigrafía y paleontología. IV Congreso Uruguayo de Geología, Actas 13:1–6

Gottschalk J, Skinner L, Waelbroeck C (2014) Hydrographic changes in the surface ocean of the sb-Antartic Atlantic linked to atmospheric CO_2 variations over the last deglaciation and Marine Isotope Stage 3. Geophys Res Abstr 16

Hessler I, Steinke S, Groeneveld J, Dupont L, Wefer G (2011) Impact of abrupt climate change in the tropical southeast Atlantic during Marine Isotope Stage (MIS) 3. Paleoceanogr. doi:10.1029/2011PA002118

Hodgson D, Verleyen E, Vyverman K, Sabbe M, Leng M, Pickering M, Keely B (2009) A geological constraint on relative sea level in Marine Isotope Stage 3 in the Larsemann Hills, Lambert Glacier region, East Antarctica (31366-33228 cal year BP). Quat Sci Rev 28:2689–2696

Hogg A, Hua Q, Blackwell P, Niu M, Buck C, Guilderson T, Heaton T, Palmer L, Reimer P, Reimer R, Turney C, Zimmerman S (2013) SHCAL 13 southern hemisphere calibration, 0–50,000 years cal BP. Radiocarbon 55:1–15

Inda H, del Puerto L (2002) Identificación taxonómica de muestras de material leñoso. Informe inédito. FC, 1–8 pp

Iriondo M, Kröhling D (2008) Cambios ambientales en la cuenca del Río Uruguay. Universidad Nacional del Litoral, Santa Fé

Jouzel J, Masson-Delmotte V, Cattani O, Dreyfus G, Falourd S, Hoffmann G, Minster B, Nouet J, Barnola JM, Chappellaz J, Fischer H, Gallet JC, Johnsen S, Leuenberger M, Loulergue L, Luethi D, Oerter H, Parrenin F, Raisbeck G, Raynaud D, Schilt A. Schwander J, Selmo E, Souchez R, Spahni R, Stauffer B, Steffensen JP, Stenni B, Stocker TF, Tison JL, Werner M, Wolff EW (2007) Orbital and millennial Antarctic climate variability over the past 800,000 years. Sci 317:793–796

Kerber L, Kinoshita A, José F, Graciano A, Oliveira E, Baffa O (2011) Electron spin resonance dating of the southern Brazilian pleistocene mammals from Touro Passo formation, and remarks on the geochronology, fauna and paleoenvironments. Quat Int 245:201–208

Long J, Stoy P (2013) Quantifying the periodicity of heinrich and dansgaard-oeschger events during marine oxygen isotope stage 3. Quat Res 79:413–423

Lopez J (2013) Early human occupation of Uruguay: radiocarbon database and archaeological implications. Quat Int 301:94–103

Lorenzo N (2009) Camélidos del Pleistoceno de Uruguay: análisis morfológico cualitativo y multivariado. Dissertation, PEDECIBA, Universidad de la República

Martínez S, Rojas A (2004) Quaternary continental mollusks from northern Uruguay: distribution and paleoecology. Quat Int 114:123–128

Martínez S, Ubilla M (2004). El Cuaternario en Uruguay. In: Veroslavsky G, Ubilla M, Martínez S (eds) Cuencas sedimentarias de Uruguay. Geología, Paleontología y Recursos Naturales. Cenozoico. DIRAC-FC, Montevideo

MEC (1989) Misión de Rescate Arqueológico de Salto Grande. Ministerio de Educación y Cultura. Montevideo 2–3:609

Morosi E, Ubilla M (2014) Preliminary report on isotopic studies ($\delta^{13}C$) in living and Late Pleistocene ungulates of Uruguay: paleoecological inferences. Abstracts 4th International Palaeontol Congress, p 757

Noriega J (1991) Un nuevo género de Furnariidae (Aves, Passeriformes) del Pleistoceno inferior-medio de la provincia de Buenos Aires, Argentina. Ameghiniana 28:317–323

Orombelli G, Maggi V, Delmonte B (2010) Quaternary stratigraphy and ice cores. Quat Int 219:55–65

Paisani JC, Pontelli ME, Osterrieth M, Lopes S, Fachin A, Guerra S, Olivera L (2014) Paleosols in low-order streams and valleys heads in the Araucaria Plateau—record of continental environmental conditions in southern Brazil at the end of MIS 3. J S Am Earth Sci 54:57–70

Panario D, Gutiérrez O (1999) The continental Uruguayan Cenozoic: an overview. Quat Int 62: 75–84

Panario D, Gutiérrez O, Sánchez L, Peel E, Oyantcabal P, Rabassa J (2014) Ancient landscapes of Uruguay. In: Rabassa J, Olliers C (eds) Gondwana landscapes in Southern South America. Springer Earth System Science, Springer, Berlin

Perea D (2008) *Nothrotherium* cf. *N. maquinense* (Xenarthra, Tardigrada) en la Formación Sopas (Pleistoceno Tardío de Uruguay). Rev Soc Uruguaya de Geología 14:2–6

Porpino K, Fernicola J, Bergqvist L (2010) Revisiting the intertropical brazilian species *Hoplophorus euphractus* (Cingulata, Glyptodontoidea) and the phylogenetic affinities of *Hoplophorus*. J Vertebr Paleontol 30:911–927

Prevosti F, Ubilla M, Perea D (2009) Large extinct canids from the Pleistocene of Uruguay: systematic, biogeographic and palaeoecological remarks. Hist Biol 21:79–89

Prosul (2009–2011) Estudo integrado do Cuaternario da Regiao Oeste do Rio Grande do Sul, Regiáo Mesopotamica e pampeana da Argentina e Noroeste do Uruguai". CNPq-Prosul 490299/2008-3. Coordenador: Dr. Ana Ribeiro (FZB)

Rabassa J, Ponce J (2013) The Heinrich and Dansgaard-Oeschger climatic events during Marine Isotopic Stage 3: searching for appropriate times for human colonization of the Americas. Quat Int 299:94–105

Ribeiro A, Scherer C (2009) Mamíferos do Pleistoceno do Rio Grande do Sul, Brasil. In: Ribeiro, Girardi, Saldanha (eds) Quaternario de Rio Grande do Sul. Integrando Conhecimentos. Monografías Sociedade Brasileira de Paleontología, Porto Alegre

Scherer C (2009) Os Camelidae Lamini (Mammalia, Artiodactyla) do Pleistoceno da America do Sul: aspectos taxonómicos e filogenéticos. Dissertation, UFGRS-IG-Pgraduacao

Stuiver M, Reimer P, Reimerl R (2103) CALIB radiocarbon calibration. http://calib.qub.ac.uk/calib/

Suárez R (2011) Arqueología durante la Transición Pleistoceno Holoceno en Uruguay: Componentes Paleoindios, organización de la tecnología lítica y movilidad de los Primeros

Americanos. Archaeopress, British Archaeological Reports (BAR) International Series 2220, Oxford

Suárez R, López J (2003) Archaeology of the Pleistocene-Holocene transition in Uruguay: an overview. Quat Int 109:65–76

Suárez R, Santos G (2010) Cazadores-recolectores tempranos, supervivencia de fauna del Pleistoceno (*Equus* y *Glyptodon*), y tecnología lítica durante el Holoceno temprano en la frontera Uruguay-Brasil. Revista de Arqueología 23:20–39

Tambussi C, Ubilla M, Acosta Hospitaleche C, Perea D (2005) Fossil records and palaeoenvironmental implications of *Chloephaga picta* (Gmelin, 1789) (Magellan Goose) and *Cariama cristata* (Linnaeus, 1766) (Seriema) from the Late Pleistocene of Uruguay. Neues Jarb Geol und Palaeontol Mh 5:257–268

Tambussi C, Acosta Hospitaleche C, Rinderknecht A, Ubilla M (2009) Parrots (Aves, Psittaciformes) in the Pleistocene of Uruguay. Ameghiniana 46:431–435

Toledo M (2011) El legado lujanense de Ameghino: revisión estratigráfica de los depósitos pleistocenos-holocenos del Valle del Río Luján en su sección tipo. Registro paleoclimático en la pampa de los estadios OIS 4 al OIS 1. Rev Asoc Geol Argentina 68:121–167

Tonni EP, Carbonari J, Huarte R (2010) Marine sediments attributed to Marine Isotope Stage 3 in the southeastern buenos aires province, Argentina. Curr Res Pleistocene 27:154–156

Ubilla M (2001) Comment on "The continental Uruguayan Cenozoic: an overview" by D. Panario and O. Gutiérrez (Quaternary International 62:75–84). Quatern Int 76–77:259–260

Ubilla M (2004) Mammalian biostratigraphy of Pleistocene fluvial deposits in northern Uruguay, South America. Proc Geologists' Assoc Lond 115:347–357

Ubilla M (2008) Postcranial morphology of the extinct caviine rodent *Microcavia criolloensis* (late Pleistocene, South America). Zoological J Linn Soc 154(4):780–806

Ubilla M, Perea D (1999) Quaternary vertebrates of Uruguay: biostratigraphic, biogeographic and climatic overview. Quat S Am Antarctic Pen 12:75–90

Ubilla M, Piñeiro G, Quintana C (1999) A new extinct species of the genus *Microcavia* (Rodentia, Caviidae) from the upper pleistocene of the northern basin of Uruguay (South America) with paleobiogeographic and paleoenvironmental comments. Stud Neotrop Fauna Environ 34: 141–149

Ubilla M, Perea D, Goso C, Lorenzo N (2004) Late Pleistocene vertebrates from northern Uruguay: tools for biostratigraphic, climatic and environmental reconstruction. Quat Int 114:129–142

Ubilla M, Perea D, Rinderknecht A, Corona A (2009) Pleistocene mammals from Uruguay: biostratigraphic, biogeographic and environmental connotations. In: Girardi S, Scherer C (eds) Ribeiro AM. Quaternario de Rio Grande do Sul. Integrando Conhecimentos. Monografías Sociedade Brasileira de Paleontología, Porto Alegre

Ubilla M, Perea D, Bond M, Rinderknecht A (2011) The first cranial remains of the Pleistocene Proterotheriid *Neolicaphrium* Frenguelli, 1921 (Mammalia, Litopterna): a comparative approach. J Vertebr Paleontol 31(1):193–201

Van Meerbeeck CJ, Renssen H, Roche DM (2009) How did Marine Isotope Stage 3 and last glacial maximum climates differ? Perspectives from equilibrium simulations. Clim Past 5: 33–51

Van Meerbeeck CJ, Renssen H, Roche DM, Wohlfarth B, Bohncke SJ, Bos J, Engels S, Helmens K, Sánchez-Goñi M, Svensson A, Vandenberghe J (2011) The nature of MIS 3 stadial-interstadial transitions in Europe: new insights from model-data comparisons. Quat Sci Rev 30:3618–3637

Verde M, Ubilla M (2002) Mammalian carnivore coprolites from the Sopas Formation (Upper Pleistocene, Lujanian Stage), Uruguay. Ichnos 9:77–80

Verde M, Ubilla M, Jiménez J, Genise J (2007) A new earthworm trace fossil from palaeosols: aestivation chambers from the late pleistocene Sopas Formation of Uruguay. Palaeogeogr Palaeoclimatol Palaeoecol 243:339–347

The Brazilian Intertropical Fauna from 60 to About 10 ka B.P.: Taxonomy, Dating, Diet, and Paleoenvironments

Mário André Trinidade Dantas and Mario Alberto Cozzuol

Abstract This chapter reviews information about the extinct fauna that lived in the Brazilian Intertropical Region (BIR) between 64 and 10 ka B.P. Data from the available literature regarding dating (^{14}C, ESR, U-series) and paleodiet reconstruction ($\delta^{13}C$) for some of taxa of the BIR are herein presented. Furthermore, paleoenvironmental reconstructions of two climatic moments are presented, one at 64 ka, and another between 27 and 10 ka B.P.

Keywords Brazilian Intertropical Region · Pleistocene megafauna · Datings · Feeding ecology

1 Introduction

The Brazilian Intertropical Region (BIR; Fig. 1) has been proposed and defined by Cartelle (1999) as a zoogeographical domain, based on the occurrence of endemic species from the Brazilian states of Goiás (GO), Minas Gerais (MG), Rio de Janeiro (RJ), Espírito Santo (ES), Bahia (BA), Sergipe (SE), Alagoas (AL), Pernambuco (PE), Rio Grande do Norte (RN), Paraíba (PB), Ceará (CE), and Piauí (PI). Mammal fossils from this region are most commonly found within "tanks" (temporary ponds), naturally formed by the accumulation of rain water (e.g., Araújo et al. 2013), or inside caves (e.g., Hubbe and Auler 2012). Fossils of medium (10–100 kg of biomass), large (100–1000 kg), and giant (more than 1000 kg) sized mammals, have been found in both areas, whereas small-sized mammals (less than

M.A.T. Dantas (✉)
Instituto Multidisciplinar em Saúde, Universidade Federal da Bahia—Campus Anísio Teixeira, Rua Rio de Contas, 58, Candeias, Vitória da Conquista, BA 45029-094, Brazil
e-mail: matdantas@yahoo.com.br

M.A. Cozzuol
Laboratório de Paleozoologia, Departamento de Biologia Geral, Universidade Federal de Minas Gerais, Av. Antônio Carlos, 6627, Belo Horizonte, MG 31270-010, Brazil
e-mail: mario.cozzuol@gmail.com

G.M. Gasparini et al. (eds.), *Marine Isotope Stage 3 in Southern South America, 60 ka B.P.–30 ka B.P.*, Springer Earth System Sciences,
DOI 10.1007/978-3-319-40000-6_12

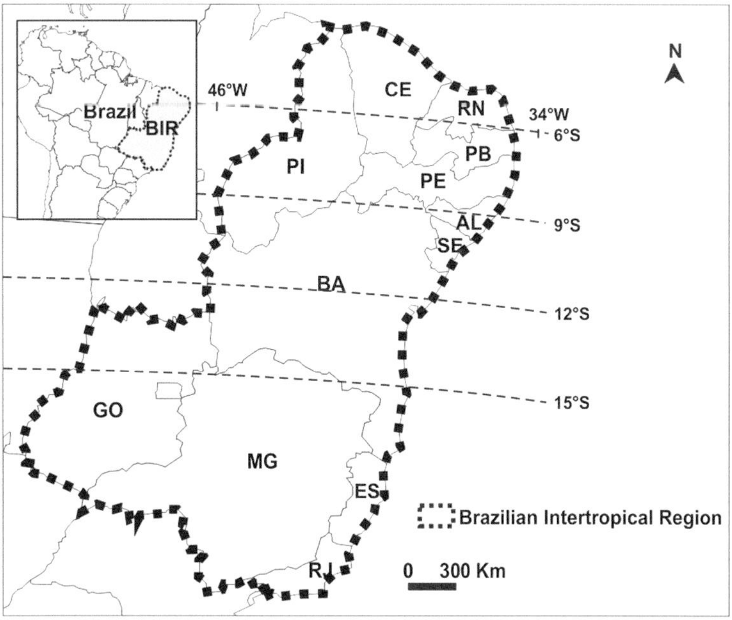

Fig. 1 Map showing the Brazilian Intertropical Region (BIR; represented by *the dotted line*; sensu Cartelle 1999)

10 kg) have mainly been found in caves. These differences can be explained by the type of fossilization taking place in each of these deposition environments.

Cartelle (1999) listed a variety of taxa occurring in the BIR, presenting details about findings from caves in the Minas Gerais and Bahia states. However, despite the well-known taxonomic knowledge of this area, dating and feeding ecology information about the taxa that lived in the BIR is scarce.

Thus, this chapter reviews the available literature in order to report: (i) an update and refinement of the information on mammal taxa occurring within BIR; (ii) a review of dating and feeding ecology information about these taxa; and (iii) commentaries, whenever possible, about the paleoenvironments in which they lived.

2 Materials and Methods

The data analyzed in this review were obtained from: (i) 08 published feeding paleoecology studies, which used isotopic carbon analysis ($\delta^{13}C$) performed on hydroxyapatite and collagen in enamel, dentine or bone (Table 2); and (ii) 13 published dating studies, performed with the Electron Spin Resonance, Carbon 14 and U-series techniques (Table 2).

Table 1 Pleistocene megafauna of the Brazilian Intertropical Region (BIR)

Taxa	BA	SE	AL	PE	PB	RN	CE	PI	GO	MG	RJ	ES
PILOSA												
Megatheriidae												
Eremotherium laurillardi (Lund, 1842)	x	x	x	x	x	x	x	x	x	x	x	x
Mylodontidae												
Mylodontinae indeterminado		x										
Glossotherium sp.					x		x		x			
Glossotherium lettsomi (Owen, 1840)	x											
Catonyx cuvieri (Lund, 1839)	x	x	x		x		x	x		x		
Valgipes bucklandi (Lund, 1839)	x					x		x		x		
Ocnotherium giganteum (Lund, 1839)	x					x				x		
Mylodonopsis ibseni Cartelle, 1991	x			x					x			
Megalonychidae				x								
Ahytherium aureum Cartelle, De Iuliis and Pujos, 2008	x											
Australonyx aquae De Iuliis, Pujos and Cartelle, 2009	x											
Nothrotheriidae												
Nothrotherium maquinense Lydekker, 1889	x				x		x			x		
CINGULATA												
Glyptodontidae												
Glyptotherium sp.	x	x		x	x	x	x	x	x	x		
Panochthus greslebini Castellanos, 1941	x			x	x	x	x	x				
Parapanocthus jaguaribensis (Moreira, 1965)	x				x	x	x					
CINGULATA												
Glyptodontidae												
Hoplophorus euphractus Lund, 1839	x			x		x		x		x		

(continued)

Table 1 (continued)

Taxa	BA	SE	AL	PE	PB	RN	CE	PI	GO	MG	RJ	ES
Dasypodidae												
Pampatherium sp.						x						
Pampatherium humboldti Ameghino, 1875	x			x				x	x	x		
Homelsina paulacoutoi (Guerra and Mahecha, 1984)	x			x	x	x	x					
Pachyarmatherium brasiliense Porpino, Berqvist and Fernicola, 2009	x	x		x		x			?	x	?	
CARNÍVORA												
Felidae			x									
Smilodon populator Lund, 1842	x	x		x	x	x	x	x	x	x		
Ursidae												
Arctotherium wingei Ameghino, 1902	x					x	x			x		
Canidae												
Protocyon troglodytes (Lund, 1838)	x				x	x	x			x		
PROBOSCIDEA												
Gomphotheriidae												
Notiomastodon platensis (Ameghino, 1888)	x	x	x	x	x	x	x	x	x	x	x	x
NOTOUNGULATA												
Toxodontidae					x							
Toxodon platensis Owen, 1840	x	x	x	x		x	x	x	x	x	x	
Piauhytherium capivarae Guérin and Faure, 2013	?			x				x				
PERISSODACTYLA												
Equidae												
Equus (*Amerhippus*) *neogaeus* Lund, 1840	x	x	x	x	x	x	x	x	x	x		
Hippidion principale (Lund, 1846)	x			x	x	x	x	x		x		
ARTIODACTYLA												
Camelidae												
Palaeolama major Liais, 1872	x	x		x	x	x	x	x	x	x		

(continued)

Table 1 (continued)

Taxa	BA	SE	AL	PE	PB	RN	CE	PI	GO	MG	RJ	ES
Cervidae						x						
Mazama gouazoubira (Fischer, 1814)	x				x	x	x		x	x		x
Ozotoceros bezoarticus (Linneus, 1758)	x				x		x		x	x		
LIPTOTERNA												
Macraucheniidae												
Xenorhinotherium bahiense Cartelle and Lessa, 1988	x		x	x	x	x		x	x	x		

Labels: confirmed presence: 'x'; unconfirmed presence: '?'. Location: *BA* Bahia; *SE* Sergipe; *AL* Alagoas; *PE* Pernambuco; *PB* Paraíba; *RN* Rio Grande do Norte; *CE* Ceará; *PI* Piauí; *GO* Goiás; *MG* Minas Gerais; *RJ* Rio de Janeiro; *ES* Espírito Santo

The current taxonomic arrangements proposed for the South American Gomphotheriidae are herein followed. Thus, the paleoecological data presented by Sánchez et al. (2004) for *Stegomastodon waringi* (Holand 1920) will be attributed to *Notiomastodon platensis* postulated by Mothé et al. (2012).

In the analyses of $\delta^{13}C$ data from hydroxyapatite and collagen in bone or dentine (e.g. MacFadden et al. 1999; Drefahl 2010; Dantas et al. 2013a; França et al. 2014), samples were chemically pretreated to eliminate the potential effects of diagenesis (secondary carbonate contamination), using the protocol described elsewhere.

The interpretation of the diet was based on the fact that most existing plants, ranging from trees and woody shrubs to grasses found on prairies and steppes at high altitudes or latitudes, use the Calvin–Benson (C_3) photosynthetic cycle. These plants present average values of $\delta^{13}C$ of −27 ‰. By contrast, the few terrestrial plants that use the Hatch-Slack (C_4) photosynthetic route are primarily tropical and subtropical grasses (Ehleringer et al. 1991; Cerling 1992). These species are typically found in open areas in warm regions subject to hydrological stress, and are able to tolerate low concentrations of CO_2. In general, C_4 plants have higher $\delta^{13}C$ values, averaging −13 ‰ (MacFadden et al. 1999; MacFadden 2005). Those plants that photosynthesize using Crassulacean Acid Metabolism (CAM), such as the succulents, present intermediate $\delta^{13}C$ values (MacFadden et al. 1999; MacFadden 2005).

Studies of modern medium- to large-sized herbivorous mammals recorded an enrichment in $\delta^{13}C$ values between 12 and 14 ‰ (13 ‰ on average) in comparison with the values recorded for the ingested vegetation (Sánchez et al. 2004). Given this, $\delta^{13}C$ values lower than −10 ‰ are typical of animals with a diet consisting exclusively of C_3 plants, while $\delta^{13}C$ values higher than −1 ‰ are consistent with a diet based on C_4 plants. Intermediate $\delta^{13}C$ values (between −10 and −1 ‰) indicate a mixed diet of C_3 and C_4 plants (MacFadden et al. 1999; MacFadden 2005).

3 Results

3.1 Taxonomy and Feeding Paleoecology

Paleoecological knowledge about the Pleistocene megafauna that lived in the BIR is mainly based on interpretations made by a few authors (e.g., Cartelle 1999). Data from ecomorphological studies have not been published yet (only as gray literature), and the few published data from carbon isotope analysis are restricted to herbivore taxa (e.g., Viana et al. 2011; Dantas et al. 2013a; França et al. 2014). The taxa registered in the BIR, as well as interpretations about their feeding ecology, are presented below. These interpretations, whenever possible, are based on analyses of fossils from the BIR, or otherwise, from other regions. The extinct medium- and small-sized herbivores were excluded from this analysis (e.g., primates, rodents, and marsupials), given the morphological similarities with the living species within these groups, what is helpful in describing their ecology. The same reasoning was applied to the other living mammal taxa, which have a generally well-described ecology (Reis et al. 2011).

3.2 Order Carnivora

Fossils of a variety of living and extinct carnivore species have been discovered in "tanks" and caves from the BIR. Extinct taxa include representatives of the three families, *Protocyon troglodytes* (Lund 1838) (Canidae), *Smilodon populator* (Lund 1842) (Felidae), and *Arctotherium wingei* (Ameghino 1902) (Ursidae), each occupying specific niches (Table 1).

P. troglodytes was a hypercarnivore species (i.e., diet composed of 70 % meat) with a body mass estimated from 16 to 37 kg. It is believed to have inhabited open fields, where it hunted in groups, feeding on medium-sized herbivores, such as species of the families Cervidae, Tayassuidae, Camelidae, and Equidae, and small-sized terrestrial sloths (Cartelle and Langguth 1999; Prevosti et al. 2005).

According to evidence collected by Prevosti and Vizcaíno (2006), the saber-toothed cat *S. populator* was probably a solitary species, with a body mass ranging from 220 to 360 kg. It has likely been specialized to predate on large-sized prey, such as the giant sloths and gomphothere species.

A. wingei was a medium-sized bear adapted to open areas and dry climates, with a body mass estimated in 43–107 kg (Soibelzon and Tarantini 2009). It was probably an omnivorous species, tending to herbivorous habits, which fed on plant soft tissues (Trajano and Ferrarezzi 1994; Soibelzon and Schubert 2011).

3.3 Order Pilosa

Sloths (extinct and living) and anteaters (Gaudin 2004) belong to this order. Pilosa (excepting the toothless anteaters) are characterized by a high degree of dental reduction and teeth that lack the enamel layer, classified as hypsodont and prismatic. The teeth of these forms of reduced and simplified dentition are generally known as molariforms, even when some of them are projected toward the front, in the position of the canines or the incisors (Paula Couto 1979). Forms of sexual dimorphism, in which the females are smaller than males, have been proposed for some Pleistocene taxa belonging to the Megatheriidae (Cartelle 1992) and Mylodontidae (Abuhid 1991 *apud* Cartelle 1999; McDonald 2006) families.

Phylogenetic analyses support the existence of nine species and four families of Pilosa during the Pleistocene inhabiting the BIR: *Eremotherium laurillardi* (Megatheriidae, Megatheriinae); *Mylodonopsis ibseni*, *Glossotherium lettsomi*, *Ocnotherium giganteum* (Mylodontidae, Mylodontinae); *Valgipes bucklandi*, *Catonyx cuvieri* (Mylodontidae, Scelidotheriinae); *Nothrotherium maquinense* (Nothrotheriidae, Nothrotheriinae); *Ahytherium aureum*, *Australonyx aquae* (Megalonychidae) (Cartelle 1991; Cartelle and De Iuliis 1995; Cartelle 1999; Gaudin 2004; Cartelle et al. 2008, 2009; De Iuliis et al. 2009).

Although there are only a few studies about the feeding paleoecology of species inhabiting the BIR, the analysis of carbon isotope ratios for some of these animals yielded important results.

Among the terrestrial sloths living in this region, *E. laurillardi* had the largest body size, with some of the specimens reaching six meters in length. Its body mass was estimated in four tons, similar to that proposed for *Megatherium americanum* Cuvier, 1796 (Fariña et al. 1998). The masticatory apparatus of *Eremotherium* and *Megatherium* is morphologically similar, indicating resembling biomechanics. These sloths had a great capacity for oral processing of food, which suggests low digestive efficiency. They had likely a powerful bite, allowing the processing of soft and fibrous types of food, and suggesting a diet composed of leaves from trees and shrubs, along with fruits (Bargo et al. 2006a; Guimarães et al. 2008).

Carbon isotope analyses for *E. laurillardi* found in the Rio Grande do Norte, Alagoas, Sergipe, and Bahia states indicate that, in the BIR, these animals had a diet based on grass and herbaceous plants (i.e., C_4 plants; $\delta^{13}C = 0.3$ and 0.91; Table 2), or a mixed diet, feeding also on leaves and fruits, of trees and shrubs (C_3 plants, $\delta^{13}C = -9.20$ to -2.06; Table 2). These species inhabited open areas or forest edges.

Fariña (1996) proposed that *E. laurillardi* might have also fed on meat, which could classify them as opportunistic omnivores, however evidence to support this hypothesis is still to be found.

According to Bargo et al. (2006a), the Mylodontidae, as opposed to the Megatheriidae, did not have a high capacity for oral processing of food; neither had they a strong bite power. Bargo et al. (2006a, b) attributed grazer habits to the Mylodontinae giant sloths *Glossotherium robustum* and *Lestodon armatus,* which would have diets based on grass and herbaceous plants. The estimated body mass of these species is about 1200–2500 kg, respectively (Fariña et al. 1998).

Two species with morphologies of the masticatory apparatus similar to the ones above mentioned were found in the BIR. *G. lettsomi* has been proposed to be a synonym of *G. robustum* (Esteban 1996 *apud* Fernicola et al. 2009), and *O.*

giganteum, which according to Cartelle (1999) is morphologically similar to *Lestodon*. Therefore, it is reasonable to expect that, given their latitudinal locations, these species would also feed on C_4 gramineae species.

Cartelle (1991) described *M. ibseni* and stated in the diagnosis that it is morphologically close to *Mylodon darwini*. Additionally, Bargo et al. (2006a, b) assigned a generalist diet to *M. darwini* and this diet is attributed tentatively to *M. ibseni* in this paper.

In the phylogenetic analysis of Scelidotheriinae proposed by Gaudin (2004) and Cartelle et al. (2009), *C. cuvieri* and *V. bucklandi* are considered morphologically close to *Scelidotherium leptocephalum*. This may suggest that *C. cuvieri* and *V. bucklandi* have shown high hypsodonty levels due to the large amount of dust particles of the soil they consumed together with the food, and also to the observed adaptations to burrowing habits. Body masses were estimated in about 500 kg (Fariña et al. 1998; Bargo et al. 2000; Vizcaíno et al. 2001). These taxa were likely browsers, feeding mainly on plant buds, fruits, or roots (Bargo 2001 *apud* Bargo et al. 2006b). Data from Pereira et al. (2013), showing values of $\delta^{13}C = -10.17$ ‰ for *V. bucklandi*, further support these assumptions (Table 2).

The only representative of the family Nothrotheriidae in the BIR is *N. maquinense*. Cartelle (1999) proposed this species should occupy a similar niche to the current arboreal sloths, feeding from the leaves found on tree tops.

Among the Megalonychidae, there are two Pleistocene species currently recognized in the BIR, *A. aureum* and *A. aquae*. According to De Iuliis et al. (2009), *A. aureum* would be more closely related to the North American taxa (e.g., *Megalonyx*), whereas *A. aquae* would be to species from the Antilles (e.g., *Megalocnus*). These species are believed to have been browsers, such as *Megalonyx* (McDonald et al. 2001), since they share similar cranial morphologies, and have the same dental formula M4/m3, with triangular molariforms and "incisor" teeth.

3.4 Order Cingulata

This order is composed of xenarthrans with a carapace of bony plates, which covers their back, sides, head top, and tail. They featured a higher number of teeth than sloths, a minimum of 28 hypsodont, rootless molariforms (Paula Couto 1979). Fossils from the BIR comprise living and extinct taxa, the latter highlighted by the giant armadillos and glyptodont.

Glyptodonts differ from giant species of armadillos (and the remaining *Dasypodidae armadillos*) by the absence of movement in the plates of their carapace, showing vertebrae fusions, and by the presence of trilobed teeth (Hoffstetter 1958; Paula Couto 1979). Two giant species of armadillos are known for the BIR, *Pampatherium humboldti* Ameghino, 1875 and *Homelsina paulacoutoi* (Guerra and Mahecha 1984), both considered as grazers of harsh vegetation (Vizcaíno 2009). Grazer glyptodonts (Vizcaíno 2009) with body masses varying from 1000 to 2000 kg (Fariña et al. 1998) were represented by the following taxa: *Panochthus*

Table 2 Review of the available dating data from 60 to ~10 ka (Carbon 14—^{14}C; Electron Spin Resonance—ESR; Thorium/Uranium—Th/U) and paleodiet reconstructions (Carbon isotopes —$\delta^{13}C$; b—bioapatite; c—collagen) for some taxa from the Brazilian Intertropical Region—BIR

Taxa	Sample	Material	Latitude	^{13}C (CO_3) ‰$_{VPDB}$	Dating technique	Reference
V. bucklandi	UGAMS 11763	Bone	05° 49′	−10.17$^{(b)}$	–	Pereira et al. (2013)
T. platensis	UGAMS 09442	Enamel	05° 52′	1.32$^{(b)}$	10.730 ± 30 (^{14}C)	Dantas et al. (2013a)
E. laurillardi	UGAMS 09436	Dentine	05° 57′	−5.22$^{(b)}$	–	Dantas et al. (2013a)
N. platensis	UGAMS 09440	Enamel	05° 57′	0.44$^{(b)}$	16.150 ± 40 (^{14}C)	Dantas et al. (2013a)
E. laurillardi	UGAMS 09435	Dentine	06° 15′	0.50$^{(b)}$	15.490 ± 40 (^{14}C)	Dantas et al. (2013a)
N. platensis	Unnumbered	Enamel	07° 11′	–	39 ± 7 (ESR)	Kinoshita et al. (2005)
N. platensis	Unnumbered	Enamel	07° 11′	–	30 ± 5 (ESR)	Kinoshita et al. (2005)
X. bahiense	Unnumbered	Enamel	07° 11′	–	39 ± 7 (ESR)	Kinoshita et al. (2005)
N. platensis	Unnumbered	Enamel	07° 45′	–	22 ± 3 (ESR)	Kinoshita et al. (2013)
T. platensis	Unnumbered	Enamel	07° 45′	–	26 ± 4 (ESR)	Kinoshita et al. (2013)
N. platensis	Unnumbered	Enamel	08° 14′	–	60 ± 9 (ESR)	Kinoshita et al. (2008)
N. platensis	Unnumbered	Enamel	08° 14′	–	63 ± 8 (ESR)	Kinoshita et al. (2008)
E. laurillardi	SM-1	Dentine	09° 22′	0.30$^{(b)}$	–	Viana et al. (2011)
N. platensis	SM-3	Enamel	09° 22′	0.00$^{(b)}$	–	Viana et al. (2011)
N. platensis	Unnumbered	Enamel	09° 22′	–	39.8 ± 1 (ESR)	Oliveira et al. (2010b)
N. platensis	Unnumbered	Enamel	09° 22′	–	10 ± 0.5 (ESR)	Oliveira et al. (2010b
T. platensis	SM-5	Enamel	09° 22′	−4.10$^{(b)}$	–	Viana et al. (2011)
P. major	Unnumbered	Enamel	09° 46′	–	38 (ESR)	Dantas et al. (2011)
T. platensis	Unnumbered	Enamel	09° 46′	–	50 (ESR)	Dantas et al. (2011)
E. laurillardi	UGAMS 09431	Dentine	09° 55′	−6.65$^{(b)}$	–	Dantas et al. (2013a)
E. laurillardi	UGAMS 09432	Dentine	09° 55′	−3.85$^{(b)}$	22.440 ± 50 (^{14}C)	Dantas et al. (2013a)
E. laurillardi	UGAMS 09433	Dentine	09° 55′	−2.45$^{(b)}$	–	Dantas et al. (2013a)

(continued)

Table 2 (continued)

Taxa	Sample	Material	Latitude	^{13}C (CO_3) ‰VPDB	Dating technique	Reference
E. laurillardi	UGAMS 13539	Dentine	09° 55′	−7.70[(b)]	10.990 ± 30 (^{14}C)	França et al. (2014)
E. laurillardi	UGAMS 13540	Dentine	09° 55′	−3.30[(b)]	11.010 ± 30 (^{14}C)	França et al. (2014)
E. laurillardi	UGAMS 13541	Dentine	09° 55′	−6.00[(b)]	9.720 ± 30 (^{14}C)	França et al. (2014)
E. laurillardi	UGAMS 13542	Dentine	09° 55′	−3.30[(b)]	9.730 ± 30 (^{14}C)	França et al. (2014)
E. laurillardi	UGAMS 13543	Dentine	09° 55′	−4.70[(b)]	11.580 ± 30 (^{14}C)	França et al. (2014)
N. platensis	UGAMS 09437	Dentine	09° 55′	0.76[(b)]	–	Dantas et al. (2013a)
N. platensis	UGAMS 09438	Enamel	09° 55′	−1.04[(b)]	13.980 ± 40 (^{14}C)	Dantas et al. (2013a)
N. platensis	UGAMS 13535	Enamel	09° 55′	−0.40[(b)]	13.380 ± 35 (^{14}C)	França et al. (2014)
N. platensis	UGAMS 13536	Enamel	09° 55′	−0.20[(b)]	16.370 ± 40 (^{14}C)	França et al. (2014)
N. platensis	UGAMS 13537	Enamel	09° 55′	−1.10[(b)]	10.440 ± 30 (^{14}C)	França et al. (2014)
N. platensis	UGAMS 13538	Enamel	09° 55′	1.30[(b)]	13.760 ± 35 (^{14}C)	França et al. (2014)
N. platensis	Unnumbered	Enamel	09° 55′	–	42 (ESR)	Dantas et al. (2011)
N. platensis	Unnumbered	Enamel	09° 55′	–	27 ± 3 (ESR)	Dantas et al. (2013b)
T. platensis	UGAMS 09446	Enamel	09° 55′	−3.68[(b)]	10.050 ± 30 (14C)	Dantas et al. (2013a)
E. laurillardi	UGAMS 09434	Dentine	10° 00′	−3.25[(b)]	–	Dantas et al. (2013a)
N. platensis	Unnumbered	Enamel	10° 00′	–	50 (ESR)	Dantas et al. (2011)
T. platensis	Unnumbered	Enamel	10° 00′	–	50 (ESR)	Dantas et al. (2011)
N. platensis	UGAMS 09439	Enamel	10° 05′	−1.86[(b)]	17.910 ± 50 (14C)	Dantas et al. (2013a)
N. platensis	UGAMS 09441	Enamel	10° 17′	−0.49[(b)]	–	Dantas et al. (2013a)
T. platensis	UGAMS 09443	Enamel	10° 17′	−1.08[(b)]	–	Dantas et al. (2013a)
T. platensis	UGAMS 09444	Dentine	10° 17′	−1.00	–	Dantas et al. (2013a)
N. maquinense	Unnumbered	Calcite	10° 18′	–	15.425 ± 491 (Th/U)	Auler et al. (2006)
N. maquinense	Unnumbered	Calcite	10° 18′	–	15.031 ± 375 (Th/U)	Auler et al. (2006)

(continued)

Table 2 (continued)

Taxa	Sample	Material	Latitude	^{13}C (CO_3) ‰$_{VPDB}$	Dating technique	Reference
T. platensis	U-96-150	Enamel	10° 21′	−5.50[(b)]	–	MacFadden (2005)
E. laurillardi	UGAMS 06136	Bone	10° 42′	−18.20[(c)]	15.770 ± 40 (14C)	Drefahl (2010)
E. (A.) neogaeus	Unnumbered	Enamel	10° 55′	1.10[(b)]	–	MacFadden et al. (1999)
E. (A.) neogaeus	Unnumbered	Enamel	10° 55′	1.70[(b)]	–	MacFadden et al. (1999)
N. platensis	Unnumbered	Enamel	10° 55′	−8.20[(b)]	–	Sánchez et al. (2004)
N. platensis	Unnumbered	Enamel	10° 55′	−5.00[(b)]	–	Sánchez et al. (2004)
T. platensis	U-96-148	Enamel	10° 55′	−12.60[(b)]	–	MacFadden (2005)
T. platensis	U-96-149	Enamel	10° 55′	−7.70[(b)]	–	MacFadden (2005)
E. laurillardi	Unnumbered	Calcite	10° 58′	–	15.000 ± 500 (Th/U)	Auler et al. (2006)
E. laurillardi	Unnumbered	Calcite	10° 58′	–	16.100 ± 3.900 (Th/U)	Auler et al. (2006)
E. laurillardi	Unnumbered	Calcite	10° 58′	–	15.800 ± 2.000 (Th/U)	Auler et al. (2006)
N. platensis	Unnumbered	Enamel	11° 32′	–	50 ± 10 (ESR)	Ribeiro et al. (2013)
T. platensis	Unnumbered	Enamel	11° 32′	–	43 ± 8 (ESR)	Ribeiro et al. (2013)
E. (A.) neogaeus	CM 11032	Enamel	12° 00′	−0.60[(b)]	–	MacFadden et al. (1999)
T. platensis	UGAMS 09445	Enamel	14° 46′	−13.24[(b)]	10.970 ± 30 (14C)	Dantas et al. (2013a)
N. platensis	Unnumbered	Enamel	19° 35′	–	64 ± 5 (ESR)	dos Avilla et al. (2013)
C. cuvieri	Unnumbered	Bone	19° 37′	–	14.030 ± 50 (14C)	Neves and Piló (2003)
C. cuvieri	Unnumbered	Bone	19° 37′	–	13.920 ± 50 (14C)	Neves and Piló (2003)
C. cuvieri	Unnumbered	Bone	19° 37′	–	9.960 ± 40 (14C)	Neves and Piló (2003)
C. cuvieri	Unnumbered	Calcite	19° 37′	–	27.1 ± 3.400 (Th/U)	Auler et al. (2006)
E. (A.) neogaeus	Unnumbered	Bone	19° 37′	–	16.900 ± 70 (14C)	Neves and Piló (2003)
E. (A.) neogaeus	Unnumbered	Bone	19° 37′	–	16.250 ± 60 (14C)	Neves and Piló (2003)
E. (A.) neogaeus	Unnumbered	Bone	19° 37′	–	16.180 ± 70 (14C)	Neves and Piló (2003)

(continued)

Table 2 (continued)

Taxa	Sample	Material	Latitude	^{13}C (CO_3) ‰VPDB	Dating technique	Reference
H. euphractus	Unnumbered	Calcite	19° 37′	–	14.849 ± 711 (Th/U)	Auler et al. (2006)
S. populator	Unnumbered	Bone	19° 37′	–	9.130 ± 150 (14C)	Neves and Piló (2003)

greslebini Castellanos, 1941, *Panocthus jaguaribensis* (Moreira 1965), *Hoplophorus euphractus* Lund (1839), and *Glyptotherium* sp. Oliveira et al. (2010a designated as *Glyptotherium* sp. for all the material previously considered as part of the *Glyptodon* genus of this region.

Additionally, *Pachyarmatherium brasilense* Porpino, Bergqvist and Fernicola (2009), which is a cingulate showing characteristics from both armadillos and glyptodonts, has also been recorded in the BIR. Downing and White (1995) assigned myrmecophagous habits to *Pachyarmatherium leiseyi* (Downing and White 1995), and it is herein believed also true for the Brazilian species *P. brasilense*.

3.5 Order Proboscidea

Only one species of Proboscidea is currently known for the BIR, *N. platensis* (Ameghino 1888). Recent studies suggest that this species lived in groups (Mothé et al. 2010), likely formed by adult females and their youngsters, and possibly other young individuals, in a similar structure to what is currently observed for living elephant populations. This species had a body mass of about four tons, and their diet consisted of grasses and shrubs (C_3 and C_4 plants; Table 2), being considered as generalists (Fariña et al. 1998; Sánchez et al. 2004; Asevedo et al. 2012).

3.6 Order Notoungulata

Two taxa of Notoungulata are recorded in the BIR: *Toxodon platensis* (Owen 1840) and *Piauhytherium capivarae* (Guerin and Faure 2013). Both species were grazers (Cartelle 1999) and are believed to have shared similar diets and body masses of about 1100 kg (Fariña et al. 1998). However, MacFadden (2005) stated that toxodonts presented a large variability in their diet, depending on the habitat. In the BIR, these species may have had more exclusive diets, mainly based on grasses and herbaceous plants, or mixed diets involving C_3 and C_4 plants, or even diets exclusively based on C_3 plants (Table 2).

3.7 Order Perissodactyla

Cartelle (1999) described two species of Perissodactyla occurring in the BIR, *Equus (Amerhippus) neogaeus* (Lund 1840) and *Hippidion principale* (Lund 1846), generally found in association, in fossiliferous outcrops (i.e., "tanks" and caves; Table 1).

Both species are considered grazers with body masses of about 300 kg (Fariña et al. 1998). However, data from carbon isotope studies, although scarce, have shown that in the low latitudes, these species likely fed predominantly on C_4 grasses, such as it has been observed in the state of Bahia (12° S, latitude) in *Equus (Amerhippus) neogaeus* [$\delta^{13}C$ from −0.6 to 1.7 ‰; MacFadden et al. (1999)]. However, in the Argentine pampas (35° S, latitude), studies have shown mixed diets, composed of C_3 and C_4 grasses, tending to a C_3 grass predomination, for *H. principale* ($\delta^{13}C$ from −12.05 to −8.08 ‰) and *Equus (Amerhippus) neogaeus* ($\delta^{13}C$ from −11.46 to −7.21 ‰) (MacFadden et al. 1999; Sánchez et al. 2006), which indicates they likely inhabited open areas. Nevertheless, Bernardes et al. (2013) suggested that these species could coexist with low trophic superposition levels, because *H. principale* might have been more selective for softer plant tissues.

3.8 Order Artiodactyla

Two fossil species of Camelidae are known for the BIR: *Palaeolama major* Liais, 1872 and *Palaeolama* sp. (Marcolino et al. 2012). *Palaeolama (Hemiauchenia) niedai* (Guérin and Faure 1999) is believed to be a junior synonym of *P. major* (Scherer 2009). Marcolino et al. (2012) reviewed the diet of this taxon, and also presented new data regarding the analysis of mummified coprolites found in association with a skeleton of *Paleolama major*, suggesting a diet composed of shrubs (C_3 plants). According to these authors, this taxon likely lived in open areas.

3.9 Order Litopterna

According to Cartelle (1999), only one litoptern species is known for the BIR: *Xenorhinotherium bahiense* (Cartelle and Lessa 1988). However, Guérin and Faure (2004) believed this species does not differ from *Macrauchenia patachonica*, assigning fossils found in the Piauí state to the latter. Given that this discussion is beyond the scope of this study, the systematic proposal of Cartelle (1999) will be considered here.

Cartelle and Lessa (1988) assigned to *X. bahiense* a diet based on grasses and herbaceous plants, what is supported by previous research about the Pleistocene flora in the region where the type specimen was found (i.e., state of Bahia), and because the specimen of the BIR was found in an aggregation with other species of grazing megamammals. Fariña et al. (1998) assigned to *M. patachonica* a body mass of about 1000 kg, which we also tentatively suggest for *X. bahiense*.

3.10 Dating Review

The South American time scale for the Pleistocene was established on the Argentine pampas region, where four land mammal ages were recognized for this epoch: the “Ensenadan,” “Bonaerian,” “Lujanian,” and “Platan” stages (Cione and Tonni 1999). The fauna from the BIR is generally assigned to the Lujanian stage (Cartelle 1999). However, available studies presenting numerical data, despite punctual, show that this fauna has been present in the area for a longer period, from, at least, 350–9 ka. This suggests that this fauna was coeval with the Bonaerian, Lujanian, and the beginning of the Platan land mammal ages.

It is recognized therefore that this fauna has lived in the BIR during the various climatic changes occurred during the Pleistocene/Holocene. The inferred ecology for these animals suggests they were adapted to open environments, such as those found today for the Caatinga and Cerrado biomes. Additionally, it is likely that during drier periods, they would have extended their distribution, whereas the opposite seems to have taken place during the more humid periods, as suggested by Cione et al. (2007).

Between 60 and about 10 ka, two distinct climatic moments are recognized for the BIR. Data from 93 to 47 ka, collected from stalagmites using $\delta^{18}O$ (Wang et al. 2004; de Barreto 2010), indicate a prolonged warmer and drier period, with short intervals of higher moisture. During this period, it is likely that the areas of the Caatinga and Cerrado would have expanded geographically and thus, connected to each other. Evidence indicates that the Caatinga already existed in this region 42 ka (De Oliveira et al. 1999; Behling et al. 2000), along with some of the megafauna taxa, such as *P. humboldti*, *H. euphractus*, *N. platensis*, and, probably, also *C. cuvieri* (Table 2), which lived during this period.

Palynological and $\delta^{18}O$ data available for the period from 40 to 10 ka are not continuous, although they seem to indicate a long period of wetness and predomination of forests, possibly forming a connection between the Amazon and Atlantic Forests (Behling et al. 2000; Auler and Smart 2001; Sifeddine et al. 2003; de Barreto 2010). For this period, there are records in the BIR of *E. laurillardi*, *C. cuvieri*, *Nothrotherium*, *T. platensis*, *N. platensis*, and *S. populator* (Table 2).

4 Discussion

4.1 Paleoenvironmental Reconstruction of the BIR at About 64 ka

Data about the feeding ecology of the megafauna from about 60 ka in the southern BIR are restricted to taphonomic and paleoecological studies of *N. platensis*, from the fossiliferous outcrop “Águas do Araxá,” in Minas Gerais. These gomphotheres inhabited a dry environment, with well-defined seasons, at about 64 ka, (Avilla et al. 2013). Fragments of conifers and grasses were found associated to teeth remains of this species, suggesting a colder and drier season; whereas barite minerals encountered in their bones are recognized as indicators of warmer seasons (Avilla et al. 2013; Dominato 2013). Palynological data from locations above 900 m a.s.l., near Araxá, corroborate this climatic environmental pattern.

4.2 *Paleoenvironmental Reconstruction for BIR at About 10–27 ka*

Published isotopic carbon ratio ($\delta^{13}C$) data are available for this period in the BIR, in relation to the following taxa: *E. laurillardi* (Lund 1842); *V. bucklandi* (Lund 1839); *N. platensis* (Ameghino 1888); *T. platensis* (Owen 1840); and *Equus* (*Amerhippus*) *neogaeus* (Lund 1840). Although the available studies are few, $\delta^{13}C$ analyses suggest *E. laurillardi* had a mixed diet, consuming large amounts of C_3 plants, exploring the edges of forests, and feeding on fruits and herbaceous species across the BIR (Table 2; Fig. 2).

N. platensis and *T. platensis* were grazers between latitudes 5° 49′ S and 6° 15′ S; whereas *V. bucklandi* lived near the edges of forests, feeding on C_3 plants (Table 2; Fig. 2). Between the latitudes 9° 22′ S and 10° 17′ S, *N. platensis* is known to have a diet exclusively based on C_4 grasses, whereas *T. platensis* had a mixed diet, although apparently favoring C_4 plants (Table 2; Fig. 2). Finally, between latitudes 10° 21′ S and 14° 46′ S, *E. neogeus* was a grazer, while *T. platensis* and *N. platensis* had mixed diets, tending to higher consumption of herbaceous C_3 plants (Table 2; Fig. 2).

These results suggest that about 27–11 ka, between the latitudes 14° S and 5° S, the BIR presented a gradual environmental change, from more open (where grasses and herbaceous plants predominated) to more forested ones. A recent

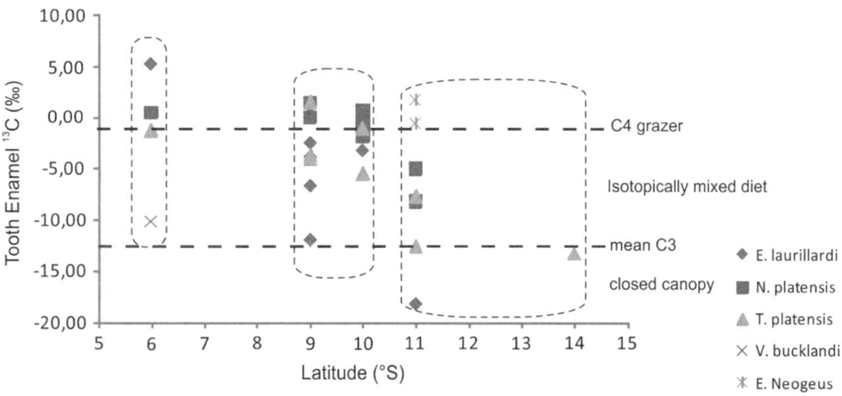

Fig. 2 Tooth enamel carbon isotope ratio ($\delta^{13}C$) values for five species in Brazilian Intertropical Region

biogeographical proposal for *N. platensis* suggests this species was adapted to dry seasonal forest environments (i.e., Caatinga, Cerrado; Dantas et al. 2013b), what may be considered as indicative that similar adaptations also happened to the remaining species of the megafauna of the region.

5 Final Remarks

This chapter represents the first step improving our knowledge about the paleoecology of the megafauna and climatic environmental patterns occurring during the Pleistocene in the BIR. Much, however, is still to be done, and it is believed that further research in this area is highly promising.

Acknowledgments To Flavia Franchini (Memorial University of Newfoundland) for the English review of the manuscript. To the anonymous reviewers which corrections and suggestions improved the quality of this manuscript.

References

Araújo-Júnior HI de, Porpino KO de, Ximenes CL, Bergqvist LP (2013) Unveiling the taphonomy of elusive natural tank deposits: a study case in the Pleistocene of northeastern Brazil. Palaeogeogr Palaeoclimatol Palaeoecol 378:52–74. doi:10.1016/j.palaeo.2013.04.001

Asevedo L, Winck GR, Mothé D, Avilla LS (2012) Ancient diet of the Pleistocene gomphothere *Notiomastodon platensis* (Mammalia, Proboscidea, Gomphotheriidae) from lowland mid-latitudes of South America: stereomicrowear and tooth calculus analyses combined. Quat Int 255:42–52. doi:10.1016/j.quaint.2011.08.037

Auler AS, Smart PL (2001) Late Quaternary paleoclimate in semiarid northeastern Brazil from U-series dating of travertine and water-table speleothems. Quat Res 55:159–167

Auler AS, Piló LB, Smart PL, Wang X, Hoffmann D, Richards DA, Edwards RL, Neves WA, Cheng H (2006) U-series dating and taphonomy of Quaternary vertebrates from Brazilian caves. Palaeogeogr Palaeoclimatol Palaeoecol 240:508–522

Avilla LS dos, Figueiredo AMG, Kinoshita A, Bertoni-Machado C, Mothé D, Asevedo L, Baffa O, Dominato VH (2013) Extinction of a gomphothere population from Southeastern Brazil: taphonomic, paleoecological and chronological remarks. Quat Int 305:85–90. doi:10.1016/j.quaint.2012.09.015

Bargo MS, Vizcaíno SF, Archuby FM, Blanco RE (2000) Limb bone proportions, strength and digging in some Lujanian (late Pleistocene—early Holocene) mylodontid ground sloths (Mammalia, Xenarthra). J Verteb Paleontol 20(3):601–610

Bargo MS, De Iuliis G, Vizcaíno SF (2006a) Hypsodonty in Pleistocene ground sloths. Acta Palaeontol Polonica 51(1):53–61

Bargo MS, Toledo N, Vizcaíno SF (2006b) Muzzle of South American Pleistocene ground sloths (Xenarthra, Tardigrada). J Morphol 267:248–263

Barreto EAS de (2010) Reconstituição da pluviosidade da Chapada Diamantina (BA) durante o Quaternário tardio através de registros isotópicos (O e C) em estalagmites. Programa de Pós-graduação em Geoquímica e Geotectônica, Universidade de São Paulo, Dissertação de Mestrado, 133 pp

Behling H, Arz HW, Tzold JRP, Wefer G (2000) Late Quaternary vegetational and climate dynamics in northeastern Brazil, inferences from marine core GeoB 3104-1. Quat Sci Rev 19:981–994
Bernardes C, Sicuro FL, Avilla LS, Pinheiro AEP (2013) Rostral reconstruction of South American hippidiform equids: new anatomical and ecomorphological inferences. Acta Palaeontol Polonica 58:669–678. doi:10.4202/app.2011.0107
Cartelle C (1991) Um novo Mylodontinae (Edentata, Xenarthra) do Pleistoceno Final da Região Intertropical brasileira. Anais da Acad Brasileira de Ciências 63:161–170
Cartelle C (1992) Edentata e megamamíferos herbívoros extintos da toca dos ossos (Ourolândia, BA). Tese de Doutorado—Universidade Federal de Minas Gerais, Belo Horizonte, 516pp
Cartelle C (1999) Pleistocene mammals of the Cerrado and Caatinga of Brazil. In: Eisenberg JF, Redford KH (eds) Mammals of the neotropics. The University of Chicago Press, Chicago, pp 27–46
Cartelle C, De Iuliis G (1995) *Eremotherium laurillardi*: the Panamerican late Pleistocene megatheriid sloth. J Vertebr Paleontol 15:830–841
Cartelle C, Langguth A (1999) *Protocyon troglodytes* (Lund): um canídeo Intertropical brasileiro. Anais da Acad Brasileira de Ciências 71:371–384
Cartelle C, Lessa G (1988) Descrição de um novo gênero e espécie de Macrauchenidae (Mammalia, Litopterna) do Pleistoceno do Brasil. Paula-Coutiana 3:3–26
Cartelle C, De Iuliis G, Pujos F (2008) A new species of Megalonychidae (Mammalia, Xenarthra) from the Quaternary of Poço Azul (Bahia, Brazil). C R Paleovol 7:335–346
Cartelle C, De Iuliis G, Ferreira RL (2009) Systematic revision of tropical brazilian Scelidotheriine sloths (Xenarthra, Mylodontoidea). J Vertebr Paleontol 29:555–566
Cerling TE (1992) Development of grasslands and savannas in East Africa during the Neogene. Palaeogeogr, Palaeoclimatol, Palaeoecol (Global and Planetary Change Section) 97:241–247
Cione AL, Tonni EP (1999) Biostratigraphy and chronological scale of upper-most Cenozoic in the Pampean Area, Argentina. Quat South Am and Antarctic Pen 12: 23–52 [In: Rabassa J, Salemme M (eds)]
Cione AL, Tonni EP, Soibelzon L (2007) Did humans cause the late Pleistocene-early Holocene mammalian extinctions in South America in a context of shrinking open areas? In: Haynes G (ed) American megafaunal extinctions at the end of the Pleistocene. Springer, Berlin, pp 125–144
Dantas MAT, Porpino KO, Bauermann SG, Prata APN, Cozzuol MA, Kinoshita A, Barbosa JHO, Baffa O (2011) Megafauna do Pleistoceno superior de Sergipe, Brasil: registros taxonômicos e cronológicos. Rev Bras Paleontol 14:311–320. doi:10.4072/rbp.2011.3.10
Dantas MAT, Dutra RP, Cherkinsky A, Fortier DC, Kamino LHY, Cozzuol MA, Ribeiro AS, Vieira FS (2013a) Paleoecology and radiocarbon dates of the Pleistocene megafauna of the Brazilian Intertropical Region. Quat Res 79:61–65. doi:10.1016/j.yqres.2012.09.006
Dantas MAT, Xavier MCT, França LM, Cozzuol MA, Ribeiro AS, Figueiredo AMG, Kinoshita A, Baffa O (2013b) A review of the time scale and potential geographic distribution of *Notiomastodon platensis* (Ameghino, 1888) in the late Pleistocene of South America. Quat Int 317:73–79. doi:10.1016/j.quaint.2013.06.031
De Iuliis G, Pujos F, Cartelle C (2009) A new ground sloth (Mammalia: Xenarthra) from the Quaternary of Brazil. CR Paleovol 8:705–715
De Oliveira PE, Barreto AMF, Suguio K (1999) Late Pleistocene/Holocene climatic and vegetational history of the Brazilian Caatinga: the fossil dunes of the middle São Francisco river. Palaeogeogr Palaeoclimatol Palaeoecol 152:319–337
Dominato VH (2013) Estudo tafonômico dos mastodontes de Araxá, Minas Gerais, Brasil. Programa de Pós-graduação em Geologia. Universidade Federal do Rio de Janeiro. Dissertação, 219p
Downing KF, White RS (1995) The cingulates (Xenarthra) of the Leisey Shell Pit local fauna (Irvingtonian), Hillsborough County, Florida. Bull Florida Mus Nat Hist 37:375–396

Drefahl M (2010) Implicações paleoambientais preliminares da análise de $\delta^{13}C$ em osso de paleomastofauna procedente de Quijingue, Bahia. Boletim de Resumos, Simpósio Brasileiro de Paleobotânica e Palinologia, Salvador, ALPP, Bahia, p 239

Ehleringer JR, Sage RF, Flanagan LB, Pearcy RW (1991) Climate change and the evolution of C 4 photosynthesis. Tree 6:95–99

Fariña RA (1996) Trophic relationships among Lujanian mammals. Evol Theor 11:125–134

Fariña RA, Vizcaíno SF, Bargo MS (1998) Body mass estimations in Lujanian (late Pleistocene-early Holocene of South America) mammal megafauna. Mastozoología Neotropical 5:87–108

Fernicola JC, Vizcaíno SF, De Iuliis G (2009) The fossil mammals collected by Charles Darwin in South America during his travels on board the HMS beagle. Revista de la Asociación Geológica Argentina 64:147–159

França LM, Dantas MAT, Bocchiglieri A, Cherkinsky A, Ribeiro AS, Bocherens H (2014) Chronology and ancient feeding ecology of two upper Pleistocene megamammals from the Brazilian Intertropical Region. Quat Sci Rev 99:78–83

Gaudin TJ (2004) Philogenetic relationships among sloths (Mammalia, Xenarthras, Tardigrada): the craniodental evidence. Zool J Linn Soc 140:255–305

Guérin C, Faure M (1999) *Palaeolama* (*Hemiauchenia*) *niedai* nov. sp., nouveau camelidae du Nordeste brésilien et sa place parmi les lamini d'Amérique du sud. Geobios 32:629–659

Guérin C, Faure M (2004) *Macrauchenia patachonica* Owen (Mammalia, Litopterna) de la région de São Raimundo Nonato (Piauí, Nordeste brésilien) et la diversité des Macraucheniidae pléistocènes. Geobios 37:516–535

Guerra CC, Mahecha GAB (1984) *Pampatherium paulacoutoi*, uma nova espécie de tatu gigante da Bahia, Brasil (Edentata, Dasypodidae). Rev BrasZoologia 2(4):229–254

Guimarães-Jr PR, Galetti M, Jordano P (2008) Seed dispersal anachronisms: rethinking the fruits extinct megafauna ate. PLoS ONE 3(3):1–13

Hoffstetter R (1958) Xenarthra. In: Piveteau J (ed) Traité de Paléontologie, Masson, Paris, pp 535–626

Hubbe A, Auler AS (2012) A large Cervidae Holocene accumulation in Eastern Brazil: an example of extreme taphonomical control in a cave environment. Int J Speleology 41:299–307. doi:10.5038/1827-806X.41.2.15

Kinoshita A, Franca AM, Almeida JAC, Figueiredo AM, Nicolucci P, Graef CFO, Baffa O (2005) ESR dates at K and X band of northeastern Brazilian megafauna. Appl Radiat Isot 62:225–229

Kinoshita A, Barreto A, Alves R, Figueiredo AM, Sarkis JES, Dias ML, Baffa O (2008) ESR dates of teeth from northeastern Brazilian megafauna. Radiat Meas 43:809–812

Kinoshita A, Mayer E, Casati R, Figueiredo AMG, Baffa O (2013) Electron Spin Ressonance dating of megafauna fossils from Lagoa dos Porcos site, Piauí, Brazil. Abstract book, EPR Biodose, Leiden, Netherlands

MacFadden BJ (2005) Diet and habitat of toxodont megaherbivores (Mammalia, Notoungulata) from the late Quaternary of South and Central America. Quat Res 64:113–124

MacFadden BJ, Cerling TE, Harris JM, Prado J (1999) Ancient latitudinal gradients of C_3/C_4 grasses interpreted from stable isotopes of New World Pleistocene horse (*Equus*) teeth. Global Ecol Biogeogr 8:137–149

Marcolino CP, Isaias RMS dos, Cozzuol MA, Cartelle C, Dantas MAT (2012). Diet of *Palaeolama major* (Camelidae) of Bahia, Brazil, inferred from coprolites. Quat Int 278:81–86. doi:10.1016/j.quaint.2012.04.002

McDonald HG (2006) Sexual dimorphism in the skull of Harlan's ground sloth. Contrib Sci 510:1–9

McDonald HG, Miller WE, Morris TH (2001) Taphonomy and significance of Jefferson's ground sloth (Xenarthra: Megalonychidae) from Utah. W N Am Nat 61:64–77

Mothé D, Avilla LS, Winck GR (2010) Population structure of the gomphothere *Stegomastodon waringi* (Mammalia: Proboscidea: Gomphotheriidae) from the Pleistocene of Brazil. Anais da Acad Bras de Ciências 82:983–996

Mothé D, Avilla LS, Cozzuol M, Winck GR (2012) Taxonomic revision of the Quaternary gomphotheres (Mammalia: Proboscidea: Gomphotheriidae) from the South American lowlands. Quaternary International 276–277:2–7. doi:10.1016/j.quaint.2011.05.018

Neves WA, Piló LB (2003) Solving Lund's dilemma: new AMS dates confirm that humans and megafauna coexisted at Lagoa Santa. Curr Res Pleistocene 20:57–60

Oliveira EV, Porpino KO, Barreto AF (2010a) On the presence of *Glyptotherium* in the Late Pleistocene of Northeastern Brazil, and the status of "*Glyptodon*" and "*Chlamydotherium*". Paleobiogeographic implications. N Jb Geol Paläont Abh 258:353–363

Oliveira LC, Kinoshita A, Barreto AMF, Figueiredo AM, Silva JLL, Baffa O (2010b) ESR dates of teeth from Brazilian megafauna. J Phys Conf Ser 249:012062. doi:10.1088/1742-6596/249/1/012062

Paula Couto C (1979) Tratado de Paleomastozoologia. Academia Brasileira de Ciências, Rio de Janeiro, p 590p

Pereira ICS, Dantas MAT, Ferreira RL (2013) Record of the giant sloth *Valgipes bucklandi* (Lund, 1839) (Tardigrada, Scelidotheriinae) in Rio Grande do Norte state, Brazil, with notes on taphonomy and paleoecology. J South Am Earth Sci 43:42–45. doi:10.1016/j.jsames.2012.11.004

Prevosti FJ, Vizcaíno SF (2006) Paleoecology of the large carnivore guild from the late Pleistocene of Argentina. Acta Palaeontolo Polonica 51:407–422

Prevosti FJ, Zurita AE, Carlini AA (2005) Biostratigraphy, systematics, and paleoecology of *Protocyon* Giebel, 1855 (Carnivora, Canidae) in South America. J South Am Earth Sci 20:5–12

Ribeiro RC, Kinoshita A, Figueiredo AMG, Carvalho IS, Baffa O (2013) Electron spin resonance dating of the late Quaternary megafauna fossils from Baixa Grande, Bahia, Brazil. Quat Int 305:91–96. doi:10.1016/j.quaint.2012.07.017

Reis NR, Peracchi AL, Pedro WA, de Lima IP (2011) Mamíferos do Brasil, 2ª edn. Nélio R dos Reis, Londrina

Sánchez B, Prado JL, Alberdi MT (2004) Feeding ecology, dispersal, and extinction of South American Pleistocene gomphotheres (Gomphotheriidae, Proboscidea). Paleobiol 30:146–161

Sánchez B, Prado JL, Alberdi MT (2006) Ancient feeding, ecology and extinction of Pleistocene horses from the Pampean Region, Argentina. Ameghiniana 43

Scherer CS (2009) Os Camelidae Lamini (Mammalia, Artiodactyla) do Pleistoceno da América do Sul: aspectos taxonômicos e filogenéticos. Tese de Doutorado-Universidade Federal do Rio Grande do Sul, Porto Alegre, 472pp

Sifeddine A, Albuquerque ALS, Ledru MPL, Turcq B, Knoppers B, Martin L, Mello WZ, Passenau H, Dominguez JML, Cordeiro RC, Abrão JJ, Bittencourt ACSP (2003) A 21,000 cal years paleoclimatic record from Caço Lake, Northern, Brazil: evidence from sedimentary and pollen analyses. Palaeogeogr Palaeoclimatol Palaeoecol 189:25–34

Soibelzon LH, Schubert BW (2011) The largest known bear, *Arctotherium angustidens*, from the early Pleistocene Pampean region of Argentina: with a discussion of size and diet trends in bears. J Paleontol 85:69–75

Soibelzon LH, Tarantini VB (2009) Estimación de la masa corporal de las especies de osos fósiles y actuales (Ursidae, Tremarctinae) de América del Sur. Rev Museo Argentino de Ciencias Naturales 11:243–254

Trajano E, Ferrarezzi H (1994) A fóssil bear from northeastern of Brazil, with a phylogenetic analysis of the South American extinct Tremarctinae (Ursidae). J Vertebr Paleontol 14:552–561

Viana MSS, Silva JLL da, Oliveira PV de, Julião MS da S (2011) Hábitos alimentares em herbívoros da megafauna pleistocênica do Nordeste do Brasil. Estudos Geológicos 21: 89–95

Vizcaíno SF (2009) The teeth of the "toothless": novelties and key innovations in the evolution of xenarthrans (Mammalia, Xenarthra). Paleobiol 35:343–366

Vizcaíno SF, Zarate M, Bargo MS, Dondas A (2001) Pleistocene large burrows in the Mar del Plata area (Buenos Aires province, Argentina) and their probable builders. Acta Paleontol Polonica 46:157–169

Wang X, Auler AS, Edwards RL, Cheng H, Cristalli PS, Smart PL, Richards DA, Shen CC (2004) Wet periods in northeastern Brazil over the past 210 kyr linked to distant climate anomalies. Nature 432:740–743

Continental Vertebrates During the Marine Isotope Stage 3 (MIS 3) in Argentina

Germán Mariano Gasparini, Esteban Soibelzon, Cecilia Deschamps, Analía Francia, Elisa Beilinson, Leopoldo Héctor Soibelzon and Eduardo Pedro Tonni

Abstract Paleontological sites in Argentina with continental vertebrates corresponding to the Marine Isotope Stage 3 (MIS 3) interval are scarce or poorly known. This situation is mainly due to the lack of absolute ages for Pleistocene fossil remains or their bearing sediments that would allow the verification of the chronology established for this interval. However, a few isolated evidences show that continental vertebrates responded to the abrupt temperature changes that characterized the MIS 3 (Heinrich colder events and Dansgaard–Oeschger warmer events). Up to date, continental vertebrate remains of this age have been found mainly in Buenos Aires province, but also in a few sites of northeastern Argentina (such as Entre Ríos, Corrientes, Formosa and Chaco provinces). In Buenos Aires province: (1) Paso Otero, in the Río Quequén Grande valley, evidence of warmer and more humid conditions were found in sediments dated in 37,800 ± 2300 radiocarbon years before present (RCYBP); (2) Mar del Sur, General Alvarado County, coastal marine sediments with continental mammals were dated in 25,700 ± 800 and 33,780 ± 1200 RCYBP; (3) Balneario Saldungaray, in the Río Sauce Grande valley, Tornquist County,

G.M. Gasparini (✉) · E. Soibelzon · C. Deschamps · A. Francia · L.H. Soibelzon
División Paleontología Vertebrados, Museo de La Plata, Facultad de Ciencias Naturales y Museo, Universidad Nacional de La Plata, Paseo del Bosque without number, CP 1900 La Plata, Buenos Aires, Argentina
e-mail: germanmgasparini@gmail.com

G.M. Gasparini · E. Soibelzon · A. Francia · E. Beilinson · L.H. Soibelzon
Consejo Nacional de Investigaciones Científicas y Técnicas (CONICET), Buenos Aires, Argentina

C. Deschamps
Comisión de Investigaciones Científicas de la provincia de Buenos Aires (CIC), La Plata, Buenos Aires, Argentina

E. Beilinson
Centro de Investigaciones Geológicas (CONICET-UNLP), Diagonal 113 N°275, CP 1900 La Plata, Buenos Aires, Argentina

E.P. Tonni
Facultad de Ciencias Naturales y Museo, Universidad Nacional de La Plata, 122 y 60, CP 1900 La Plata, Argentina

G.M. Gasparini et al. (eds.), *Marine Isotope Stage 3 in Southern South America, 60 ka B.P.–30 ka B.P.*, Springer Earth System Sciences,
DOI 10.1007/978-3-319-40000-6_13

gastropods associated with mammal remains were dated in 32,300 ± 1800 and 27,500 ± 670 RCYBP; (4) Los Pozos, Marcos Paz County, sediments dated between 29,000 and 33,000 RCYBP are associated with remains of mammals, birds, reptiles and amphibians; (5) San Pedro, San Pedro County, sediments bearing vertebrate fauna have two OSL datings of 37,626 ± 4198 and 41,554 ± 3756 years B.P. (YBP). In Entre Ríos province, Río Ensenada valley, Diamante Department, some levels of the Tezanos Pinto Formation with OSL datings between 9000 and 35,000 YBP yielded remains of grazer megamammals and other taxa characteristic of the modern Patagonian Domain. In the province of Corrientes, Arroyo Toropí, Bella Vista, vertebrate remains dated with OSL from 36,000 to 52,000 YBP show a clear taxonomic change in response to climatic fluctuations. In Formosa province, Río Bermejo, Villa Escolar sediments of the Fortín Tres Pozos Formation, bearing vertebrate fauna have an OSL age of 58,160 ± 4390 YBP. In the province of Chaco, Charata locality, gastropods associated with mammal remains were dated between 22,000 and 27,000 RCYBP. A larger amount of absolute datings of the bearing sediments and especially taxon dates are needed to determine more accurately the response of the fauna to the climate changes characteristic of MIS 3.

Keywords Chronology · Mammals · Paleontology · OSL · Paleoclimate · Late Pleistocene · Radiocarbon dating

1 Introduction

The Dansgaard/Oeschger (D/O; warmer events) and Heinrich (H; colder events) cycles are recorded in many marine and continental sites worldwide. However, the information is strongly biased to the Northern Hemisphere. The marine records are mainly from the North Atlantic, and the continental ones, from western North America, Europe, and China (Voelker 2002; Hessler et al. 2010; Kanner et al. 2012). Continental biological *proxies* are scarce and belong almost exclusively to palynological information about the then existing flora.

In Argentina, and especially in the Pampean Region, numerous sites with Pleistocene–Holocene vertebrates particularly of the last glacial period (110,000–12,000 YBP) are well known and have been extensively studied. However, due to the lack of numerical ages of Pleistocene fossil remains or their bearing sediments it is difficult to assign them the global chronology established for Marine Isotope Stage 3 (MIS 3). Thus, the sites that can be attributed to this interval with certainty are quite scarce or poorly known. Up to date, continental vertebrate remains of this age have been identified mainly in localities of the Buenos Aires province, but also in a few sites of northeastern Argentina such as Entre Ríos, Corrientes, Formosa, and Chaco provinces (Fig. 1).

Mammals in particular are especially sensitive to climate change and even more those populations that inhabit the extremes of the geographical distribution of the species (Millien et al. 2006, and literature therein). In this regard, Buenos Aires

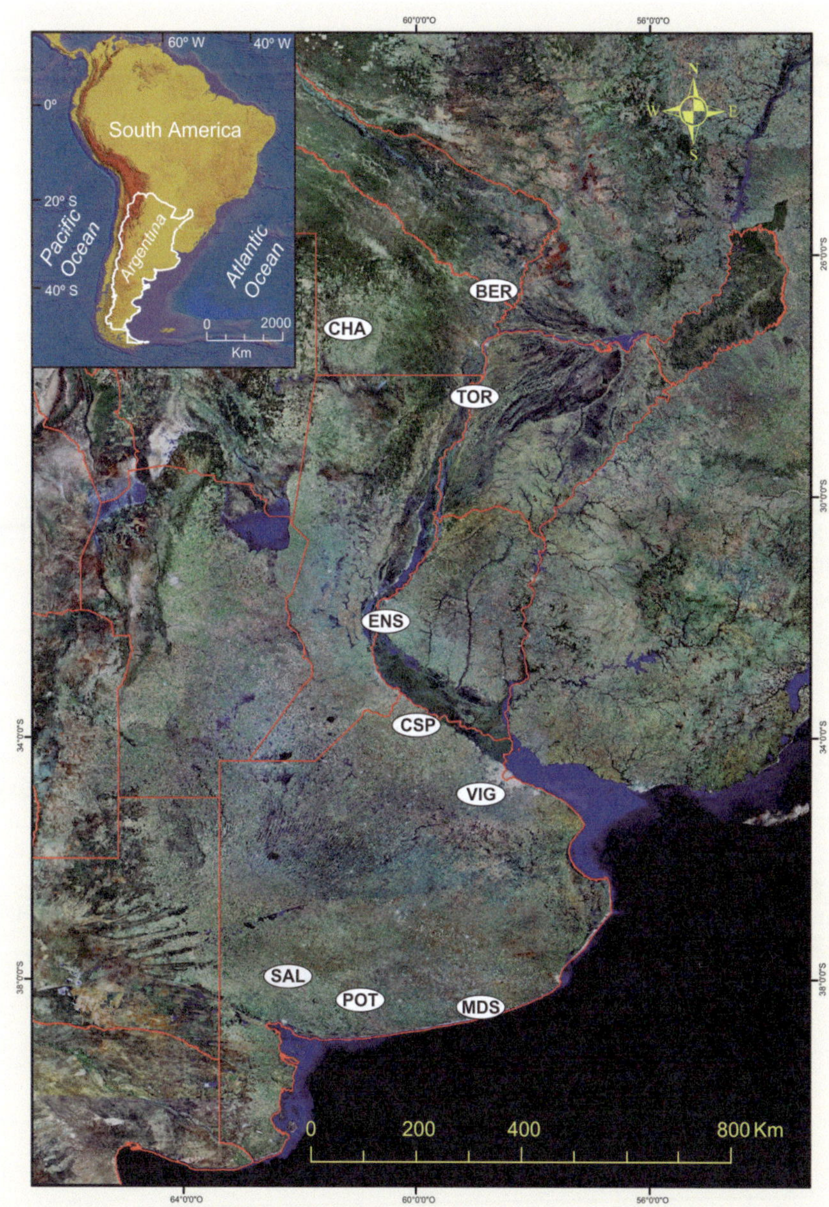
60°0'0"O
56°0'0"O
60° W
40° W
0°
South America
20° S
Pacific Ocean
Argentina
Atlantic Ocean
40° S
0
2000
Km
N
W
E
S
26°0'0"S
BER
CHA
TOR
30°0'0"S
ENS
CSP
34°0'0"S
VIG
SAL
POT
MDS
38°0'0"S
0
200
400
800 Km
64°0'0"O
60°0'0"O
56°0'0"O

◀ **Fig. 1** Geographic location of continental paleontological sites of Argentina, involved in the temporal lapse of MIS 3. *BER* Río Bermejo cliffs, near the Villa Escolar locality, Laishi Department (Formosa province); *CHA* Sitio 71, Charata locality (Chaco province); *CSP* Reserva Paleontológica Campo Spósito, Bajo del Tala, Río Baradero, San Pedro County (Buenos Aires province); *ENS* Río Ensenada, near the locality of Diamante, Diamante Department (Entre Ríos province); *MDS* Mar del Sur, General Alvarado County (Buenos Aires province); *POT* Paso Otero, in the Río Quequén Grande valley, Lobería County (Buenos Aires province); *SAL* Balneario Saldungaray in the Río Sauce Grande valley, Tornquist County (Buenos Aires province); *TOR* Arroyo (i.e., creek) Toropí, near Bella Vista locality, Bella Vista Department (Corrientes province); *VIG* Nicolás Vignona III Quarry, at the southern margin of the Río Matanza, Los Pozos locality, Marcos Paz County (Buenos Aires province)

province (included in the Pampean Region) is a good example of an ecotone between faunas of the Brazilian (in Argentina fossils have been recovered in Entre Ríos, Corrientes, Formosa and Chaco provinces) and Patagonian subregions (sensu Hershkovitz 1958; Cione et al. 2015), and consequently an interesting place to study the reaction of the fauna to environmental changes.

The composition of South American vertebrate assemblages was already established at the beginning of the Pleistocene (ca. 2.6 Ma), after the strong influence of legions of mammals entering through the Panama corridor during the major phase of the Great American Biotic Interchange (GABI) (see Cione et al. 2015; Vucetich et al. 2015). Thus, along the Pleistocene, the faunal changes were determined by their different reactions to climate change. On the one hand, the species often formed non-analog assemblages (the association of species that are today allopatric), typical of this period (Bell et al. 2004). On the other hand, megamammals and large carnivores became eventually extinct. A few isolated evidences suggest that continental vertebrates responded to the abrupt temperature changes, at millennia scale, that characterized the MIS 3 with H and D/O events.

In this contribution the faunal information of the Argentine continental paleontological sites involved in the MIS 3 interval has been gathered. The dynamics of the vertebrate fauna are also discussed here from the standpoint of the sudden changes in temperature.

1.1 Argentine Sites and Vertebrate Records During MIS 3

1. Buenos Aires province

   1.a Balneario Saldungaray (SAL) in the Río Sauce Grande valley, Tornquist County, Sierras Australes (38° 12′ 15″S and 61°46′ 06″W; Figs. 1 and 2): Arenoso Medio Member of the Agua Blanca Formation (Rabassa 1989; =Upper section of the San José Sequence in Zavala and Quattrocchio 2001). A single vertebrate remains was found in this unit, which represents an indeterminate species of the chinchillid rodent *Lagostomus*.

   The basal levels of this unit yielded abundant gastropods of the species *Plagiodontes patagonicus*, *Austroborus dorbignyi,* and *Discoleus aguirrei*,

Fig. 2 Balneario Saldungaray (SAL) in the Río Sauce Grande valley, Tornquist County. **a** General view; **b** detail of the section 200 m downstream. Photograph by Cecilia M. Deschamps

being dominant *P. patagonicus*, followed by the other two. These three species currently inhabit the area and are usually associated, but the most frequent and dominant species is *D. aguirrei*, followed by *P. patagonicus* and *A. dorbigny*. One dating of 32,300 ± 1800 radiocarbon years before present (RCYBP; Figini et al. 1989; LP115) was obtained on valves of *P. patagonicus*. Another dating on the valves of the same species from the same lithostratigraphic unit in this locality yielded 27,500 ± 670 RCYBP (Figini et al. 1989; Rabassa 1989; LP2859) (see Table 1).

Table 1 Argentine sites with the OSL, TL and radiocarbon dates corresponding to the MIS 3 with vertebrate remains

Provinces	Paleontological site	Dates	References
Buenos Aires	Río Sauce Grande, Balneario Saldungaray, Tornquist County	27,500 ± 670 and 32,300 ± 1800 RCYBP	Figini et al. (1989), Rabassa (1989)
	Río Quequén Grande, Paso Otero, Lobería County	37,800 ± 2300 RCYBP	Prado et al. (1987), Pomi and Tonni (2011), Cenizo et al. (2015)
	Mar del Sur, General Alvarado County	25,700 ± 800 and 33,780 ± 1200 RCYBP	Tonni et al. (2010)
	"Nicolás Vignona III" Quarry, southern margin of Río Matanza, Los Pozos locality, Marcos Paz county	29,070 ± 1420/31,040 ± 740/31,950 ± 830/32,070 ± 1210/32,580 ± 1520 RCYBP	Gasparini et al. (2013), Beilinson et al. en preparation
	"Reserva Paleontológica Campo Spósito", Bajo Del Tala, Río Baradero, San Pedro County	OSL 37,626 ± 4198 and 41,554 ± 3756 years B.P.	Prado and Alberdi (2012), Aguilar (2013)
Entre Ríos	Río Ensenada, Diamante locality, Diamante Department	TL 35,890 ± 1030 and 31,690 ± 1620 years (Lower Member) B.P. TL 8150 ± 400 and 9390 ± 630 years (Upper Member) B.P. OSL 33,000 years B.P. (alluvial faces)	Kröhling (1999) Kröhling (1999) Ferrero (2013), Ferrero et al. (2015), Brunetto et al. (2015)
Corrientes	Arroyo Toropí, Bella Vista locality, Bella Vista Department	OSL 36,000 and 52,000 years B.P.	Tonni et al. (2005), Francia et al. (2012), Francia (2014)
Formosa	Río Bermejo cliffs, Villa Escolar locality, Laishi Department	OSL 58,160 ± 4390 years BP	Zurita et al. (2009)
Chaco	Sitio 71, Charata locality, Chacabuco Department	22,600 ± 380, 24,010 ± 430 and 26,630 ± 370 RCYBP	Gasparini et al. (2015)

This site was included in this paper because of the radiocarbon datings, but from the geomorphological point of view the unit bearing the dated valves was considered older, and was correlated with the Upper section of the San José Sequence assigned to the Middle Pleistocene (see details in Zavala and Quattrocchio 2001; Deschamps 2003, 2005; Verzi et al. 2004). The vertebrate remains so far found do not contribute to elucidate this issue.

Fig. 3 Paso Otero (POT), in the Río Quequén Grande valley, Lobería County. Photograph by Eduardo P. Tonni

1.b Paso Otero (POT), in the Río Quequén Grande valley, Lobería County (38° 11′ 48″S and 59° 06′ 56″W; Figs. 1 and 3): La Chumbiada Member of the Luján Formation.

According to Prado et al. (1987), the following mammals are recorded in the La Chumbiada Member (cited as "sector pardo" of the Guerrero Member). Xenarthrans: *Glyptodon* sp., *Doedicurus clavicaudatus*, *Panochthus* sp., *Eutatus seguini*; Perissodactyla: *Equus* (*A.*) *neogaeus*; Artiodactyla: *Lama guanicoe*, *L. gracilis*, Cervidae (Odocoileinae) indet.; Carnivora: *Dusicyon avus*, *Lycalopex gymnocercus*; Rodentia: *Lagostomus maximus*, *Dolichotis patagonum*, *Lundomys* sp.

Cenizo et al. (2015) described a bird assemblage from Paso Otero. This assemblage included 22 taxa of different avian families, associated with aquatic, semiacquatic and wading (i.e., Anatidae, Rallidae, Podicipedidae), and terrestrial habits (i.e., Tinamidae, Falconidae, Strigidae, Furnariidae).

Sediments of the La Chumbiada Member cropping out at Paso Otero are dated in 37,800 ± 2300 RCYBP (on the gastropod *Chilina fluminea*: LP1928; Pomi and Tonni 2011; Cenizo et al. 2015) (see Table 1).

Another locality, which unfortunately yielded no vertebrate remains, has two datings that helped to constrain the age of the end of the deposition of the La Chumbiada Member and the beginning of the Guerrero Member of the Luján Formation. This locality is Arroyo Tapalqué (Olavarría County, 36° 52′ 44″S–60° 18′ 38″W; Fig. 4). La Chumbiada Member is dated in 29,150 ± 800 RCYBB (LP 268), and the base of the Guerrero Member in 21,040 ±450 RCYBP (LP396), both on valves of the gastropod *Heleobia parchappei* (Figini et al. 1998).

Fig. 4 Arroyo Tapalqué, Olavarría County. The *red circle* indicates the place where the La Chumbiada and Guerrero members of the Luján Formation were dated. Photograph by Eduardo P. Tonni

1.c Mar del Sur (MDS), General Alvarado County (38° 20′ 55″S and 57° 59′ 28″W; Figs. 1 and 5).

Coastal marine sediments with continental mammals were dated in 25,700 ± 800 and 33,780 ± 1200 RCYBP (Tonni et al. 2010) (see Table 1).

This still unnamed stratigraphic unit, about 1 m thick, is composed of green yellowish silty-clayey sands, slightly compacted. It contains isolated osteoderms of the extinct megamammal *Glyptodon* sp. (Tonni et al. 2010) and shells in life position of *Tagelus plebeius,* a euhaline bivalve species which is a common inhabitant of estuaries and coastal lagoons of Argentina. The shells of *T. plebeius* are also associated with shells of *Heleobia australis*, a gastropod of wide range of salinity (mesohaline to euhaline, ca. 10–35 ‰).

1.d Nicolás Vignona III Quarry (VIG), at the southern margin of the Río Matanza, Los Pozos locality, Marcos Paz County (34° 54′ 40.4″S and 58° 42′ 11.9″W; Figs. 1 and 6).

The sedimentary succession starts with laminated siltstones and fine sandstones of a gray-greenish coloration and a high participation of *Helobia australis* and *Diplodon* sp. Shells of both species were dated in 32,070 ± 1210 and 31,040 ± 740 RCYBP (LP2602, LP2665). These deposits are overlain by brown sandstones with trough cross-stratification and paleosol development. They are associated with remains of mammals, birds, reptiles, and amphibians (see Table 1). Finally, the uppermost 2 m are composed of light brown sandy siltstones with abundant *Helobia australis* valves that yielded ages of 32,580 ± 1520 and 29,070 ± 1420 RCYBP.

Fig. 5 Mar del Sur (MDS), General Alvarado County. Photograph by Esteban Soibelzon

The fauna exhumed from VIG shows taxa mainly adapted to open or semiopen and arid or semiarid environments (e.g. *Panochthus*; *Doedicurus*; *Glyptodon*; *Eutatus*; *Megatherium*; *Lestodon*; *Notiomastodon*; *Macrauchenia*; *Lestodelphys* cf. *L. juga*; *Toxodon* and *Hippidion*). In addition, there are extant taxa at the same stratigraphic level (e.g., *Chaetophractus*, *Ctenomys*, *Dolichotis*, *Lagostomus*, *Lama*, and *Tayassu pecari*). The bearing level was dated on specimens of the bivalve *Ostrea* sp. (LP2729) giving an age of 31,950 ± 830 RCYBP.

Similarly to what happens in Balneario Saldungaray, the Nicolás Vignona III Quarry site was included in this paper because of the radiocarbon datings. From the sedimentological and geomorphological points of view, the deposits bearing the dated valves are considered older, and might be correlated with the MIS 5e transgression (Upper Pleistocene). The vertebrate remains so far found do not contribute to elucidate this issue.

1.e Reserva Paleontológica Campo Spósito (CSP), Bajo del Tala, Río Baradero, San Pedro County (33° 44′ 34″S and 59° 36′ 6″W; Figs. 1 and 7).

The vertebrate remains housed at the Museo Paleontológico "Fray Manuel de Torres" (San Pedro) include the following taxa: *Megatherium americanum*, *Notiomastodon platensis*, *Macrauchenia patachonica*, *Morenelaphus* sp.,

Fig. 6 **a–b** General views of the Nicolás Vignona III Quarry (VIG), at the southern margin of the Río Matanza, Los Pozos locality, Marcos Paz County. **c** detail of the level with *Helobia australis*. Photographs by Esteban Soibelzon

Antifer sp., *Toxodon platensis*, *Lestodon armatus*, *Panochthus tuberculatus*, *D. clavicaudatus*, *Glyptodon* sp., *Equus (A.) neogaeus*, *Hippidion principale* and *Hemiauchenia* sp. The same taxa at a generic level were recognized for this site by Aguilar (2013).

Sediments bearing this vertebrate fauna have two OSL datings of 37,626 ± 4,198 and 41,554 ± 3,756 years BP (Prado and Alberdi 2012) (see Table 1).

2. Entre Ríos province

   Río Ensenada (ENS), near the locality of Diamante, Diamante Department (32° 04′ 10″S and 60° 38′ 17″W; Figs. 1 and 8).

   The Tezanos Pinto Formation is the typical Late Pleistocene–Early Holocene loess unit of the northeastern Pampean Region. It includes two members, which are mainly developed at the divides. The Lower Member is dated by TL in 35,890 ± 1,030 and 31,690 ± 1,620 YBP (Kröhling 1999). The Upper Member yielded TL ages of 8,150 ± 400 and 9,390 ± 630 YBP (Kröhling

Fig. 7 Reserva Paleontológica Campo Spósito (CSP), Bajo del Tala, Río Baradero, San Pedro County. **a** View of access to the outcrop; **b** detail of the exposed units. Photographs by José Luis Aguilar

1999) (see Table 1). Within the valleys, the Tezanos Pinto Formation is represented by alluvial and fluvial facies. One OSL dating for the alluvial facies gave an age of 33,000 years (Ferrero 2009) and the fossil record is restricted to *Smilodon populator* (Ferrero 2013; Ferrero et al. 2015; Brunetto et al. 2015). For the valley of the Río Ensenada, Ferrero (2009) and Ferrero and Noriega (2009) described mammal taxa traditionally related to dry and cold climatic conditions, which are widely distributed in the Late Pleistocene of other areas of Argentina (Gasparini et al. 2011) as follows: *E. seguini*, *Glyptodon reticulatus*, *P. tuberculatus*, *L. gymnocercus* [cited as *Dusicyon gymnocercus*], *S. populator*,

Fig. 8 Río Ensenada (ENS), Diamante Department. **a** General view of the locality; **b** detail of the main exposure. Photographs by Brenda Ferrero

M. patachonica, *T. platensis*, *Equus (A.) neogaeus*, *H. principale*, *Antifer ultra*, *Morenelaphus brachyceros*, *Hemiauchenia paradoxa*, *L. guanicoe*, and *L. gracilis*. However, they are found in the Salto Ander Egg Formation, in association with interglacial fauna (Ferrero 2013; Ferrero et al. 2015; Brunetto et al. 2015).

3. Corrientes province
   Arroyo (i.e., creek) Toropí (TOR), near Bella Vista locality, Bella Vista Department (28° 30′ 27″S and 59° 02′ 43″W; Figs. 1 and 9).
   The Toropí/Yupoí Formation includes an interesting assemblage of fossil vertebrates in which mammals are the most diverse and frequent, whereas reptiles are comparatively scarce. The bearing levels were dated by OSL yielding ages between ca. 52,000–36,000 YBP (Tonni et al. 2005; Francia et al. 2012) (see Table 1).
   Francia (2014) and Francia et al. (2015) reported the following vertebrates: Mammals: *N. platensis*, *Equus* (*A.*) cf. *E.* (*A.*) *neogaeus*, *Hippidion* sp., *Morenelaphus lujanensis*, *Hippocamelus sulcatus*, cf. *Mazama* sp., *T. pecari*, *Tayassu* sp., *Neolicaphrium recens*, *Hemiachenia paradoxa*, *T. platensis*, *Chaetophractus villosus*, *Euphractus* aff. *E. sexcinctus*, *Propraopus sulcatus*, *Holmesina paulacoutoi*, *Pampatherium typum*, *Neosclerocalyptus paskoensis*, *Neosclerocalyptus* sp., *P. tuberculatus*, *Glyptodon* sp., *Scelidotherium* sp., *Scelidodon* sp., *Galea* aff. *G. tixiensis*, *D. patagonum*, *S. populator*, and *Pantera onca*; Reptiles: *Chelonoidis lutzae*, and *Boa constrictor*.
4. Formosa province
   Río Bermejo cliffs (BER), near Villa Escolar locality, Laishi Department (26° 36′ S and 58° 40′ W; Figs. 1 and 10).
   The sediments cropping out at the Río Bermejo, can reach a thickness of 8-9 meters; the sediments bearing vertebrate fauna were considered by Zurita et al. (2009) as the Fortín Tres Pozos Formation. However, Iriondo (2010) assigned its lower portion to the Río Bermejo Formation and the upper one to the La Fidelidad Formation (Zurita et al. 2014).
   Zurita et al. (2009: 277) described the fauna as formed by "Pampean Patagonian elements". This fauna includes remains assigned to *Glyptodon* sp., *Neosclerocalyptus* cf. *N. paskoensis*, *P. typum*, *Pampatherium* sp., *Megatherium* sp., cf. *Morenelaphus*, cf. *H. paradoxa*, and *Toxodon* sp.
   At the lower third of the Río Bermejo Formation (sensu Iriondo 2010), sediments bearing vertebrate fauna have an OSL date that indicates an age of 58,160 ± 4,390 YBP (UIC2108BL; Zurita et al. 2009) (see Table 1). In addition, ^{14}C dating in the middle section of the Río Bermejo Formation indicated an approximate age of 9,500 YBP (Zurita et al. 2011, 2014).
5. Chaco province
   The paleontological site named as Sitio 71, Charata locality (CHA), Chacabuco Department (27° 11′ 60″ S and 61° 10′ 48″ W; Figs. 1 and 11) comprises alluvial sandy silts ("*grandes abanicos aluviales*", "large alluvial fans", sensu Iriondo et al., 2000) reworked by aeolian processes. Two paleosol levels were identified at the exposed profile, corresponding to stabilization moments of the landscape.

Fig. 9 Arroyo Toropí (TOR), Bella Vista Department. **a** General view of the locality; **b** detail of the exposed units. Photographs by Jorge Carrillo Briceño and Analía Francia

Sediments overlying the upper paleosol correspond to Level 1; Level 2 was deposited between both paleosols, and Level 3 is below the lower paleosol. Level 2 was dated in 22,600 ± 380 and 24,010 ± 430 RCYBP (*Pomacea* sp.: LP 3141 and LP 3188, respectively); a similar specimen of Level 3 was dated in 26,630 ± 370 RCYBP (LP 3142). The dates are stratigraphically consistent.

The faunal assemblage found at levels 2 and 3 correspond to *Glyptodon* sp., *Neosclerocalyptus* sp., *Equus* (*A.*) sp., and *Toxodon* sp. At level 3, a skull and mandible belonging to the tayassuid *Catagonus* sp. were recorded (Gasparini et al. 2015).

Fig. 10 Río Bermejo cliffs (BER), near Villa Escolar locality, Laishi Department. **a** View of the Río Bermejo; **b** detail of the cliffs. Photographs by Alfredo Zurita

Fig. 11 Paleontological site named as Sitio 71, Charata locality (CHA). Photograph by Guillermo Lamenza

2 Discussion

Data obtained from the studied sites allow some inferences to be made regarding paleoenvironments and paleoclimates prevailing in the interval corresponding to MIS 3, on the basis of the response of vertebrates to climate change and environmental requirements.

In Buenos Aires province, specifically in the assemblage at Paso Otero, the record of *Lundomys* sp. (Sigmodontinae, Mammalia) has been used by Teta and Pardiñas (2006; see also Pomi and Tonni 2011) to refer the bearing level to a warmer period. In addition, the signs of termite activity observed on the bones of this assemblage confirm such environmental conditions since termites are predominantly a tropical order of insects (Pomi and Tonni 2011). However, together with these evidences, xeric mammals such as *Lama gracilis* and *Lestodelphys* sp. were found. The association of species with different ecological requirements suggests that the deposit may represent an averaged time during which very rapid climate changes occurred.

According to Cenizo et al. (2015), the ecological requirements of the avian taxa reported for Paso Otero indicate the existence of open grasslands with the presence of freshwater and permanent ponds, similar to the environment found today in southeastern Buenos Aires province (e.g., "Area Interserrana Bonaerense"). Consequently, the high similarities in ecological requirements and species diversity of the assemblages of Paso Otero and the avifauna currently living in the area, suggest similar climatic conditions to the present ones.

In Mar del Sur, Tonni et al. (2010) provided new evidence about the abrupt warming that occurred during MIS 3. According to the geological data at this site, sea-level rise accompanied the climatic change.

Gasparini et al. (2013) described for Nicolás Vignogna III Quarry the first record of *T. pecari* in the central-northern region of Buenos Aires province. The genus was recorded in the Middle Pleistocene (Bonaerian) of southern Buenos Aires province (Río Sauce Grande, Bajo San José; Deschamps 2005) in sediments correlated with MIS 11, a very warm period although not necessarily wetter in this area (Verzi et al. 2004). During the Late Pleistocene, *T. pecari* was distributed southern to its current range, probably evidencing different paleoenvironmental conditions. The genus *Tayassu* by itself is insufficient to infer the prevailing environmental conditions, because of its wide current geographical distribution and broad ecological tolerance (Menegaz and Ortíz Jaureguizar 1995, Gasparini and Zurita 2005, Gasparini 2013). However, it has not been recorded together with faunas of cold climate. Besides, other recorded taxa which include living species such as *Lestodelphys*, *Chaetophractus*, *Ctenomys*, *Dolichotis*, *Lagostomus*, and *Lama* are characteristic of arid and semiarid areas. Today *T. pecari* is not sympatric with the species of *Dolichotis* and *Lestodelphys*.

In Entre Ríos province, Río Ensenada, the faunal assemblage is relevant because it includes taxa of Late Pleistocene megamammals typical of the Pampean Region and two species of Camelidae (one of them, extinct). The geographic distribution of living species (*L. guanicoe*) excludes the area of the Argentine Mesopotamia (Entre Ríos, Corrientes and Misiones provinces) being characteristic of arid and semiarid areas.

In Corrientes province, the faunal assemblage found in Arroyo Toropí is characterized by taxa with clearly different ecological requirements, to which it must be added the presence of taxa that are not currently present in this area; moreover, they live in distant geographical areas (Francia 2014). The record of climate-sensitive taxa such as *D. patagonum*, and *B. constrictor* suggests that the sequence was deposited under less humid environmental conditions than the present ones. In this sense, Scillato-Yané et al. (1998) inferred that the annual rainfall must have been lower than 700 mm, which strongly contrasts with the present situation with annual rainfall exceeding 1300 mm (Corrientes meteorological station, for decades 1941–1950 and 1981–990). However, the record of some elements with tropical or subtropical affinities, both extinct (e.g., *H. paulacoutoi*) and living ones (e.g., *P. onca*), suggests that some episodes of higher humidity also occurred during the deposition of the sequence.

In Formosa province, the mastofauna assemblage recorded at the Río Bermejo cliffs, near Villa Escolar (together with those previously known from the south and central eastern Chaco) includes taxa adapted to open and relatively cold environments (Zurita et al. 2009). This is in agreement with the conditions of deposition of the sediments that form the alluvial fan of the Río Bermejo. These authors also stated that *"esta paleofauna resulta, desde un punto de vista taxonómico, más afín a aquella registrada en el Pleistoceno tardío de la región Pampeana que a la conocida para la región Mesopotámica de Argentina, en donde se observa una "mezcla" de taxones típicamente pampeanos con otros de origen brasílico, indicadores de ambientes más húmedos y cálidos…"* ("*this paleofauna is, from a taxonomic point of view, closer to that recorded in the Late Pleistocene of the Pampean Region than to that one known for the Mesopotamian Region in which there is a mix of typically Pampean taxa with Brazilian ones, indicators of more humid and warm environments…*") (Zurita et al. 2009: 285).

In Chaco province, the faunal assemblage of Sitio 71 was found at levels 2 and 3; the tayassuid *Catagonus* was found in the latter. This genus has a set of morphological features that suggests adaptations to cursorial habits on dry and relatively open environments. The bearing sediments and the particularities of the faunal assemblages, as well as the chronological data, allow confirming that in this area of Argentina, arid and semiarid conditions, with scarce or absent vegetation cover were developed during the last part of MIS 3 and part of MIS 2. These environmental conditions allowed the settlement of megamammals adapted to open environments.

Taking into account the above-mentioned, the assemblages of vertebrates of the few sites that can be attributed to the MIS 3 suggest that they responded to the sudden changes of temperature that characterized this interval. In this sense, Paso Otero, Nicolás Vignona III Quarry (taking into account the comments on its age), and Arroyo Toropí are particularly informative. In these localities, taxa with different ecological requirements were found within a single lithostratigraphic unit. These lithostratigraphic units represent an average sample in terms of geological time, and not a single paleoclimatic event. Thus, the vertebrate assemblages of these units represent successive biocenoses instead of a single one. In summary, this scenario could be explained as a taphocoenosis produced by the association of taxa with different ecological requirements that lived in different intervals of time, being their stratigraphic association the result of the averaged time represented in the deposit. These mixed-up vertebrate assemblages would also suggest that the climatic changes and the resulting climatic episodes would have been fast, intense, and short.

In consequence the "non-analogous or disharmonious assemblages" (see Semken 1974; Lundelius 1983; Graham 1985; Graham and Mead 1987; Bell et al. 2004; Morgan and Emslie 2010) could be the result of rapid faunal changes without resolution in the fossil record.

A larger amount of absolute datings of the bearing sediments and especially taxon dates are needed to determine more accurately the faunal response to climate change.

Acknowledgments The authors thank the Facultad de Ciencias Naturales y Museo (UNLP) and the Consejo Nacional de Investigaciones Científicas y Técnicas (CONICET). This manuscript was partially funded by PICT 2010-0804 and PIP 0496.

References

Aguilar JL (2013) Fósiles de la región pampeana. Hallazgos paleontológicos en San Pedro. Universitas, Editorial Científica Universitaria, Córdoba, 178pp

Bell CJ, Lundelius EL, Barnosky AD, Graham RW, Lindsay EH, Ruez DR, Semken HA, Webb SD, Zakrzewski RJ (2004) The Blancan, Irvingtonian, and Rancholabrean mammal ages. In: Woodburne MO (ed) Late Cretaceous and Cenozoic mammals of North America, p.: 232–314. New York, Columbia

Brunetto E, Ferrero BS, Noriega JI (2015) Late Pleistocene lithostratigraphy and sequences in the Southwestern (Argentina): evidences of the last interglacial stage. J S Am Earth Sci 58: 111–128

Cenizo M, Agnolin F, Pomi L (2015) A New Pleistocene bird assemblage from the Southern Pampas (Buenos Aires, Argentina). Palaeogeogr Palaeoclimatol Palaeoecol 420:65–81

Cione AL, Gasparini GM, Soibelzon E, Soibelzon LH, Tonni EP (2015) The Great American biotic interchange. a South American perspective. In: Rabassa J, Lohmann G, Notholt J, Mysak LA, Unnithan V (eds) Springer Brief Monographies in Earth System Sciences. South America and the Southern Hemisphere. Springer, Berlin, 97pp. doi:10.1007/978-94-017-9792-4

Deschamps CM (2003) Estratigrafía y paleoambientes del Cenozoico en el sur de la Provincia de Buenos Aires. El aporte de los vertebrados. Doctoral Thesis (unpublished), Facultad de Ciencias Naturales y Museo, Universidad Nacional de La Plata, La Plata, 317pp

Deschamps CM (2005) Late Cenozoic mammal biochronostratigraphy in southwestern Buenos Aires Province, Argentina. Ameghiniana 42(4):733–750 (Buenos Aires)

Ferrero BS (2009) Mamíferos del Cuaternario de la provincia de Entre Ríos, Argentina: Diversidad y evolución. Aspectos bioestratigráficos y paleozoogeográficos de una fauna particular. Doctoral Thesis (unpublished), Facultad de Ciencias Naturales y Museo, Universidad Nacional de La Plata, La Plata, 425pp

Ferrero BS (2013) Los vertebrados del Cuaternario de la provincia de Entre Ríos. II Simposio del Plioceno y Pleistoceno del Centro y Norte de. Ameghiniana Suplemento 50(6):R11 (Buenos Aires)

Ferrero BS, Noriega J (2009) La paleontología de vertebrados en el Cuaternario de la provincia de Entre Ríos (Argentina): estado actual y perspectivas. In: Ribeiro AM, Bauermann SG, Scherer CS (eds) Quaternário do Rio Grande Do Sul. Integrando conhecimentos, vol 1. Monografías, Sociedad Brasileira de Paleontología, pp 207–215

Ferrero BS, Noriega JI, Brunetto B (2015) Bioestratigrafía y geocronología de la Formación Salto Ander Egg (Pleistoceno tardío). XXIX Jornadas de Paleontología de Vertebrados. Diamante, Entre Ríos

Figini A, Rabassa J, Tonni EP, Huarte R, Gómez G, Carbonari J, Zubiaga A (1989) Dataciones radiocarbónicas de gasterópodos terrestres en sedimentos del Pleistoceno superior y Holoceno del valle del río Sauce Grande, provincia de Buenos Aires. I Jornadas Geológicas Bonaerenses Actas, pp 809–824. Tandil, 1989

Figini A, Huarte R, Carbonari J, Tonni EP (1998) Edades C-14 en un perfil de arroyo Tapalqué, provincia de Buenos Aires, Argentina. Contribución a la cronología de acontecimientos faunístico-ambientales. X Congreso Latinoamericano de Geología y VI Congreso Nacional de Geología Económica, Actas 1:27–31. (Buenos Aires, 1998)

Francia A (2014) Vertebrados cuaternarios de Corrientes, paleoambientes, paleoclimas y diversidad. Un análisis comparativo con faunas continentales equivalentes del Cono Sur de América del Sur. Doctoral Thesis (unpublished), Facultad de Ciencias Exactas y Naturales y Agrimensura, Universidad Nacional del Nordeste, Corrientes, 369pp

Francia A, Carlini AA, Zurita AE, Miño-Boilini AR, Kruck W (2012) Cronología de las unidades litoestratigráficas aflorantes en el Arroyo Toropí, Provincia de Corrientes, y los registros paleofaunísticos. Reunión de Comunicaciones Científicas y Tecnológicas, Corrientes. On-line, CB-033 (http//www.unne.edu.ar)

Francia A, Zurita AE, Carlini AA (2015) How Marine Isotope Stage 3 (MIS 3) is reflected in northern faunal assemblage of Argentina: the Xenarthra Cingulata case. Quat Int 377(2015): 126–139. doi:10.1016/j.quaint.2015.03.012

Gasparini GM (2013) Records and Stratigraphical Ranges of South American Tayassuidae (Mammalia, Artiodactyla). J Mamm Evol 20(1):57–68. doi:10.1007/s10914-011-9172-z

Gasparini GM, Zurita AE (2005) Primer registro fósil de *Tayassu pecari* (Link) (Mammalia, Artiodactyla) en la Argentina. Ameghiniana 42(2):473–480 (Buenos Aires)

Gasparini GM, Rodríguez S, Soibelzon LH, Beilinson E, Soibelzon E, Missagia RV (2013) *Tayassu pecari* (Link, 1795) (Mammalia, Cetartiodactyla): comments on its South American fossil record, taxonomy and paleobiogeography. Hist Biol ID: 858247. doi:10.1080/08912963.2013.858247

Gasparini GM, Lamenza GN, Ruella A, Tonni EP, González O (2015) Cronología y fauna del sitio 71 (Charata, provincia del Chaco, Argentina). III Simposio del Mioceno-Pleistoceno del Centro y Norte de Argentina. Corrientes, Argentina

Graham RW (1985) Response of mammalian communities to environmental changes during the late Quaternary. In: Diamond J, Case TJ (eds) Community ecology. Harper and Row, New York, pp 300–313

Graham RW, Mead JI (1987) Environmental fluctuations and evolution of mammalian faunas during the last deglaciation in North America. In: The geology of North America, vol K-3, North America and adjacent oceans during the last deglaciation, pp 371–402

Hershkovitz P (1958) A geographic classification of Neotropical Mammals. FieldZool 36:581–620

Hessler I, Dupont L, Bonnefille R, Behling H, González C, Helmens KF, Hooghiemstra H, Lebamba J, Ledru MP, Lézine AM, Maley J, Marret F, Vincens A (2010) Millennial-scale changes in vegetation records from tropical Africa and South America during the last glacial. Quat Sci Rev 29:2882–2899

Iriondo MH (2010) Geología del Cuaternario en Argentina. Editorial Moglia, Corrientes, Argentina, 437pp

Iriondo MH, Colombo F, Kröhling D (2000) El abanico aluvial del Pilcomayo, Chaco (Argentina-Bolivia-Paraguay): características y significado sedimentario. Geogaceta 28:79–82

Kanner LC, Burns SJ, Cheng H, Edwards RL (2012) High-latitude forcing of the South American summer monsoon during the last glacial. Science 335(6068):570–573

Kröhling DM (1999) Sedimentary maps of loessic units outcropping in North Pampa, Argentina. In: Iriondo M (ed) South American Loess and related topics. Quat Int 62:49–55

Lundelius E (1983) Climatic implications of Late Pleistocene and Holocene faunal associations in Australia. Alcheringa 7:125–149

Menegaz AN, Ortíz Jaureguizar E (1995) Los Artiodáctilo. In: Alberdi MT, Leone G, Tonni EP (eds) Evolución biológica y climática de la región Pampeana durante los últimos cinco millones de años. Un ensayo de correlación con el Mediterráneo occidental. Monografías del Museo Nacional de Ciencia Naturales, Consejo Superior de Investigaciones Científicas, Madrid, 15, pp 311–337

Millien V, Lyons SK, Olson L, Smith FA, Wilson AB, Yom-Tov Y (2006) Ecotypic variation in the context of global climate change: revisiting the rules. Ecol Lett 9:853–869

Morgan GS, Emslie SE (2010) Tropical and western influences in vertebrate faunas from the Pliocene and Pleistocene of Florida. Quat Int 217:143–158

Pomi LH, Tonni EP (2011) Termite traces on bones from the Late Pleistocene of Argentina. Ichnos 18:166–171

Prado JL, Alberdi MT (2012) Equidos y gonfoterios del Pleistoceno tardío de San Pedro, provincia de Buenos Aires, Argentina. Estud Geol 68(2):261–276

Prado J, Menegaz A, Tonni EP, Salemme M (1987) Los mamíferos de la fauna local Paso Otero (Pleistoceno tardío) provincia de Buenos Aires. Aspectos paleoclimáticos y bioestratigráficos. Ameghiniana 24:217–233 (Buenos Aires)

Rabassa J (1989) Geología de los depósitos del Pleistoceno Superior y Holoceno en las cabeceras del río Sauce Grande, provincia de Buenos Aires. 1° Jornadas Geológicas Bonaerenses Actas 765–790. Tandil 1989

Scillato-Yané GJ, Tonni EP, Carlini AA, Noriega JI (1998) Nuevos Hallazgos de Mamíferos del Cuaternario en el Arroyo Toropí, Corrientes, Argentina. Aspectos Bioestratigráficos, Paleoambientales y Paleozoogeograficos. X Congreso Latinoamericano de Geología y VI Congreso Nacional de Geología Económica Actas I:263–268

Semken HA (1974) Micromammal distribution and migration during the Holocene. In: Abstracts 3rd Annual Meeting American Quaternary Association, p 25

Teta P, Pardiñas UFJ (2006) Pleistocene record of the marsh rat of the Genus *Lundomys* in Southern South America: Paleoclimatic significance. Curr Res Pleistocene 23:202–204

Tonni EP, Carlini AA, Zurita AE, Frechen M, Gasparini GM, Budziak D, Kruck W (2005) Cronología y Bioestratigrafía de la Unidades del Pleistoceno aflorantes en el Arroyo Toropí, provincia de Corrientes, Argentina. 19°Congreso Brasileiro de Paleontología y 6°Congreso Latino-Americano de Paleontologia. On-line abstracts volume

Tonni EP, Carbonari JE, Huarte R (2010) Marine sediments attributed to Marine Isotope Stage 3 in the southeastern Buenos Aires Province, Argentina. Geosciences 154–156

Verzi D, Deschamps C, Tonni EP (2004) Biostratigraphic and palaeoclimatic meaning of the Middle Pleistocene South American rodent *Ctenomys kraglievichi* (Caviomorpha, Octodontidae). Palaeogeogr Palaeoclimatol Palaeoecol 212:315–329

Voelker A (2002) Global distribution of centennial-scale records for marine isotope stage (MIS) 3: a database. Quat Sci Rev 21:1185–1212

Vucetich MG, Arnal M, Deschamps CM, Pérez ME, Vieytes EC (2015) A brief history of caviomorph rodents as told by the fossil record. In: Vassallo A, Antonucci D (eds) Biology of caviomorph rodents; diversity and evolution. Sociedad Argentina para el estudio de los Mamíferos, pp 11–62

Zavala CA, Quattrocchio ME (2001) Estratigrafía y evolución geológica del río Sauce Grande (Cuaternario), provincia de Buenos Aires. Rev Asoc Geol Argentina 56:25–37

Zurita AE, Miño-Boilini AR, Carlini AA, Iriondo M, Alcaraz MA (2009) Paleontología del Chaco Oriental. Una nueva localidad con mamíferos fósiles pleistocenos en el río Bermejo (Formosa, Argentina). Rev Mex Ciencias Geológicas 26:277–288

Zurita AE, Rodríguez-Bualó SM, Bogan S, Miño-Boilini AR, Alcaraz MA, Lutz AI, Friedrichs J (2011) A latest Pleistocene-Early Holocene palaeofaunal association in Northern Argentina. V Congreso Latinoamericano de Paleontología de Vertebrados, Libro de resúmenes: 048

Zurita AE, Miño-Boilini AR, Francia A, Erra G, Alcaraz MA, Carlini AA, Lutz AI, Friedrichs J (2014) Paleontología y cronología del Cuaternario de las provincias de Corrientes y Formosa, Argentina. Acta geológica lilloana 26 (1):75–86 (San Miguel de Tucumán, Argentina)

Marine Isotope Stage 3 (MIS 3) Versus Marine Isotope Stage 5 (MIS 5) Fossiliferous Marine Deposits from Uruguay

Alejandra Rojas and Sergio Martínez

Abstract Uruguay has several marine deposits of undoubtedly Late Pleistocene age, but there is conflicting evidence when comparing ages obtained by different methods. While ^{14}C datings suggest younger ages (related to Marine Isotope Stage 3—MIS 3), OSL, where available, indicate older times (related to Marine Isotope Stage 5—MIS 5). The analysis of the abundant molluscan fauna and the presence of extralimital warm water taxa points to a higher than present water temperature for the Uruguayan coast. The referred discrepancies are discussed and a MIS 5 age is preferred according to all available evidence.

Keywords Marine Isotope Stage 3 (MIS 3) · Marine Isotope Stage 5 (MIS 5) · Late Pleistocene · Molluscs · Bivalves · Gastropods · Paleoecology · Paleobiogeography · Extralimital species · Uruguay

1 Introduction

The Pleistocene Epoch was a time of global climatic and sea level changes that had a profound impact in the marine and continental realms, influencing the configuration of our present-day biota. The information for the reconstruction of these oscillations comes from different sources that include ice cores and a wide spectrum of continental and marine records. It is widely recognized the existence throughout the Pleistocene of numerous glacial–interglacial cycles characterized by cold (stadials) and warm (interestadials) lapses (e.g., Emiliani 1955; Shackleton 1969; Crowley and North 1991; Winograd et al. 1997). In recent years and especially for the Late Pleistocene, a great improvement in the knowledge of the timing and

A. Rojas (✉) · S. Martínez
Departamento de Paleontología, Instituto de Ciencias Geológicas, Facultad de Ciencias, Universidad de la República, Iguá 4225, CP 11400 Montevideo, Uruguay
e-mail: alejandra@fcien.edu.uy

S. Martínez
e-mail: smart@fcien.edu.uy

G.M. Gasparini et al. (eds.), *Marine Isotope Stage 3 in Southern South America, 60 ka B.P.–30 ka B.P.*, Springer Earth System Sciences,
DOI 10.1007/978-3-319-40000-6_14

magnitude of these climatic changes in both hemispheres comes from ice cores from Greenland and Antarctica (e.g., Petit et al. 1999; North Greenland Ice Core Project Members 2004; EPICA Community Members 2004, 2006; Jouzel et al. 2007; Orombelli et al. 2010). These high resolution sources of information allowed the recognition of the pronounced and abrupt Dansgaard–Oeschger (D–O) warming events in Greenland and the more gradual Antarctic Isotope Maxima (AIM) in Antarctica (EPICA Community Members 2006; Huber et al. 2006) superimposed to the traditional isotopic stages. These millenial-scale changes were also recognized in other records (e.g., Barker et al. 2009).

The Marine Isotope Stage 5 (MIS 5) had an average duration between 130 and 71 ka B.P. and was a lapse of warm conditions with the warmest peak considered to be the Last Interglacial or MIS 5e (Zubakov and Borzenkova 1990; Winograd et al. 1997; Petit et al. 1999; Shackleton et al. 2003; Jouzel et al. 2007). The Last Interstadial or Marine Isotopic Stage 3 (MIS 3) had an average duration between 60 and 27 ka B.P. and was characterized by overall lower temperatures than present and than MIS 5 and by the record of numerous D–O events (Huber et al. 2006; Van Meerbeeck et al. 2009, 2011; Buiron et al. 2012; Long and Stoy 2013).

One of the sources of information that have contributed to the reconstruction of the environmental changes occurred in our near past come from littoral fossiliferous assemblages around the world. Pleistocene marine assemblages have been studied from the Pacific coast of North America (e.g., Valentine and Jablonski 1993; Roy et al. 1995; Powell et al. 2004), Mexico (De Diego-Forbis et al. 2004), Atlantic islands at different latitudes (Muhs et al. 2002; Ávila et al. 2009), Pacific Islands (Muhs et al. 2002), Japan (Amano 1994; Kitamura and Ubukata 2003), Australasia (Murray-Wallace and Belperio 1991; Murray-Wallace et al. 2000; Murray-Wallace 2002), Antarctica (Berkman et al. 1998), and Europe (Zazo et al. 2003; Nielsen et al. 2006; Garilli 2011). In South America, Pleistocene marine deposits are known from the Pacific coast (e.g., Ortlieb et al. 1990; Rivadeneira and Carmona 2008), Beagle Channel (Rabassa et al. 2009) and Atlantic coast, mostly from Argentina (see Aguirre and Whatley 1995; Isla et al. 2000; Aguirre 2003; Aguirre et al. 2011; Gordillo and Isla 2011; Charó et al. 2013, 2014) and in a lesser extent from Brazil (e.g., Lopes and Simone 2012; Lopes et al. 2013).

In Uruguay, undoubtedly Pleistocene fossil assemblages were first recognized by Martínez et al. (2001) in the Puerto de Nueva Palmira (western Uruguay, Colonia County) and La Coronilla (eastern Uruguay, Rocha County). Subsequently, Rojas (2007) provided a new analysis of the molluscan content of these two assemblages and included another one located at Zagarzazú (Colonia County). More recently, Martínez et al. (2013) provided a biogeographic analysis of Pleistocene and Holocene molluscan faunas of the Southwestern Atlantic, including Brazilian, Argentinean, and Uruguayan data.

After the first characterization of the Pleistocene molluscan assemblages from Uruguay provided by Martínez et al. (2001), new data on the faunal composition, biogeographic inferences, and geochronological context have become available. Thus, the aims of this contribution are (a) to update the paleontological content of the Late Pleistocene fossil assemblages from Uruguay, (b) to provide a paleoecologic and biogeographic analysis of the faunal elements that are useful for paleoenvironmental reconstruction, and (c) to discuss these results in the light of the MIS 3 versus MIS 5 temporal and climatic scenario.

2 Geographical Setting

The Late Pleistocene fossiliferous deposits found so far in Uruguay are three, two of them located in the western coast, and the remaining in the eastern coast (Fig. 1). The Puerto de Nueva Palmira outcrop can nowadays be found in the area occupied by the port of the city of Nueva Palmira. About 12 km to the south the Zagarzazú site crops out in the abrasion platform of the beach. Finally, the La Coronilla fossil assemblage crops out in the abrasion platform of the homonymous beach in the Rocha County (eastern Uruguay). The Zagarzazú and La Coronilla deposits are frequently covered by sand.

The study area comprises the Uruguayan coastal waters currently dominated by the fluviomarine gradient of the Río de la Plata and Atlantic Ocean. The Río de la Plata estuary between Argentina and Uruguay receives from the west, freshwater

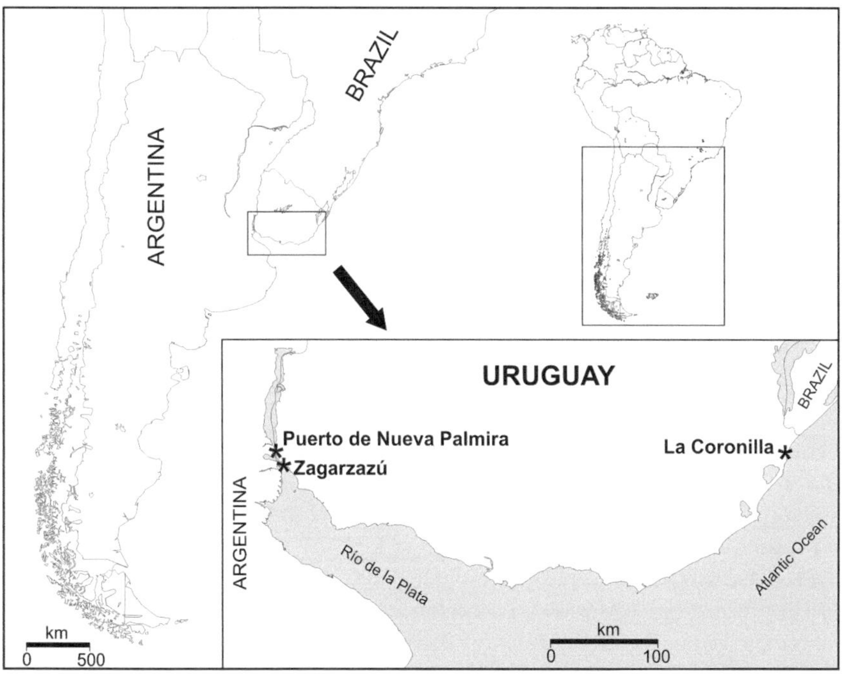

Fig. 1 Geographic location of the Late Pleistocene fossil assemblages of Puerto de Nueva Palmira, Zagarzazú, and La Coronilla

and sediment discharge coming from the Paraná and Uruguay rivers and from the east, the marine waters of the Atlantic Ocean (Urien 1972). It is divided by a submerged shoal (Barra del Indio) into an inner fluvial system and an outer mixohaline brackish system (Mianzan et al. 2001). The biogeography of the living biota responds to this gradient with a west to east dominance of freshwater, estuarine, and marine taxa, respectively (Masello and Menafra 1998; Giberto et al. 2004; Giménez et al. 2005; Brazeiro et al. 2006). The Río de la Plata estuary is a highly variable environment at different timescales and the boundaries and position of the salinity zones vary according to wind balance, river discharge, season, and phenomena such as El Niño Southern Oscillation (ENSO) (e.g., Guerrero et al. 1997; Acha et al. 2008; Möller et al. 2008; Nagy et al. 2008).

The large-scale dynamics of the Southwestern Atlantic Ocean are dominated by the northward flowing Malvinas (Falkland) Current and the southward flowing Brazilian Current. The former transports cold subantarctic waters along the Argentinean shelf whereas the latter carries tropical and subtropical waters along the continental margin of South America (Olson et al. 1988; Piola et al. 2000; Odebrecht and Castello 2001). Both currents meet approximately between 25° S and 45° S at the Brazil–Malvinas Confluence Zone, producing a complex oceanographic area with highly variable physicochemical and biological attributes on the shelf and slope. The temperature gradient developed by these water masses outline the current malacological provinces of the Southwestern Atlantic: the warm water Brazilian Province, the confluence Argentinean Province, and the cold Magallanic Province (e.g., Scarabino 1977; Briggs 1995; Martínez and del Río 2002; Martínez et al. 2013). Studies in Brazil, Argentina, and Uruguay showed that the faunal composition of the Argentinean Province includes a combination of species of warm-temperate and cool-temperate affinities, besides endemic ones. In this scenario, the Río de la Plata and its freshwater discharge acts as an ecological barrier and represents a broad ecotone between the southern and northern areas (Masello and Menafra 1998; Scarabino et al. 2006a, b).

3 Geological Setting

The Quaternary marine deposits of Uruguay have their origin in the transgressive–regressive events characteristic of this period of global-scale climatic oscillations. The fossiliferous deposits have been correlated with adjacent Atlantic units from Argentina and Brazil by various authors (Goñi and Hoffstetter 1964; Forti-Esteves 1974; Martínez 1990; Aguirre and Whatley 1995).

From the lithostratigraphic viewpoint they are included in the Chuy and/or Villa Soriano formations (Goñi and Hoffstetter 1964; Goso 1972). The Villa Soriano Formation represents the fossiliferous deposits that Caorsi and Goñi (1958) named "Arcillas grises del Vizcaíno" which were later formalized as a lithostratigraphic unit by Goso (1972). These deposits crop out along a narrow stripe parallel to the present coastline of Uruguay, from the Río Negro to the Merín Lagoon margins.

Lithologically, this unit has wide grain-size variability, from clay to medium sands and occasionally gravel and pebbles (Preciozzi et al. 1988). Most authors consider as a characteristic of this unit the abundant fossiliferous content (e.g., Serra 1943; Caorsi and Goñi 1958); thus, in a sort of circular reasoning, most fossils have been attributed to this unit. The age of the Villa Soriano Formation has been a matter of discussion. Bossi (1966) and Preciozzi et al. (1988) considered it as of Late Pleistocene–Holocene age, meanwhile Figueiras (1962), Bossi et al. (1975), and Sprechmann (1978) considered it only of Holocene age. Martínez et al. (2006) assigned to the Villa Soriano Formation the fossiliferous deposits radiocarbon dated as Holocene. This unit has been correlated with the "Querandinense" deposits from Argentina (Goñi and Hoffstetter 1964; Aguirre and Whatley 1995) and with the Patos Group in Rio Grande do Sul State, Brazil (Martínez 1988).

The Chuy Formation was defined by Delaney (1963), although the first published mention corresponds to Goñi and Hoffstetter (1964). This unit was originally described as composed by yellow reddish medium sands. Later on, other lithological types, not included in its definition, were added to its characterization. According to Navarro (1990), the Chuy Formation extends from the Río Santa Lucía basin to the east of Uruguay. Its scarce fossil remains were reported from wells by Sprechmann (1978). The stratigraphic relationships of this unit indicate that it belongs to the Pleistocene (Goñi and Hoffstetter 1964; Goso 1972; Sprechmann 1978).

The poor definition of these lithostratigraphic units, their wide and overlapping lithological characterization, and the use of nonlithological criteria to recognize them, hinders their identification in the field and consequently the placement of fossil remains (see Martínez and Ubilla 2004; Martínez et al. 2013, for further details). Thus, the assignment of the Late Pleistocene fossiliferous deposits studied here to the Villa Soriano Formation or the Chuy Formation is problematic and not useful for practical purposes at the moment.

4 Description of the Outcrops and Previous Work

4.1 Puerto de Nueva Palmira

The fossiliferous deposit is located at around 12 m of altitude and is approximately 2 m thick (Figs. 2 and 3a–c). The lithology is medium to coarse sand with embedded clasts which maximum diameter can reach approximately 4 cm. Shells are mostly randomly distributed and densely packed. Bivalves are frequently disarticulated, although specimens with articulated valves rarely occur. Abrasion and fragmentation are very common, revealing some degree of local transport. The assemblage likely represents the accumulation of shells in multiple high energy events in a proximal environment influenced by waves (Martínez et al. 2001; Rojas 2007).

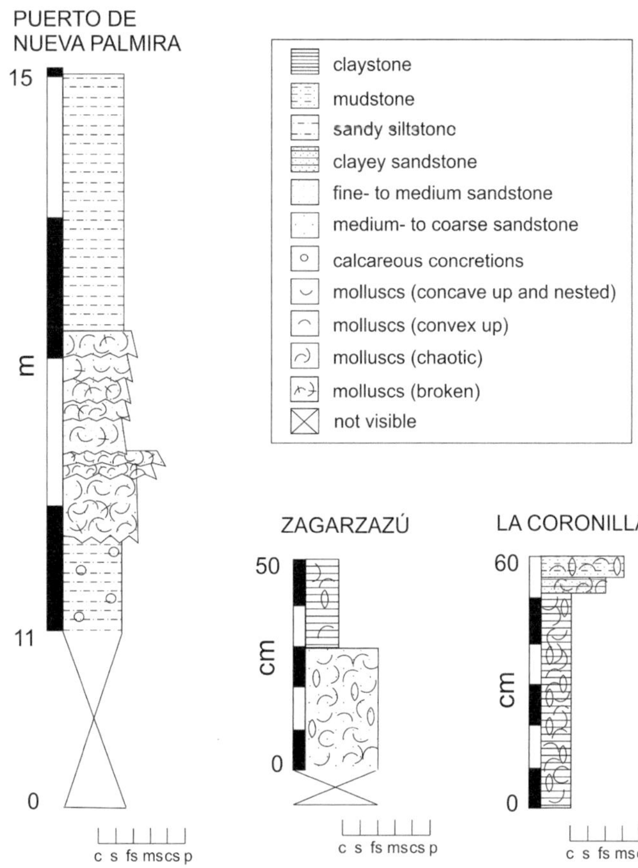

Fig. 2 Stratigraphic sections of the studied deposits

The Puerto de Nueva Palmira fossil assemblage is one of the most mentioned in the literature concerning the Quaternary marine deposits from Uruguay. Since the first reports about this deposit, some physical changes in the area have occurred. For example, the port was built with the consequent modification of the original landscape. Teisseire (1927, 1928) reported the molluscan fauna of a deposit which he refers as located on the cliff of the cemetery of Nueva Palmira (formerly nearby the port) with an altitude of 7–10 m. This author mentions the presence of *Anomalocardia brasiliana*, *Bullia cochlidium*, *Cardium muricatum*, *Mactra isabelleana*, *Pitar rostratum*, *Thais haemastoma, Bullia deformis,* and *Fissuridea patagonica*. The present authors believe that this deposit studied by Teisseire and subsequent authors may be equivalent to the deposit referred here, because Teisseire (1927), Kraglievich (1928), and Fontana (1930) mentioned that in the cliff of the cemetery a new port (a Free Trade Zone) was being constructed at that time. The last author corrected the altitude assigned by Teisseire (1928) to 15 m and

Fig. 3 Photographs of the fossiliferous Pleistocene outcrops. **a–c** Puerto de Nueva Palmira; **d–f** Zagarzazú; **g–i** La Coronilla. Images (**f**) and (**i**) show *Tagelus plebeius* in life position

reported the presence, at approximately 6 m a.s.l., of varied fossil shells with gastropods that needed to be studied. He also commented (as Kraglievich 1928: Fig. 6) about the abundant representation of *Corbula mactroide* (sic). Both authors based on this bivalve, correlated this deposit with the "Querandinense" units. After that, Frenguelli (1930) referred to the deposit of the Puerto de Nueva Palmira, which he assigned to the "Belgranense" stage (Pleistocene) based upon its nature, position, aspect, and paleontological content. This author reported the presence of *Littoridina charruana*, *Bulla striata*, *Bittium varium*, *Purpura haemastoma*, *Bullia globulosa*, *Ostrea spreta*, *Ostrea puelchana*, *A. brasiliana,* and *M. isabelleana*. Regarding the presence of *Corbula mactroides* in this deposit, Frenguelli (1930) suggested that Kraglievich (1928) had misidentified this species with *M. isabelleana* which is very abundant there. But Kraglievich (1932) reaffirmed that the species found was *Corbula mactroides*. Later on, Roselli (1939) mentioned this deposit and commented on the diverse opinions regarding its assignment to the "Belgranense" or "Querandinense" episodes. Serra (1943) presented a stratigraphic section of the outcrop and made comments about the most abundant fossils, *M. isabelleana* and *Purpura haemastoma*. Calcaterra (1971) expressed doubts on the proposed age for this deposit based upon its elevation (more than 12 m a.s.l.) and the thickness of the overlying sediments. He also listed the presence of *M. isabelleana*, *T. haemastoma*, *Arca* sp., *Crassostrea* sp., and *Buccinanops* sp. Francis (1975) referred to the coastal cliff of Nueva Palmira and, besides listing some molluscan species, he made reference to the historical controversy between Kraglievich and Frenguelli on the presence or absence of *Erodona mactroides* in this outcrop, which he was not able to find. He was surprised about the elevation of the deposit in relation to others typically attributed to the "Querandinense". Roselli (1976) thoroughly described the deposit and stuck to the interpretation proposed by Frenguelli (1930) concerning the Late Pleistocene age of the deposit. This author provided a list of taxa that includes *Mactra* (*Mactratoma*) *isabelleana*, *A. brasiliana*, *Thais* (*Stramonita*) *haemastoma*, *Buccinanops globulosus*, *Buccinanops cochlidium*, *Buccinanops deformis*, *Plicatula spondyloidea*, *Ostrea spreta*, *Arca bisulcata*, *Ostrea puelchana*, *Tagelus* (*Tagelus*) *gibbus*, *Diodora patagonica*, *Urosalpinx rushi*, *Littoridina australis*, *Nassarium* sp., *Brachidontes dominguensis*, *Trachycardium muricatum*, *Balanus* sp., *Pitar* (*Pitar*) *rostrata*, *Clamys* sp., *Glycymeris longior*, *Siphonaria* (*Pachysiphonaria*) *lessoni*, *Crepidula dilatada* ¿var. patagonica?, *Calliostoma* sp., *Erodona mactroides*, *Mactra* sp., ¿*Olivancillaria*?, ¿*Chione*?, *Drupa* (*Drupa*) *necocheana,* and two terrestrial gastropods. Afterward, the paleontology, geochronology, and paleoenvironmental conditions of the locality were studied by Martínez et al. (2001). Supplementary information may be found in Rojas (2007).

4.2 Zagarzazú

This fossiliferous deposit is located in the abrasion platform of the Zagarzazú resort approximately 0.5–1 m above mean sea level (a.m.s.l.; Figs. 2 and 3d–f). Its exposure depends on the littoral dynamics, and sometimes it may be covered by sand or water. It is composed of fine sands at the base and green clays at the top. Fossil shells are present in both levels. The fossil preservation is variable, but mostly good. Bivalve shells in life position are frequent (Fig. 3f) and complete, whole specimens with delicate shells can be found, although disarticulated and broken remains do occur as well. Few shells showing abrasion were found. The depositional environment is inferred as proximal, under low energy conditions (Rojas 2007).

The scientific knowledge of the Zagarzazú assemblage is very recent. Verde (2003) and Rojas (2003) presented preliminary reports about the ichnofossils and molluscan content. Rojas (2007) did a more extensive analysis of the molluscan fauna, paleoenvironmental conditions, and available radiometric dating.

4.3 La Coronilla

This deposit is located at the abrasion platform of the La Coronilla beach, 0.5–1 m a.s.l. (Figs. 2 and 3g–i). Similarly to the Zagarzazú locality, its exposure depends on the littoral dynamics as it is frequently covered by sand or water. It is lithologically composed of a greenish-grayish sandy claystone in which fossils are well preserved. Taxa with delicate shells occur. It is frequent the preservation of shells in life position (Fig. 3i). Fragmentation of the shells is common but not abrasion. These features and the lack of shell size sorting suggest a predominantly autochthonous–parautochthonous assemblage without significant transport (Rojas 2007; Rojas et al. 2014). Shell fragmentation can be related to the activity of biological agents. The depositional environment is inferred as proximal and mostly protected from wave action.

Previous works regarding the La Coronilla fossil fauna can be found in Martínez et al. (2001), who studied the paleontology, geochronology, and paleoenvironments. Additional information can be found in Rojas (2007). Lorenzo and Verde (2004) and Rojas et al. (2014) presented ichnological information, and Rojas and Urteaga (2011) dealt with the record of chitons in this assemblage.

5 Absolute Ages

Bracco and Ures (1998), Martínez et al. (2001), Rojas (2007), and Rojas (2010) published absolute dates for the Uruguayan Late Pleistocene assemblages, based upon conventional ^{14}C and OSL dating. These data are compiled in Table 1.

Table 1 Absolute ages available for the Late Pleistocene fossil assemblages from Uruguay

Locality	Dating method	Lab. number	Taxon	Age obtained	Source
Puerto de Nueva Palmira	^{14}C conventional	LP-738	*Mactra isabelleana*	31,000 ± 1200	Martínez et al. (2001)
Puerto de Nueva Palmira	^{14}C conventional	LP-730	*Anomalocardia brasiliana*	34,600 ± 2000	Martínez et al. (2001)
La Coronilla 1	^{14}C conventional	LP-884	*Ostrea equestris*	29,500 ± 600	Martínez et al. (2001)
La Coronilla 2	^{14}C conventional	LP-824	*Mactra isabelleana*	35,500 ± 1900	Martínez et al. (2001)
Zagarzazú	^{14}C conventional	LP-1466	*Tagelus plebeius* (life position)	35,500 ± 1900	Rojas (2007)
Puerto de Nueva Palmira (PNPL01)	OSL	UIC2632	–	80,680 ± 5500	Rojas (2010)
Zagarzazú (ZZZL01)	OSL	UIC2633	–	88,355 ± 7070	Rojas (2010)

5.1 ^{14}C Conventional

Bracco and Ures (1998) provided ^{14}C conventional ages for the Nueva Palmira site without locating the samples in a stratigraphic column or indicating the beds from which they were collected; albeit they mentioned an elevation of 8–10 m a.s.l. They mentioned two radiocarbon ages of 31,900 ± 700 years B.P. (URU 0081) obtained from nonidentified shells and another age of 35,300 ± 1,150 years B.P. (URU 0087) from carbonate clasts included in the deposit. Martínez et al. (2001) provided ^{14}C conventional ages obtained from infaunal bivalves of 31,000 ± 1,200 years B.P. (LP-738) on *M. isabelleana* and 34,600 ± 2,000 years B.P. (LP-730) on *A. brasiliana*. The X-ray diffraction analyses applied to evaluate the consistency of the samples gave reliable results for these ages.

For the La Coronilla deposit, Martínez et al. (2001) obtained ages of 29,500 ± 600 years B.P. (LP-884) on *Ostrea equestris* and 35,500 ± 1,900 years B.P. on *M. isabelleana*.

For Zagarzazú, Rojas (2007) obtained a ^{14}C conventional age of 35,500 ± 1,900 years B.P. (LP-1466) on the infaunal bivalve *Tagelus plebeius*. The specimens used for dating were collected in life position.

The datings reported by Martínez et al. (2001) and Rojas (2007) were performed at the Laboratorio de Tritio y Carbono of the Universidad de La Plata (LATYR). All these ages were considered as minimum ages; thus, fossil assemblages are thought to be older than their measured radiocarbon age (Martínez et al. 2001; Rojas 2007).

5.2 OSL Ages

In order to extend the temporal range of numerical dating and to obtain an independent source of ages for the Pleistocene deposits, the present authors have applied the Optically Stimulated Luminescense (OSL) dating technique to sandy fossiliferous deposits. As suitable deposits for this method, the Puerto de Nueva Palmira and Zargarzazú beds were selected because of their sandy lithology. Although OSL dating has been better developed in continental environments such as aeolian sediments, this dating technique has also been used for littoral marine deposits (e.g., Mallinson et al. 2008; Simms et al. 2009; Suguio et al. 2011). From a methodological point of view, sampling was made using PVC tubes wrapped with thick black tape in order to prevent light reaching the sample. Tubes were dug horizontally into the profile and samples were then extracted. The extremes of each tube were discarded in the laboratory to use only the central part of the sample. For each sample, ages were obtained on quartz grains of 150–250 μm in the Luminescence Dating Research Laboratory, Department of Earth and Environmental Sciences of the University of Illinois at Chicago. Results were provided by Rojas (2010). Ages of 80,680 ± 5,500 years B.P. and 88,355 ± 7,070 years B.P. were obtained for the

Puerto de Nueva Palmira and Zagarzazú localities, respectively. Additional data of the OSL ages can be found in Appendix.

6 Fossil Composition of the Late Pleistocene Assemblages

The Late Pleistocene fossil assemblages from Uruguay preserve a diverse and abundant bivalve and gastropod fauna (Martínez et al. 2001; Rojas 2007). In Tables 2 and 3 the present authors forwarded an update of these molluscan taxa based upon the works of Martínez et al. (2001) and Rojas (2007), due to their explicit description of sampling method employed and the availability of the specimens collected.

Table 2 Gastropod taxa found in the Late Pleistocene fossil assemblages of Uruguay

	Locality		
	Puerto de Nueva Palmira	Zagarzazú	La Coronilla
Acteocina candei	x	x	x
Bittiolum varium	x		x
Boonea jadisi			x
Boonea seminuda			x
Bostrycapulus odites	x		x
Buccinanops globulosus	x	x	
Bulla occidentalis	x		
Costoanachis sertulariarum			x
Crepidula plana			x
Crepidula protea			x
Crepidula sp.		x	
Cylichnella bidentata	x	x	x
Diodora patagonica	x		x
Epitonium albidum		x	
Epitonium georgettinum		x	
Melanella sp.			x
Fargoa bushiana			x
Finella dubia	x	x	x
Heleobia sp.	x	x	x
Iselica anomala			x
Littoraria flava		x	
Lottia subrugosa	x		
Lucapinella henseli			x

(continued)

Table 2 (continued)

	Locality		
	Puerto de Nueva Palmira	Zagarzazú	La Coronilla
Marshallora sp.	x		
Olivella tehuelcha			x
Olivella sp.	x		x
Parvanachis spp.		x	x
Siphonaria lessoni	x		
Stramonita haemastoma	x	x	
Tegula patagonica			x
Turbonilla multicostata			x
Turbonilla uruguayensis			x
Turbonilla spp.		x	x
Urosalpinx haneti			x
Vitrinellidae indet.			x
Zidona dufresnei			x

Table 3 Bivalve taxa found in the Late Pleistocene fossil assemblages of Uruguay

	Locality		
	Puerto de Nueva Palmira	Zagarzazú	La Coronilla
Abra uruguayensis			x
Adrana patagonica			x
Aequipecten tehuelchus			x
Anatina anatina			x
Anomalocardia brasiliana	x	x	x
Atrina seminuda			x
Brachidontes sp.	x	x	x
Bushia rushii	x	x	x
Cardiomya sp.			x
Corbula caribaea	x	x	x
Corbula lyoni		x	
Crassinella lunulata			x
Cyclinella tenuis			x
Cyrtopleura sp.	x	x	
Donax sp.		x	
Ennucula puelcha			x
Ennucula uruguayensis			x
Erodona mactroides	x	x	
Ervilia concentrica			x
Gastrochaena sp.			x
Gouldia cerina	x		x

(continued)

Table 3 (continued)

	Locality		
	Puerto de Nueva Palmira	Zagarzazú	La Coronilla
Laevicardium sp.			x
Limaria sp.			x
Lunarca ovalis	x		x
Macoma constricta	x	x	
Macoma uruguayensis		x	x
Mactra isabelleana	x	x	x
Merisca martinicensis			x
Musculus sp.			x
Mytilus edulis			x
Chione subrostrata	x		x
Noetia bisulcata			x
Nucula semiornata	x	x	x
Ostrea equestris	x	x	x
Phlyctiderma semiaspera	x		x
Pitar rostratus	x		x
Pitar palmeri			x
Plicatula gibbosa	x		x
Scapharca brasiliana			x
Semele proficua			x
Sphenia fragilis			x
Tagelus plebeius	x	x	x
Tellina gibber	x	x	x
Trachycardium muricatum			x
Veneridae indet.			x

Other molluscs, such as polyplacophorans, have been recorded only for the La Coronilla site (Rojas and Urteaga 2011). The chiton species found in this deposit are *Chaetopleura angulata*, *C. isabellei*, *C. asperrima*, and *Ischnochiton striolatus*.

All specimens referenced in Martínez et al. (2001), Rojas (2007), and Rojas and Urteaga (2011) are housed in the Colección Paleontológica at the Facultad de Ciencias, Universidad de la República (FCDPI), Montevideo. Some of the specimens listed by Martínez et al. (2001) and Rojas (2007) were revised in order to improve their taxonomic assignment. For instance, specimens originally assigned to *Chrysallida gemmulosa* are reclassified herein as *Fargoa bushiana*. Specimens referenced as *Turbonilla americana* by Martínez et al. (2001) are reclassified as *Turbonilla multicostata* and those classified as *Heleobia australis* are considered as *Heleobia* sp., due to the difficult identification of the different species of this genus using only their shells. Specimens assigned by Martínez et al. (2001) to *Clausinella gayi* are now thought to be another yet unidentified venerid. Although the species

Ennucula uruguayensis has been synonymized with *E. puelcha*, the present authors maintained it as a valid species, based on the observations reported by Scarabino et al. (2006b). *Crepidula* sp., listed by Martínez et al. (2001) from Puerto de Nueva Palmira is herein identified as *Bostrycapulus odites*. *Parvanachis isabellei* from Martínez et al. (2001) and *Parvanachis paessleri* from Rojas (2007) are presented here as *Parvanachis* spp. because the assignment of these specimens to a particular species of *Parvanachis* deserves caution and further comparisons.

In addition, occasional remains of other non-molluscan taxa such as balanids, decapods, ostracods, bryozoans, serpulids, corals, echinoid, and fishes have been found (Rojas 2007). Bioerosion traces on shells reported by Verde (2003) for the Zagarzazú locality include *Entobia* isp., *Gastrochaenolites* isp., and *Meandropolydora* isp. Lorenzo and Verde (2004) found *Entobia* isp., *Caulostrepsis taenicola,* and *Oichnus paraboloides* on molluscan shells from Puerto de Nueva Palmira and La Coronilla localities. Recently, Rojas et al. (2014) reported the first gastropod drill hole on a fossil chiton plate. This corresponds to the ichnospecies *Oichnus simplex* placed on an intermediate valve of *C. angulata* found in the La Coronilla site. Verde (2003) recorded ichnofossils in soft sediments from the Zagarzazú locality. These structures correspond to *Ophiomorpha nodosa* and *Thalassinoides* isp., along with other yet unidentified bivalve ichnofossils.

7 Analysis of the Molluscan Fauna

A total of 85 molluscan taxa (36 gastropods, 45 bivalves, and 4 polyplacophorans) have been recorded so far in the Late Pleistocene fossil assemblages of Uruguay. Ongoing research suggests that the molluscan diversity is even higher. La Coronilla locality has the richest assemblage, recording 70 molluscan taxa. It is followed by the Puerto de Nueva Palmira and Zagarzazú deposits that record 32 taxa and 27 taxa, respectively.

According to the available information (Clavijo et al. 2005; Martínez et al. 2006; Rojas 2007; Rojas and Urteaga 2011) the gastropods *Bittiolum varium*, *Melanella* sp., *F. bushiana*, *Iselica anomala*, *Littoraria flava*, *Lucapinella henseli*, *T. multicostata* and the Vitrinellidae specimens, the bivalves *Anatina anatina*, *Atrina seminuda*, *Cardiomya* sp., *Cyclinella tenuis*, *E. uruguayensis*, *Ervilia concentrica*, *Gastrochaena* sp., *Gouldia cerina*, *Laevicardium* sp., *Limaria* sp., *Macoma constricta*, *Musculus* sp., *Pitar palmeri,* and *Merisca martinicensis* have been exclusively recorded in Pleistocene deposits and are absent from the Holocene molluscan assemblages.

All recorded molluscs are extant and live along the western coast of the Atlantic Ocean. Only *Siphonaria lessoni* is a cold water species influenced by the Malvinas Current (Scarabino 1977) which represents approximately 1 % of the taxa recorded in the Pleistocene assemblages. Meanwhile, in the recent Uruguayan coast

Fig. 4 Extralimital warm water gastropods from the Pleistocene fossil assemblages from Uruguay. **a** *Littoraria flava* (FCDPI 3948); **b** *Bittiolum varium* (FCDPI 6661); **c** *Fargoa bushiana* (FCDPI 4274); **d** *Bulla occidentalis* (FCDPI 3060). Scale bar **a**–**c** 1 mm; **d** 5 mm

approximately 19 % of the molluscan fauna are considered to be cold water species (Sicardi 1967; Sprechmann 1978).

The molluscs found in the Late Pleistocene assemblages also record 17 taxa that show a retraction in their recent biogeographic range (Figs. 4, 5, and 6). The gastropods *B. varium*, *Bulla occidentalis*, *F. bushiana,* and *L. flava*, the bivalves *A. anatina*, *A. brasiliana*, *C. tenuis*, *E. concentrica*, *G. cerina*, *Laevicardium* sp.,

Fig. 5 Extralimital warm water bivalves from the Pleistocene fossil assemblages from Uruguay. **a** *Scapharca brasiliana* (FCDPI 4202); **b** *Ervilia concentrica* (FCDPI 4217); **c** *Macoma constricta* (FCDPI 4145); **d** *Limaria* sp. (FCDPI 2881); **e** *Laevicardium* sp. (FCDPI 4204); **f** *Pitar palmeri* (FCDPI 4222); Scale bar: **a**, **c**, **e**, **f** 5 mm; **b**, **d** 1 mm

Limaria sp., *M. constricta*, *Chione subrostrata*, *P. palmeri*, *Scapharca brasiliana,* and *M. martinicensis,* and the chiton *I. striolatus* have their southernmost boundary of distribution in Brazilian waters. Most of these taxa were recorded exclusively at the La Coronilla assemblage. From the precedent group, only *A. brasiliana* has been recorded in all Pleistocene deposits studied. *G. cerina*, *N. subrostrata,* and *B. varium* were recorded both at La Coronilla and Puerto de Nueva Palmira localities. For the moment, *B. occidentalis* has been only recorded at Puerto de Nueva Palmira, *L. flava* only in Zagarzazú and *M. constricta* is shared by Zagarzazú and Puerto de Nueva Palmira localities.

Fig. 6 Extralimital warm water bivalves from the Pleistocene assemblages from Uruguay. **a** *Anatina anatina* (FCDPI 4192); **b** *Gouldia cerina* (FCDPI 4220); **c** *Chione subrostrata* (FCDPI 4206); **d** *Anomalocardia brasiliana* (FCDPI 4205); **e** *Cyclinella tenuis* (FCDPI 4221); **f** *Merisca martinicensis* (FCDPI 4717);. Scale bar: **a** 10 mm; **b** 1 mm; **c–f** 5 mm

8 Discussion

Several lines of evidence must be considered in order to assign the Uruguayan Late Pleistocene fossil assemblages either to MIS 3 or MIS 5. These lines are: (a) the numerical ages obtained for the deposits, (b) sea level and deposit elevation above mean sea level, (c) the environmental conditions inferred from the identified fossils.

(a) Numerical ages

The standard radiocarbon ages obtained for the three Pleistocene fossil assemblages fall within the scope of MIS 3. These ages were considered by Martínez et al. (2001) and Rojas (2007) as minimum ages. Due to the warmer water affinities of the molluscan fauna, these authors considered that the real age of the assemblages would fall within MIS 5e. In relation to the ages reported by Martínez et al. (2001), Tonni et al. (2010) argued that the ages reported for the Puerto de Nueva Palmira and the La Coronilla deposits were finite ages and statistically distinguishable from the dating limit of the radiocarbon dating method. Thus, Tonni et al. (2010) based their opinion only on the numerical ages, but not taking into account the paleoecological information, and considered that the Uruguayan Pleistocene deposits are representative of the final period of MIS 3.

However, the OSL ages recently obtained for the Puerto de Nueva Palmira and Zagarzazú localities are significantly older compared to those obtained by ^{14}C, falling within MIS 5a. The OSL ages, obtained from coastal deposits elsewhere, have shown to be an accurate dating method (e.g., Mallinson et al. 2008; Simms et al. 2009; Suguio et al. 2011) that indicates the time lapse occurred since the last exposure of quartz sand to sunlight. In fact, for instance, Simms et al. (2009) assigned on the basis of OSL dating to MIS 5a deposits from the northwestern Gulf of Mexico which had previously radiocarbon dated as of MIS 3.

(b) Sea level and deposits elevation

During the Late Pleistocene, sea level above or at its present position has been widely recognized for MIS 5e up to approximately 9 m a.s.l. (e.g., Stirling et al. 1998; McCulloch and Esat 2000; Shackleton 2000; Murray-Wallace 2002; Waelbroeck et al. 2002; Hearty et al. 2007; Schellmann and Radtke 2004; Siddall et al. 2007; Alley et al. 2010; Muhs et al. 2011; Dutton and Lambeck 2012). Coastal deposits of this substage usually crop out several meters above present sea level.

Regarding the MIS 5a substage, considerable controversy exists about the position of sea level in that moment. The estimates for eustatic sea level altitude range approximately from +3 (or even +6 and +10) to −30 m, relative to modern sea level (Ludwig et al. 1996; Muhs et al. 2002; Potter and Lambeck 2003; Potter et al. 2004; Wehmiller et al. 2004; Dorale et al. 2010; Abad et al. 2013; Medina-Elizalde 2013). The elevation of the MIS 5a deposits is very variable depending on the studied sites and their tectonic setting (see Coyne et al. 2007, and references therein). Although they often represent submerged outcrops, for example, Muhs et al. (2002) and Wehmiller et al. (2004) report MIS 5e and 5a deposits at similar low elevations.

There is a certain consensus that MIS 3 interestadial sea level was well below its present position, with estimates from approximately −80 to −20 m (Yokoyama et al. 2001; Murray-Wallace 2002; Siddall et al. 2003, 2008). The elevation of MIS 3 deposits around the world is variable, but they are mostly found below present sea level although they have been also reported up to 8 m a.s.l. (see Hodgson et al. 2009).

Considering the sea level scenarios throughout the Late Pleistocene and taking into account that the studied deposits are considered to be in a mostly tectonically stable area, the Pleistocene assemblages from Uruguay are likely to have been deposited during a sea level stand higher than the present one, such as the inferred for MIS 5. Regarding the height of the Uruguayan deposits, the La Coronilla and Zagarzazú deposits are located at about 0.5–1 m a.s.l., whereas the Puerto de Nueva Palmira site is placed about 12–13 m a.s.l. (Martínez et al. 2001; Rojas 2007). These elevation differences are notorious and deserve an explanation, especially considering that the Puerto de Nueva Palmira and Zagarzazú deposits are separated by only about 12 km.

Taphonomy and biological indicators suggest that the Zagarzazú assemblage was deposited in shallow waters protected from waves and probably close to the coastline, due to the presence of the *Thalassinoides–Ophiomorpha* association (see Verde 2003). The Puerto de Nueva Palmira assemblage reflects the deposition during high energy events that could represent accumulation in the more proximal foreshore. Thus, the water table could have been thicker in Zagarzazú and thinner in Puerto de Nueva Palmira deposit, explaining at least in part the height difference between both deposits. Like the Zagarzazú assemblage, the La Coronilla one is also inferred to have been deposited in a shallower water environment protected from waves. Thus the coastline at that time would have been situated inland, possibly reducing the elevation difference of the deposits. However, even considering the depth of deposition, the Puerto de Nueva Palmira assemblage is placed at an important altitude. Although there is still no published reference to neotectonics in the Uruguayan coast, Martínez et al. (2001) considered likely that neotectonics had resulted in a local uplift of the Puerto de Nueva Palmira deposit, which remains to be proved. Another possibility to take into account is that the Uruguayan Pleistocene deposits have different ages, as suggested by the available datings. Consequently, they may have been deposited during different sea-level position scenarios. For instance, the Puerto de Nueva Palmira assemblage could have been deposited during MIS 5e and the Zagarzazú deposit during MIS 5a.

(c) Paleoenvironmental inferences from the molluscan assemblages

Another line of evidence that should be considered in order to infer the most probable age of the Late Pleistocene invertebrate assemblages from Uruguay is the paleoenvironmental information provided by the fauna. As it has been mentioned above, only one species (*S. lessoni*) recorded in the Puerto de Nueva Palmira assemblage is considered to be a cold water taxon, which currently lives in rocky shores of the Uruguayan coast (e.g., Scarabino et al. 2006a; Brazeiro et al. 2006). However, the Late Pleistocene assemblages include 17 extralimital warm water molluscan taxa that currently do not reach the Uruguayan coast (Fig. 7). These molluscs represent 20 % of the taxa recorded providing a strong environmental signal. The former presence of these taxa in the Uruguayan coast points to the existence of higher temperatures than present (Martínez et al. 2001; Rojas 2007; Rojas and Urteaga 2011). The majority of the warm water molluscs recorded has their southernmost distribution boundary in Santa Catarina, Brazil. Thus, it can be inferred that the temperature regime of the Uruguayan coast when these assemblages lived could have been similar to those found today in that region of the Brazilian coast. Considering the Late Pleistocene period, the only time interval in which similar or higher temperatures than present are inferred is the substage MIS 5e. For example, taking into account Southern Hemisphere data, temperatures about 3 °C (Petit et al. 1999) or +4 to +5 °C higher than present were reconstructed from Antarctic ice cores (EPICA 2006; Jouzel et al. 2007). Bianchi and Gersonde (2002) and Cortese and Abelmann (2002) inferred +2 to +3 °C higher than present temperatures based on diatoms and radiolarians from marine sediment records in the Atlantic sector of the Southern Ocean. Also, a global dataset of ice, marine, and terrestrial sequences suggests that global temperatures were on average 1.5 °C higher than today, also showing a strong latitudinal temperature gradient with greater warming at higher latitudes (Turney and Jones 2010).

Additionally, MIS 5c fossil assemblages worldwide commonly record extralimital warm water species. This is the case for Pacific and western Atlantic islands like Hawaii and Bermuda but also around North America and Greenland (see Muhs et al. 2002), for Mediterranean and eastern Atlantic islands assemblages (e.g., Zazo et al. 2003, 2013; Garilli 2011; Montesinos et al. 2014). Most studies considering shifting geographic ranges of species invoke weaker cold currents and stronger warm currents to explain biogeographic changes shown by calid water taxa during the warmer MIS 5e (e.g., Muhs et al. 2002, among others). Similarly, a stronger and southward reaching warm Brazilian Current has been invoked to explain the presence of warm water molluscs in the Late Pleistocene deposits from Uruguay (Martínez et al. 2001; Rojas 2007). Also, the same mechanism has been proposed for the finding of northward displaced species in Holocene Argentinean assemblages, especially those from the province of Buenos Aires (Aguirre 1991, 1993; Aguirre and Farinati 2000; Aguirre et al. 2005).

The environmental and paleobiogeographical information derived from the Late Pleistocene fossil assemblages from Uruguay fits in the environmental scenario reconstructed for MIS 5e from multiple lines of evidence. However, some issues remain to be considered. For example, the La Coronilla assemblage contains 12 extralimital warm water molluscan taxa, whereas the Puerto de Nueva Palmira records six taxa and the Zagarzazú assemblage records three taxa. As it was explained before, different depositional conditions are inferred for these three assemblages. Thus, taphonomic processes may have played a role in the preservation potential of the fauna. Another variable that must be considered is salinity. The La Coronilla assemblage is located in the Atlantic sector of the Uruguayan coast; thus, fully marine conditions are to be expected. However, the Puerto de Nueva Palmira and Zagarzazú deposits are in the western coast of Uruguay, where freshwater environments develop today. Although the salinity requirements of the taxa recorded in these assemblages suggest higher salinities in that area, a western freshwater discharge was supposed to be already working then (e.g., Potter 1998; Iriondo 2004; Veroslavsky and Ubilla 2007), lowering salinity to some extent. Thus, this difference with the La Coronilla assemblage may be playing a role in the lower species richness of the western deposits. There is still another possibility to consider. Maybe the three assemblages are not coetaneous, as suggested by the available OSL ages, implying deposition under rather different environmental conditions. Meanwhile the warmer water character of the La Coronilla fossil assemblage probably represents a MIS 5e deposit, the Puerto de Nueva Palmira, and Zagarzazú assemblages may be likely representing the occurrence of MIS 5a deposits.

Globally, there are fewer coastal records comprising MIS 5a and as mentioned above, there is conflicting evidence about the sea-level position and temperature regimes of this substage relative to the present conditions. Some studies of tectonically stable areas from the Northern Hemisphere suggest that sea level during

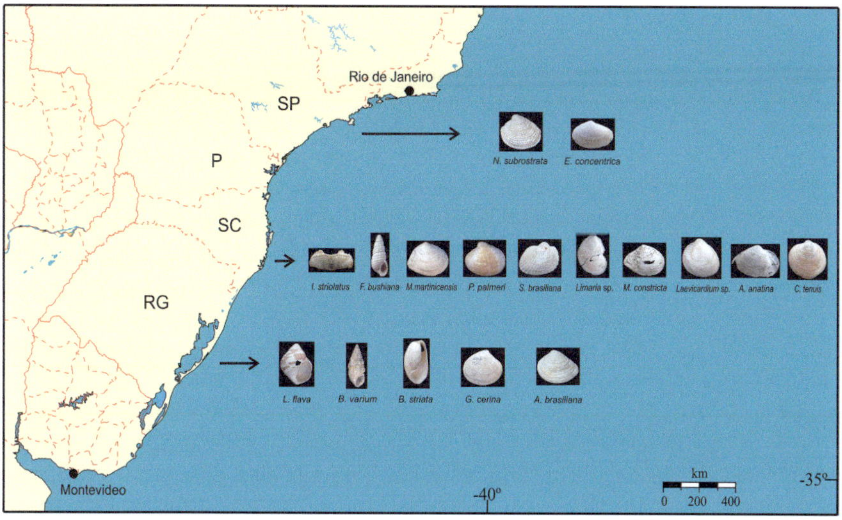

Fig. 7 Current southernmost limit of geographic distribution of the extralimital warm water molluscs found in the Late Pleistocene assemblages from Uruguay. *RG* Rio Grande do Sul; *SC* Santa Catarina; *P* Parana; *SP* São Paulo. Biogeographic data were taken from Ríos (1994, 2009), Scarabino and Zaffaroni (2004), Amaral et al. (2006), Pimenta et al. (2009), Rosenberg (2009)

MIS 5a was above the present one, with temperature regimes similar to present conditions explained by the increased melting of ice sheets (e.g., Ludwig et al. 1996; Muhs et al. 2002; Dorale et al. 2010). Some studies of pollen and diatoms from Baffin Island, Canada, show that climate was warmer than the Holocene around 80 ka B.P., implying the almost complete deglaciation of the Laurentide ice sheet (Miller et al. 1999). Regarding the Southern Hemisphere data, controversies are recorded by pollen and beetles (e.g., Burge and Shulmeister 2007; Fletcher and Thomas 2010) from Tasmania and New Zealand.

Considering the OSL ages obtained for Puerto de Nueva Palmira and Zagarzazú deposits and the fact that they record less extralimital warm water molluscs than the fossil assemblage of La Coronilla, these western assemblages could represent deposits of the less understood MIS 5a substage.

Ongoing research on the Late Pleistocene deposits from Uruguay suggests that the knowledge of the molluscan composition of the assemblages can be improved and richness seems to be higher than that reported by previous studies (Martínez et al. 2001; Rojas 2007; Martínez et al. 2013). Along with the paleoecological interpretations, new geochronological data including AMS and other dating methods such as U/Th are needed to achieve a higher resolution in the timing of the environmental and faunal changes occurred in the Uruguayan coast during the last thousand years.

9 Conclusions

Uruguay has three undoubtedly Late Pleistocene localities with marine fossiliferous deposits; the western Puerto de Nueva Palmira, Zagarzazú assemblages and the eastern La Coronilla assemblage. Conflicting ages were obtained from different dating methods. Whereas standard radiocarbon ages are related to MIS 3, OSL datings from the western deposits are older and related to MIS 5a. The paleoecological and biogeographical analyses of the abundant molluscan fauna point to a higher than present water temperature for the Uruguayan coast, due to the presence of a significant number of extralimital warm water taxa. These taxa that currently live in Brazilian waters are well represented in the La Coronilla assemblage and have a lesser representation in the Puerto de Nueva Palmira and Zagarzazú assemblages. Higher temperatures than present and widespread biogeographic changes of marine fauna have been widely recognized for MIS 5e (Last Interglacial) assemblages. Thus, all available evidence suggest that the Late Pleistocene assemblages from Uruguay were deposited during MIS 5, thus the radiocarbon dating indicating MIS 3 times should be considered as minimum ages. The La Coronilla assemblage most likely represents a MIS 5e assemblage, meanwhile Puerto de Nueva Palmira and Zagarzazú deposits could have been originated during the MIS 5a substage, as suggested by the available OSL ages and the lower representation of extralimital northern molluscs.

Acknowledgments We acknowledge the invitation by Eduardo P. Tonni and Jorge Rabassa to attend the "Simposio multidisciplinario, El estadio isotópico 3 en Argentina y el sur de América del Sur: 60.000 a 25.000 años atrás", at La Plata, Argentina, from which this publication has derived. Germán M. Gasparini kindly helped with editorial issues. We thank Fabrizio Scarabino for providing samples of the La Coronilla assemblage. Juan Carlos Zaffaroni and Alexandre Dias Pimenta helped with micromollusc determinations. Mariano Verde assisted during the fieldwork and Mariana Demicheli aided in the preparation of samples and picking of specimens. Martín Ubilla helped during OSL sampling and provided useful bibliography. We also thank the personnel at Museo Lucas Rosselli and Administración Nacional de Puertos (ANP) of Nueva Palmira for allowing the access to the Puerto de Nueva Palmira deposit. This is a contribution to "Programa de Desarrollo de las Ciencias Básicas" (PEDECIBA-Biología), Comisión Sectorial de Investigación Científica (CSIC), and Agencia Nacional de Investigación e Innovación (ANII/FCE2007_034 and ANII/FCE2007_058) for providing funding for this research.

Appendix: Data of the OSL ages obtained for Puerto de Nueva Palmira and Zagarzazú

Sample code	Lab number	Equivalent dose (Grays)[a]	U (ppm)[b]	Th (ppm)[b]	K_2O (%)[b]	H_2O (%)[c]	Cosmic dose (mGrays/year)[d]	Total dose (mGrays/year)	OSL age (year)[e]
PNPLO1	UIC2632	73.58 ± 4.78	0.6 ± 0.1	1.5 ± 0.1	0.64 ± 0.01	10 ± 3	0.016 ± 0.002	0.91 ± 0.04	80,680 ± 5,500
ZZZLO1	UIC2633	124.53 ± 8.16	0.7 ± 0.1	2.1 ± 0.1	1.34 ± 0.01	20 ± 5	0.016 ± 0.002	1.41 ± 0.06	88,355 ± 7,070

[a]Equivalent dose determined by the multiple aliquot regenerative dose method under blue (470 nm) excitation (Jain et al. 2003). Blue emissions are measured with 3-mm-thick Schott BG-39 and one, 3-mm-thick Corning 7–59 glass filters that blocks >90 % luminescence emitted below 390 nm and above 490 nm in front of the photomultiplier tube. The coarse-grained (150–250 μm) quartz fraction is analyzed

[b]U ,Th, and K20 determined by ICP-MS at Activation Laboratory Ltd., Ontario

[c]Average water content estimated from particle size characteristics assuming periodic wetting in the vadose zone

[d]Cosmic dose rate component from Prescott and Hutton (1994) based on latitude, longitude, elevation, and burial depth of samples

[e]All errors are at one sigma and ages are calculated from AD 2010. Analyses preformed by Luminescence Dating Research Laboratory, Dept. of Earth & Environmental Sciences, Univ. of Illinois-Chicago

References

Abad M, Rodríguez-Vidal J, Aboumaria K, Zaghloul MN, Cáceres LM, Ruiz F, Martínez-Aguirre A, Izquierdo T, Chamorro S (2013) Evidence of MIS 5 sea-level highstands in Gebel Mousa coast (Strait of Gibraltar, North of Africa). Geomorph 182:133–146

Acha EM, Mianzan H, Guerrero R, Carreto J, Giberto D, Montoya N, Carignan M (2008) An overview of physical and ecological processes in the Rio de la Plata. Est Cont Shelf Res 28:1579–1588

Aguirre ML (1991) Holocene macrobenthic molluscan associations from northeastern Buenos Aires Province, Argentina. Quat South Am Antarct Peninsula 7:161–195

Aguirre ML (1993) Palaeobiogeography of the Holocene molluscan fauna from northeastern Buenos Aires Province, Argentina: its relation to coastal evolution and sea level changes. Palaeogeogr Palaeoclimatol Palaeoecol 102:1–26

Aguirre ML (2003) Late Pleistocene and Holocene palaeoenvironments in Golfo San Jorge, Patagonia: molluscan evidence. Mar Geol 194:3–30

Aguirre ML, Farinati EA (2000) Moluscos del Cuaternario marino de la Argentina. Bol Acad Nac Cienc Córdoba 64:235–333

Aguirre M, Whatley RC (1995) Late quaternary marginal marine deposits and palaeoenvironments from northeastern Buenos Aires Province, Argentina: a review. Quat Sci Rev 14:223–254

Aguirre ML, Donato M, Richiano S, Farinati EA (2011) Pleistocene and Holocene interglacial molluscan assemblages from Patagonia and Bonaerensian littoral (Argentina, SW Atlantic): palaeobiodiversity and palaeobiogeography. Palaeogeogr Palaeoclimatol Palaeoecol 308:277–292

Aguirre ML, Negro Sirch Y, Richiano S (2005) Late Quaternary molluscan assemblages from the coastal area of Bahía Bustamante (Patagonia, Argentina): paleoecology and paleoenvironments. J South Am Earth Sci 20:13–32

Alley RB, Andrews JT, Brigham-Grette J, Clarke GKC, Cuffey KM, Fitzpatrick JJ, Funder S, Marshall SJ, Miller GH, Mitrovica JX, Muhs DR, Otto-Bliesner BL, Polyak L, White JWC (2010) History of the Greenland Ice Sheet: paleoclimatic insights. Quat Sci Rev 29:1728–1756

Amano K (1994) An attempt to estimate the surface temperature of the Japan Sea in the Early Pleistocene by using a molluscan assemblage. Palaeogeogr Palaeoclimatol Palaeoecol 108:369–378

Amaral ACZ, Rizzo AE, Arruda EP (2006) Manual de identificação dos invertebrados marinhos da Região Sudeste-Sul do Brasil, vol 1. Edusp, Sao Paulo

Ávila SP, Madeira P, Zazo C, Kroh A, Kirby M, Marques da Silva C, Cachão M, de Frias Martins AM (2009) Palaeoecology of the Pleistocene (MIS 5.5) outcrops of Santa Maria Island (Azores) in a complex oceanic tectonic setting. Palaeogeogr Palaeoclimatol Palaeoecol 274:18–31

Barker S, Diz P, Vautravers MJ, Pike J, Knorr G, Hall IR, Broecker WS (2009) Interhemispheric Atlantic seesaw response during the last deglaciation. Nature 457:1097–1102

Berkman PA, Andrews JT, Bjorck S, Colhoun EA, Emslie SD, Goodwin ID, Hall BL, Hart CP, Hirakawa K, Igarashi A, Ingolfsson O, Lopez-Martinez J, Lyons WB, Mabin MCG, Quilty PG, Taviani M, Yoshida Y (1998) Circum-Antarctic coastal environmental shifts during the Late Quaternary reflected by emerged marine deposits. Ant Sci 10(3):345–362

Bianchi C, Gersonde R (2002) The Southern Ocean surface between Marine Isotope Stages 6 and 5d: shape and timing of climate changes. Palaeogeogr Palaeoclimatol Palaeoecol 187:151–177

Bossi J (1966) Geología del Uruguay. Departamento de Publicaciones de la Universidad de la República, Montevideo, Uruguay

Bossi J, Ferrando LA, Fernández N, Elizalde G, Morales H, Ledesma JJ, Carballo E, Medina E, Ford I, Montaña JR (1975) Carta Geológica del Uruguay. Escala 1:1.000.000. Dirección de Suelos y Fertilizantes, Ministerio de Agricultura y Pesca, Montevideo

Bracco R, Ures C (1998) Las variaciones del nivel del mar y el desarrollo de las culturas prehistóricas en el Uruguay. Actas Congr Uruguayo Geol 2:16–20

Brazeiro A, Borthagaray AI, Giménez L (2006) Patrones geográficos de diversidad bentónica en el litoral rocoso de Uruguay. In: Menafra R, Rodríguez L, Scarabino F, Conde D (eds) Bases para la conservación y el manejo de la costa uruguaya. Vida Silvestre, Montevideo
Briggs JC (1995) Global biogeography. Developments in palaeontology and stratigraphy, vol 14. Elsevier, Amsterdam
Buiron D, Stenni B, Chappellaz J, Landais A, Baumgartner M, Bonazza M, Capron E, Frezzotti M, Kageyama M, Lemieux-Dudon B, Masson-Delmotte V, Parrenin F, Schilt A, Selmo E, Severi M, Swingedouw D, Udisti R (2012) Regional imprints of millennial variability during the MIS 3 period around Antarctica. Quat Sci Rev 48:99–112
Burge PI, Shulmeister J (2007) An MIS 5a to MIS 4 (or early MIS 3) environmental and climatic reconstruction from the northwest South Island, New Zealand, using beetle fossils. J Quat Sci 22:501–516
Calcaterra A (1971) Algunas observaciones sobre la fauna de moluscos de la Formación Querandina (Deptos de Colonia y Soriano, Uruguay). Com Soc Malac Uruguay 3(20):79–82
Caorsi J, Goñi J (1958) Geología Uruguaya, vol 37. Boletín Instituto Geológico Uruguayo
Charó MP, Fucks EE, Gordillo S (2013) Moluscos marinos bentónicos del Cuaternario de Bahía Anegada (sur de Buenos Aires, Argentina): variaciones faunísticas en el Pleistoceno tardío y Holoceno. Rev Mex Cienc Geol 30(2):404–416
Charó MP, Gordillo S, Fucks EE, Giaconi LM (2014) Late Quaternary molluscs from the northern San Matías Gulf (Northern Patagonia, Argentina), southwestern Atlantic: faunistic changes and paleoenvironmental interpretation. Quatern Int 352:26–47
Clavijo C, Scarabino F, Rojas A, Martínez S (2005) Lista sistemática de los moluscos marinos y estuarinos del Cuaternario del Uruguay. Com Soc Malac Uruguay 9(88):381–411
Cortese G, Abelmann A (2002) Radiolarian-based paleotemperatures during the last 160 kyr at ODP Site 1089 (Southern Ocean, Atlantic Sector). Palaeogeogr Palaeoclimatol Palaeoecol 182:259–286
Coyne MK, Jones B, Ford D (2007) Highstands during Marine Isotope Stage 5: evidence from the Ironshore Formation of Grand Cayman. Br West Indies Quat Sci Rev 26:536–559
Crowley TJ, North GR (1991) Paleoclimatology. Oxford University Press Inc, New York
De Diego-Forbis T, Douglas R, Gorsline D, Nava-Sanchez E, Mack L, Banner J (2004) Late Pleistocene (last interglacial) terrace deposits, Bahia Coyote, Baja California Sur, Mexico. Quatern Int 120:29–40
Delaney PJC (1963) Fisiografia e geología de superficie da Planicie Costeira do Rio Grande do Sul. Dissertation, Universidade de Sao Paulo
Dorale JA, Onac BP, Fornos JJ, Gines J, Gines A, Tuccimei P, Peate DW (2010) Sea-level highstand 81,000 years ago in Mallorca. Science 327:860–863
Dutton A, Lambeck K (2012) Ice volume and sea level during the last interglacial. Science 337:216–219
Emiliani C (1955) Pleistocene temperatures. J Geol 63:538–578
EPICA Community Members (2004) Eight glacial cycles from an Antarctic ice core. Nature 429:623–628
EPICA Community Members (2006) One-to-one coupling of glacial climate variability in Greenland and Antarctica. Nature 444:195–198
Figueiras A (1962) Sobre nuevos hallazgos de moluscos subfósiles de la Transgresión Querandina. Com Soc Malac Uruguay 1(3):53–68
Fletcher MS, Thomas I (2010) A quantitative Late Quaternary temperature reconstruction from western Tasmania, Australia. Quat Sci Rev 29:2351–2361
Fontana MA (1930) Memoria de la excursión científica a Nueva Palmira: Abril de 1927. Rev Soc Am Arq 4:119–178
Forti Esteves IR (1974) Bioestratigrafia e palaeoecologia (Mollusca) do Quaternario da Planicie costeira do Rio Grande do Sul (Brasil). Annais do Congresso Brasileiro de Geologia 28:133–149
Francis JC (1975) Esquema bioestratigráfico regional de la República Oriental de Uruguay. Actas 1er Congr Arg Paleont Biostr 2:539–568

Frenguelli J (1930) Apuntes de Geología Uruguaya. Bol Inst Geol Perf 11:1–47
Garilli V (2011) Mediterranean quaternary interglacial molluscan assemblages: palaeobiogeographical and palaeoceanographical responses to climate change. Palaeogeogr Palaeoclimatol Palaeoecol 312:98–114
Giberto DA, Bremeca CS, Acha EM, Mianzan H (2004) Large-scale spatial patterns of benthic assemblages in the SW Atlantic: the Río de la Plata estuary and adjacent shelf waters. Est Coast Shelf Sci 61:1–13
Giménez L, Borthagaray AI, Rodríguez M, Brazeiro A, Dimitriadis C (2005) Scale-dependent patterns of macrofaunal distribution in soft-sediment intertidal habitats along a large-scale estuarine gradient. Helgol Mar Res 59:224–236
Goñi JC, Hoffstetter R (1964) Uruguay. Lexique stratigraphique. international, 5 Amerique Latine, 9a. Centre National de la Recherche Scientifique, Paris
Gordillo S, Isla FI (2011) Faunistic changes between the Middle/Late Pleistocene and the Holocene on the Atlantic coast of Tierra del Fuego: molluscan evidence. Quat Int 233:101–112
Goso H (1972) Cuaternario. Technical report, Programa de Estudio y Levantamiento de Suelos, Ministerio de Ganadería y Agricultura, Montevideo
Guerrero RA, Acha EM, Framiñan MB, Lasta CA (1997) Physical oceanography of the Río de la Plata Estuary, Argentina. Cont Shelf Res 17(7):727–742
Hearty PJ, Hollin JT, Neumann AC, O'Leary MJ, McCulloch M (2007) Global sea-level fluctuations during the Last Interglaciation (MIS 5e). Quat Sci Revs 26:2090–2112
Hodgson DA, Verleyen E, Vyverman W, Sabbe K, Leng MJ, Pickering MD, Keely BJ (2009) A geological constraint on relative sea level in Marine Isotope Stage 3 in the Larsemann Hills, Lambert Glacier region, East Antarctica (31366–33228 cal yr BP). Quat Sci Rev 28:2689–2696
Huber C, Leuenberger M, Spahni R, Flückiger J, Schwander J, Stocker TF, Johnsen S, Landais A, Jouzel J (2006) Isotope calibrated Greenland temperature record over Marine Isotope Stage 3 and its relation to CH_4. Earth Planet Sci Lett 243:504–519
Iriondo M (2004) The littoral complex at the Paraná River mouth. Quat Int 114:143–154
Isla FI, Rutter NW, Schnack EJ, Zarate MA (2000) La transgresión Belgranense en Buenos Aires. Una revisión a cien años de su definición. Rev Cuaternario Cienc Ambientales 4:3–14
Jain M, Bøtter-Jensen L, Singhvi AK (2003) Dose evaluation using multiple-aliquot quartz OSL: test of methods and a new protocol for improved accuracy and precision. Radiat Meas 37 (1):67–80
Jouzel J, Masson-Delmotte V, Cattani O, Dreyfus G, Falourd S, Hoffmann G, Minster B, Nouet J, Barnola JM, Chappellaz J, Fischer H, Gallet JC, Johnsen S, Leuenberger M, Loulergue L, Luethi D, Oerter H, Parrenin F, Raisbeck G, Raynaud D, Schilt A, Schwander J, Selmo E, Souchez R, Spahni R, Stauffer B, Steffensen JP, Stenni B, Stocker TF, Tison JL, Werner M, Wolff EW (2007) Orbital and millennial Antarctic climate variability over the past 800,000 years. Science 317:793–796
Kitamura A, Ubukata T (2003) The sequence of local recolonization of warm-water marine molluscan species during a deglacial warming climate phase: a case study from the early Pleistocene of the Sea of Japan. Palaeogeogr Palaeoclimatol Palaeoecol 199:83–94
Kraglievich L (1928) Apuntes sobre la Geología y Paleontología de la República Oriental del Uruguay. Rev Am Arq 2:5–61
Kraglievich L (1932) Nuevos apuntes para la geología y paleontología uruguayas. Anales Mus Hist Nat Montvideo 2(3):257–320
Long J, Stoy P (2013) Quantifying the periodicity of Heinrich and Dansgaard-Oeschger events during Marine Oxygen Isotope Stage 3. Quat Res 79:413–423
Lopes RP, Simone LRL (2012) New fossil records of Pleistocene marine mollusks in southern Brazil. Rev Bras Paleontol 15(1):49–56
Lopes RP, Simone LRL, Dillenburg SR, Schultz CL, Pereira JC (2013) A Middle Pleistocene marine molluscan assemblage from the southern coastal plain of Rio Grande do Sul State, Brazil. Rev Bras Paleontol 16(3):343–360

Lorenzo N, Verde M (2004) Estructuras de bioerosión en moluscos marinos de la Formación Villa Soriano (Pleistoceno tardío—Holoceno) de Uruguay. Rev Bras Paleontol 7:319–328
Ludwig KR, Muhs DR, Simmons KR, Halley RB, Shinn EA (1996) Sea-level records at 80 ka from tectonically stable platforms: Florida and Bermuda. Geology 24(3):211–214
Mallinson D, Burdette K, Mahan S, Brook G (2008) Optically stimulated lumninescense age controls on late Pleistocene and Holocene coastal lithosomes, North Carolina, USA. Quat Res 69:97–109
Martínez S (1988) Moluscos fósiles holocenos de la margen occidental de la Laguna Merín (Formación Villa Soriano, Uruguay). Com Soc Malacol Uruguay 6:409–418
Martínez S (1990) Taphonomy and Palaeoecology of Holocene mollusks from the western margin of the Merin Lagoon (Villa Soriano Fm., Uruguay). Quat South Am Antarct Peninsula 7:121–135
Martínez S, Ubilla M (2004) El Cuaternario de Uruguay. In: Veroslavsky G, Ubilla M, Martínez S (eds) Cuencas sedimentarias de Uruguay: Geología, palaeontología y recursos naturales, Cenozoico. DIRAC, Montevideo
Martínez S, del Río CJ (2002) Late Miocene molluscs from the Southwestern Atlantic Ocean (Argentina and Uruguay): a palaeobiogeographic analysis. Palaeogeogr Palaeoclimatol Palaeoecol 188:167–187
Martínez S, del Río C, Rojas A (2013) Biogeography of the quaternary molluscs of the Southwestern Atlantic Ocean. Springer Briefs in Earth System Sciences, South America and the southern hemisphere
Martínez S, Rojas A, Ubilla M, Verde M, Perea D, Piñeiro G (2006) Molluscan assemblages from the marine Holocene of Uruguay: composition, geochronology, and paleoenvironmental signals. Amegh 43(2):385–397
Martínez S, Ubilla M, Verde M, Perea D, Rojas A, Guérèquiz R, Piñeiro G (2001) Paleoecology and geochronology of Uruguayan coastal marine Pleistocene deposits. Quat Res 55:246–254
Masello A, Menafra R (1998) Comunidades macrobentónicas de la zona costera uruguaya y áreas adyacentes. In: Wells PG, Daborn GR (Eds) El Río de la Plata—Una revisión ambiental. Un informe del Proyecto ECOPLATA. Dalhousie University, Halifax
McCulloch MT, Esat T (2000) The coral record of last interglacial sea levels and sea surface temperatures. Chem Geol 169:107–129
Medina-Elizalde M (2013) A global compilation of coral sea-level benchmarks: implications and new challenges. Earth Planet Sci Lett 362:310–318
Mianzan H, Lasta C, Acha M, Guerrero R, Macchi G, Bremen C (2001) The Río de la Plata Estuary, Argentina-Uruguay. In: Seeliger U, Kjerfve B (eds) Coastal marine ecosystems of Latin America. Springer-Verlag, Berlin
Miller GH, Mode WN, Wolfe AP, Sauer PE, Bennike O, Forman SL, Short SK, Staford TW Jr (1999) Stratifed interglacial lacustrine sediments from Baffin Island, Arctic Canada: chronology and paleoenvironmental implications. Quat Sci Rev 18:789–810
Möller OO Jr, Piola AR, Freitas AC, Campos EJD (2008) The effects of river discharge and seasonal winds on the shelf off southeastern South America. Cont Shelf Res 28:1607–1624
Montesinos M, Ramos AJG, Lomoschitz A, Coca J, Redondo A, Betancort JF, Meco J (2014) Extralimital Senegalese species during Marine Isotope Stages 5.5 and 11in the Canary Islands (29° N): Sea surface temperature estimates. Palaeogeogr Palaeoclimatol Palaeoecol 410:153–163
Muhs DR, Simmons KR, Randall Schumann R, Halley RB (2011) Sea-level history of the past two interglacial periods: new evidence from U-series dating of reef corals from south Florida. Quat Sci Rev 30:570–590
Muhs DR, Simmons KR, Steinke B (2002) Timing and warmth of the Last Interglacial period: new U-series evidence from Hawaii and Bermuda and a new fossil compilation for North America. Quat Sci Revs 21:1355–1383
Murray-Wallace CV (2002) Pleistocene coastal stratigraphy, sea-level highstands and neotectonism of the southern Australian passive continental margin—a review. J Quat Sci 17(5–6): 469–489
Murray-Wallace CV, Belperio AP (1991) The last interglacial shoreline in Australia—a review. Quat Sci Rev 10(5):441–461

Murray-Wallace CV, Beu AG, Kendrick GW, Brown LJ, Belperio AP, Sherwood JE (2000) Palaeoclimatic implications of the occurrence of the arcoid bivalve *Anadara trapezia* (Deshayes) in the Quaternary of Australasia. Quat Sci Rev 19(6):559–590

Nagy GJ, Severov DN, Pshennikov VA, De los Santos M, Lagomarsino JJ, Sans K, Morozov EG (2008) Rio de la Plata estuarine system: relationship between river flow and frontal variability. Adv Space Res 41:1876–1881

Navarro R (1990) A brief review on the knowledge on the Quaternary of the Southern and Southeastern Coast of Uruguay. Quat South Am Antarct Peninsula 7:107–117

Nielsen JK, Hanken N-M, Nielsen JK, Hansen KS (2006) Biostratigraphy and palaeoecology of the marine Pleistocene of Rhodes, Greece: Scleractinia, Serpulidae, Mollusca and Brachiopoda. Bull Geosci 81(3):173–196

North Greenland Ice Core Project Members (2004) High-resolution record of Northern Hemisphere climate extending into the last interglacial period. Nature 431:147–151

Odebrecht C, Castello JP (2001) The convergence ecosystem in the Southwest Atlantic. In: Seeliger U, Kjerfve B (eds) Coastal marine ecosystems of Latin America. Springer, Berlin

Olson DB, Podestá GP, Evans RH, Brown OB (1988) Temporal variations in the separation of Brazil and Malvinas Currents. Deep Sea Res 35(12):1971–1990

Orombelli G, Maggi V, Delmonte B (2010) Quaternary stratigraphy and ice cores. Quat Int 219:55–65

Ortlieb L, Devries T, Diaz A (1990) Ocurrencia de *Chione broggi* (Pilsbry & Olsson, 1943) (Pelecypoda) en depositos litorales cuaternarios del sur del Perú: implicaciones paleoceanográficas. Bol Soc Geol Perú 81:127–134

Petit J-R, Jouzel J, Raynaud D, Barkov NI, Barnola JM, Basile I, Bender M, Chappellaz J, Davis M, Delaygue G, Delmotte M, Kotlyakov VM, Legrand M, Lipenkov VY, Lorius C, Pepin L, Ritz C, Saltzman E, Stievenard M (1999) Climate and atmospheric history of the past 420000 years from the Vostok ice core, Antarctica. Nature 399:429–436

Pimenta AD, Absalão RS, Miyaji C (2009) A taxonomic review of the genera *Boonea*, *Chrysallida*, *Parthenina*, *Ivara*, *Fargoa*, *Mumiola*, *Odostomella* and *Trabecula* (Gastropoda, Pyramidellidae, Odostomiinae) from Brazil. Zootaxa 2049:39–66

Piola AR, Campos EJD, Möller OO Jr, Charo M, Martínez C (2000) Subtropical shelf off eastern South America. J Geophys Res 105:6565–6578

Potter E-K, Lambeck K (2003) Reconciliation of sea-level observations in the Western North Atlantic during the last glacial cycle. Earth Planet Sci Lett 217:171–181

Potter E-K, Esat TM, Schellmann G, Radtke U, Lambeck K, McCulloch MT (2004) Suborbital-period sea-level oscillations during marine isotope substages 5a and 5c. Earth Planet Sci Lett 225:191–204

Potter PE (1998) The Mesozoic and Cenozoic paleodrainage of South America: a natural history. J South Am Earth Sci 10:331–344

Powell CL II, Grant LB, Conkling SW (2004) Paleoecologic Analysis and Age of a Late Pleistocene Fossil Assemblage at a Locality in Newport Beach, Upper Newport Bay, Orange County, California. The Veliger 47(3):171–180

Preciozzi F, Spoturno J, Heinzen W, Rossi P (1988) Memoria explicativa de la Carta Geológica del Uruguay a la escala 1: 500.000. DINAMIGE, Montevideo

Prescott JR, Hutton JT (1994) Cosmic ray contributions to dose rates for luminescence and ESR dating. Radiat Meas 23:497–500

Rabassa J, Gordillo S, Ocampo C, Rivas Hurtado P (2009) The southernmost evidence for an interglacial transgression (Sangamon?) in South America. First record of upraised Pleistocene marine deposits in Isla Navarino (Beagle Channel, Southern Chile). Geol Acta 6(3):251–258

Ríos E (1994) Seashells of Brazil. Editora Fundaçao Universidade do Rio Grande, Rio Grande

Ríos E (2009) Compendium of Brazilian sea shells. Evangraf, Rio Grande

Rivadeneira MM, Carmona ER (2008) A Late Pleistocene macro-benthic assemblage in Caleta Patillos, northern Chile: paleoecological and paleobiogeographical interpretations. Rev Geol Chile 35:163–173

Rojas A (2003) Moluscos marinos cuaternarios del Balneario Zagarzazú, Departamento de Colonia, Uruguay. Pub Esp Soc Zool Uruguay 1:83
Rojas A (2007) Moluscos de aguas cálidas del Cuaternario Marino del Uruguay. Dissertation, Universidad de la República
Rojas A (2010) Primeras edades obtenidas por datación por luminiscencia (OSL) en depósitos marinos fosilíferos del Pleistoceno tardío de Uruguay. X Congr Argent Paleontol y Bioestratigr y VII Congr Lat Paleontol 1:205
Rojas A, Urteaga D (2011) Late Pleistocene and Holocene chitons (Mollusca, Polyplacophora) from Uruguay: palaeobiogeography and palaeoenvironmental reconstruction in mid latitudes of the Southwestern Atlantic. Geobios 44:377–386
Rojas A, Verde M, Urteaga D, Scarabino F, Martínez S (2014) The first predatory drillhole on a fossil chiton plate: an occasional prey item or an erroneous attack? Palaios 29:414–419
Roselli FL (1939) Apuntes de Geología y Paleontología uruguayas. Bol Soc Am Cs Nat "Kraglievich-Fontana" 1(2):29–102
Roselli FL (1976) Contribución al estudio de la Geo Paleontología Departamentos de Colonia y Soriano (República Oriental del Uruguay). Imprenta Cooperativa, Montevideo
Rosenberg G (2009) Malacolog 4.1.1: a database of Western Atlantic Marine Mollusca. [WWW database (version 4.1.1)]. http://www.malacolog.org/. Accessed 26 Nov 2014
Roy K, Jablonski D, Valentine JW (1995) Thermally anomalous assemblages revisited: patterns in the extraprovincial latitudinal range shifts of Pleistocene marine mollusks. Geology 23 (12):1071–1074
Scarabino F, Zaffaroni JC (2004) Estatus faunístico de veinte especies de moluscos citadas para aguas uruguayas. Com Zool Mus Nac Hist Nat Antr 13(202):1–15
Scarabino F, Zaffaroni JC, Carranza A, Clavijo C, Nin M (2006a) Gasterópodos marinos y estuarinos de la costa uruguaya: faunística, distribución, taxonomía y conservación. In: Menafra R, Rodríguez L, Scarabino F, Conde D (eds) Bases para la conservación y el manejo de la costa uruguaya. Vida Silvestre, Montevideo
Scarabino F, Zaffaroni JC, Clavijo C, Carranza A, Nin M (2006b) Bivalvos marinos y estuarinos de la costa uruguaya: faunística, distribución, taxonomía y conservación. In: Menafra R, Rodríguez L, Scarabino F, Conde D (eds) Bases para la conservación y el manejo de la costa uruguaya. Vida Silvestre, Montevideo
Scarabino V (1977) Moluscos del Golfo San Matías (Provincia de Río Negro, República Argentina). Inventario y claves para su identificación. Com Soc Malacol Uruguay 4:177–285
Schellmann G, Radtke U (2004) A revised morpho- and chronostratigraphy of the Late and Middle Pleistocene coral reef terraces on Southern Barbados (West Indies). Earth Sci Rev 64:157–187
Serra N (1943) Memoria explicativa del mapa geológico del Departamento de Colonia. Bol Inst Geol Uruguay 30
Shackleton NJ (1969) The Last Interglacial in the Marine and Terrestrial Records. Proc R Soc Lond B Biol Sci 174(1034):135–154
Shackleton NJ (2000) The 100,000 year Ice Age cycle identified and found to lag temperature, carbon dioxide, and orbital eccentricity. Science 289:1897–1902
Shackleton NJ, Sánchez Goñi MF, Pailler D, Lancelot Y (2003) Marine isotope substage 5e and the Eemian Interglacial. Global Planet Change 36:151–155
Sicardi OE (1967) La influencia de las Corrientes marinas sobre la malacofauna uruguaya. Com Soc Malacol Uruguay 2(12):49–60
Siddall M, Rohling EJ, Almogi-Labin A, Hemleben Ch, Meischner D, Schmelzer I, Smeed DA (2003) Sea-level fluctuations during the last glacial cycle. Nature 423:853–858
Siddall M, Rohling EJ, Thompson WG, Waelbroeck C (2008) Marine isotope stage 3 sea level fluctuations: data synthesis and new outlook. Rev Geophys 46:RG4003
Siddall M, Chappell J, Potter E-K (2007) Eustatic sea level during past interglacials. In: Sirocko F, Claussen M, Sanchez-Goni MF, Litt T (Eds) The climate of past interglacials. Developments in Quaternary Science Series No. 7. Elsevier, Amsterdam

Simms AR, DeWitt R, Rodríguez AB, Lambeck K, Anderson JB (2009) Revisiting marine isotope stage 3 and 5a (MIS3-5a) sea levels within the northwestern Gulf of Mexico. Global Planet Change 66:100–111

Sprechmann P (1978) The paleoecology and paleogeography of the Uruguayan coastal area during the Neogene and Quaternary. Zitt 4:3–72

Stirling CH, Esat TM, Lambeck K, McCulloch MT (1998) Timing and duration of the Last Interglacial: evidence for a restricted interval of widespread coral reef growth. Earth Planet Sci Lett 160:745–762

Suguio K, Bezerra FHR, Barreto AMF (2011) Luminescence dated Late Pleistocene wave-built terraces in northeastern Brazil. Acad Bras Cienc 83(3):907–920

Teisseire A (1927) Expedición a los departamentos de Colonia y Soriano. Rev Amigos Arqueol 1:47–61

Teisseire A (1928) Contribución al estudio de la geología y la paleontología de la República Oriental del Uruguay. Región de Colonia. Anales Univ 122(series 37):319–469

Tonni EP, Carbonari JE, Huarte RA (2010) Marine Sediments Attributed to Marine Isotope Stage 3 in the Southeastern Buenos Aires Province, Argentina. Curr Res Pleist 27:154–156

Turney CSM, Jones RT (2010) Does the Agulhas Current amplify global temperatures during super-interglacials? J Quat Sci 25(6):839–843

Urien CM (1972) Río de la Plate estuary environments. In: Nelson BW (ed) Environmental framework of coastal plain estuaries, Boulder; Geol Soc Am Mem 133:213–234

Valentine JW, Jablonski D (1993) Fossil communities: compositional variation at many time scales. In: Ricklefs RE, Schluter D (eds) Species diversity in ecological communities: Historical and geographical perspectives. University of Chicago Press, Chicago

Van Meerbeeck CJ, Renssen H, Roche DM (2009) How did Marine Isotope Stage 3 and Last Glacial Maximum climates differ? Perspectives from equilibrium simulations. Clim Past 5:33–51

Van Meerbeeck CJ, Renssen H, Roche DM, Wohlfarth B, Bohncke SJP, Bos JAA, Engels S, Helmens KF, Sánchez-Goñi MF, Svensson A, Vandenberghe J (2011) The nature of MIS 3 stadial e interstadial transitions in Europe: new insights from model data comparisons. Quat Sci Rev 30:3618–3637

Verde M (2003) Marine invertebrate ichnofossils from the Quaternary of Uruguay. Pub Esp Soc Zool Uruguay 1:94

Veroslavsky G, Ubilla M (2007) A "snapshot" of the evolution of the Uruguay River (Del Plata Basin): the Salto depositional sequence (Pleistocene, Uruguay, South America). Quat Sci Rev 26:2913–2923

Waelbroeck C, Labeyrie L, Michel E, Duplessy JC, McManus JF, Lambeck K, Balbon E, Labracherie M (2002) Sea-level and deep water temperature changes derived from benthic foraminifera isotopic records. Quat Sci Rev 21:295–305

Wehmiller JF, Simmons KR, Cheng H, Edwards L, Martin-McNaughton J, York LL, Krantz DE, Shen C (2004) Uranium-series coral ages from the US Atlantic Coastal Plain—The "80 ka problem" revisited. Quat Int 120:3–14

Winograd IJ, Landwehr JM, Ludwig KR, Coplen TB, Riggs AC (1997) Duration and structure of the past four interglaciations. Quat Res 48:141–154

Yokoyama Y, Esat TM, Lambeck K (2001) Coupled climate and sea-level changes deduced from Huon Peninsula coral terraces of the last ice age. Earth Planet Sci Lett 193:579–587

Zazo C, Goy JL, Dabrio CJ, Bardají T, Hillaire-Marcel C, Ghaleb B, González-Delgado JA, Soler V (2003) Pleistocene raised marine terraces of the Spanish Mediterranean and Atlantic coasts: records of coastal, sea-level highstands and climate changes. Mar Geol 194(1–2):103–133

Zazo C, Goy JL, Dabrio CJ, Lario J, González-Delgado JA, Bardají T, Hillaire-Marcel C, Cabero A, Ghaleb B, Borja F, Silva PG, Roquero E, Soler V (2013) Retracing the Quaternary history of sea-level changes in the Spanish Mediterranean-Atlantic coasts: geomorphological and sedimentological approach. Geomorphology 196:36–49

Zubakov V, Borzenkova I (1990) Global paleoclimate of the Late Cenozoic. Elsevier, New York

Vegetation and Climate in Southern South America during Marine Isotope Stage 3 (MIS 3): an Overview of Existing Terrestrial Pollen Records

Ana María Borromei and Lorena Laura Musotto

Abstract Data from terrestrial pollen records in the Chilean sector show that Marine Oxygen Isotope Stage 3 (MIS 3) was characterized by the Heinrich (stadials)/Dansgaard–Oeschger (interstadials) oscillations. At central Chile (32–38° S) an open beech/podocarp woodland was apparently established during the last ice age under cold and humid climates, where nowadays grows a semiarid, broad sclerophyllous vegetation, while episodes of aridity with rise of temperature were indicated by expansion of chenopods–amaranths. At the Southern Lake District and northern Isla Grande de Chiloé (40°–42° 30′ S), the stadial events were characterized by higher amounts of grasses indicative of the Subantarctic Parkland vegetation. This vegetation implied summer temperatures of ~6 °C (~8 °C below present). The interstadials were represented by expansion of the Valdivian-North Patagonian Evergreen Forest-Subantarctic Evergreen Forest implying summer temperatures of ~12 °C. In the Argentine sector, the steppe environmental conditions prevailed. Interstadial conditions are pointed out at 39° S, in NW Patagonia. Meanwhile, in southern Patagonia at 51°–52° S, and Tierra del Fuego at 54° S, the climatic conditions during MIS 3 are interpreted as colder and drier than today.

Keywords MIS 3 terrestrial pollen records · Palaeoenvironments · Palaeoclimates · Southern South America

1 Introduction

General knowledge of the Marine Oxygen Isotope Stage 3 (MIS 3, 59–29 cal ka B.P.) comes mainly from climate records distributed in the northern hemisphere, whereas there is limited palaeoclimatic information covering this period from the southern

A.M. Borromei (✉) · L.L. Musotto
Departamento de Geología, Instituto Geológico del Sur (INGEOSUR—CONICET), Universidad Nacional del Sur, San Juan 670, B8000ICN Bahía Blanca, Argentina
e-mail: borromei@criba.edu.ar

L.L. Musotto
e-mail: loremusotto@criba.edu.ar

G.M. Gasparini et al. (eds.), *Marine Isotope Stage 3 in Southern South America, 60 ka B.P.–30 ka B.P.*, Springer Earth System Sciences,
DOI 10.1007/978-3-319-40000-6_15

hemisphere (Voelker and participants 2002). According to Kaplan et al. (2008), this is due not only to the poor or lack of preservation of relevant deposits, but also to the ca. 35 ^{14}C ka limit of radiocarbon dating. Of the Late Pleistocene palynological terrestrial sites at the southern extremity of the South American continent, few of them extend earlier than the Last Glacial Maximum (LGM) and those that predate the LGM are, for the most part, discontinuous. The terrestrial pollen records during MIS 3 are mainly situated in the Chilean sector: central Chile (32–38° S) and Southern Lake District-northern Isla Grande de Chiloé (40°–42° 30′ S). On the Argentine sector, the pollen data come from a few sequences located at the Province of Neuquén (39° S), southeastern Province of Santa Cruz (51–52° S) and Isla Grande de Tierra del Fuego (54° S). High-resolution pollen data are also available from a marine sediment core (ODP site 1233) located at 41° S west of the Chilean coast (Heusser et al. 2006).

The aim of this chapter is to present an overview of the published pollen datasets from southern South America covering the MIS 3 (Table 1). Most of the sections showed in the table, cover different intervals of the MIS 3. Only three localities (the Laguna de Tagua Tagua, Fundo Nueva Braunau and Taiquemó sites) from Chile, and one locality (Laguna Potrok Aike site) from Argentina, comprise an uninterrupted succession of this time span.

2 Modern Climate and Vegetation

2.1 Chilean Sector

2.1.1 Central Chile (32–38° S)

The Mediterranean climate region of central Chile is today hotter and drier in summer and cooler and relatively wetter in winter as a consequence of the seasonal, latitudinal change in the position of the Polar Front. Located at the boundary of the polar and subtropical maritime air masses, the front in winter (July) is positioned at relatively lower latitudes than in summer (January). As a result, cyclonic storms from the south, part of the westerly wind regime, penetrate the region during the colder months of the year (June–August). Climatic data from stations at lower altitudes indicate mean temperatures of approximately 18–21 °C in summer and around 8–12 °C in winter; total annual precipitation is between 100 and 800 mm with 84–90 % falling during autumn and winter (Miller 1976). Conditions, however, vary inland from the ocean. Cooler air, moving inland from the Pacific Ocean, orographically loses its moisture when crossing the Cordillera de la Costa, leaving the interior valleys as relatively drier areas.

Adapted to a regional, semiarid climate with summer drought, broad sclerophyllous woodland or *matorral* vegetation (Heusser 1990, 1994) covers much of the Cordillera de la Costa, Valle Central, and lower slopes of the Andes (Fig. 1). The woodland covers mountain slopes to a maximum altitude of 1600 m. The latitudinal and altitudinal relationships are with semiarid thorn shrub-succulent

Table 1 Marine oxygen isotope stage 3 (MIS 3) terrestrial pollen records from Chilean and Argentine sector

Site no.	Site name	Latitude (S)	Longitude (W)	References
Chilean sector				
1	Laguna de Tagua Tagua	34° 30′	71° 10′	Heusser (1983, 1990, 1994, 2003), Valero-Garcés et al. (2005)
2	Rupanco	–	–	Heusser (1981)
3	Puerto Octay road cut	40° 58′	72° 53′	Heusser (1981), Heusser et al. (1999)
4	Puerto Varas longitudinal	–	–	Heusser (1981)
5	Punta Penas	–	–	Heusser (1981)
6	Punta Pelluco	–	–	Heusser (1981)
7	Frutillar Bajo	41° 08′	73° 01′	Heusser et al. (1999)
8	Fundo Nueva Braunau	41° 17′	73° 04′	Heusser et al. (2000), Heusser (2003)
9	Canal Tenglo	41° 28′	72° 58′	Heusser et al. (1999)
10	Río Negro	42° 03′	73° 50′	Villagrán (1990)
11	Taiquemó	42° 17′	73° 60′	Heusser et al. (1999), Heusser (2003), Heusser and Heusser (2006)
12	Teguaco	42° 17′	73° 35′	Heusser et al. (1999)
13	Delcahue	42° 20′	73° 39′	Heusser et al. (1999)
14	Pidpid	42° 24′	73° 46′	Villagrán (1985, 1990)
Argentine sector				
15	Bajada de Rahue	39° 22′	70° 56′	Markgraf et al. (1986)
16	Laguna Potrok Aike	51° 58′	70° 23′	Recasens et al. (2012)
17	Magallanes maar	52° 07′	69° 16′	Corbella et al. (2000)
18	Lago Fagnano	54° 33′	67° 19′	Bujalesky et al. (1997)
19	Lago Fagnano 1	54° 55′	67° 17′	Ponce et al. (2014)

vegetation, established to the north, and lowland deciduous beech forest, lying mostly to the south. Woodland typically contains *Cryptocarya alba* (Lauraceae), the arboreal species *Schinus latifolius* (Anacardiaceae), and *Peumus boldus* (Monimiaceae). Along the cloud-covered coast, *Beilschmiedia miersii* (Lauraceae), and in drier sectors, *Lithraea caustica* (Anacardiaceae), *Quillaja saponaria,* and *Kageneckia oblonga* (Rosaceae) are important constituents. A large sector of the drier interior between the Cordillera de la Costa and the Andes ranges at the border of broad sclerophyllous woodland is occupied by steppe-scrub or *espinal*. The shrubs *Trevoa trinervis* (Rhamnaceae), *Colliguaya odorifera* (Euphorbiaceae) and *Cestrum parqui* (Solanaceae) are featured here, mixed with grasses and composites.

Stands of southern beech (*Nothofagus*) of limited size occasionally occur in the mountains in different parts of the broad sclerophyllous formation. The *Lowland Deciduous Beech Forest* develops below 1650 m a.s.l. It includes the thermophilic and drought resistant species such as *N. obliqua*, *N. alpina*, and *N. glauca* (Heusser 1981). The high Andean deciduous beech forest comprises cold resistant species such as *N. antarctica* and *N. pumilio* at 1600–1800 m a.s.l. The latter exhibits higher moisture requirements than *N. antarctica*. The more oceanic evergreen beech (*N. dombeyi*) forest in the Andes reaches the northern end of its range close to 34° 30′ S at 600–1000 m a.s.l. Podocarp, notably the Andean *Prumnopitys andina*, does not range within the limits of the formation and grows at mountain altitudes (1000–1100 m a.s.l.), principally in the zone of lowland deciduous forest.

2.1.2 Southern Lake District–Northern Isla Grande de Chiloé (40°–42° 30′ S)

The region lies across the seasonal swing of the Polar Front where air masses of the subpolar westerlies and the subtropical Pacific Ocean anticyclone are in contact. Consequently, cyclonic storms associated with frontal systems of the westerlies and accompanied by strong wind and heavy precipitation, are frequent (Miller 1976). Oceanic (with no dry season) climate in this region, is dictated by cyclonic conditions in the belt of westerlies coupled with topography and cold offshore Humboldt Current. Average annual precipitation is commonly 1500–2000 mm, increasing to 4000 mm in the coastal cordillera and over 5000 mm in the Andes mountains. Temperature in January (summer) averages 14–16 °C and in July (winter) 7–8 °C (Heusser 1981).

Four distinctive forest zones are recognized between sea level and the alpine tundra (Schmithüsen 1956) (Fig. 1). The *Valdivian Evergreen Forest* reaches ca. 400 m and descends as a zone southward to the shore of the Lago Llanquihue and to sea level along the Seno Reloncaví and the Golfo de Ancud. Communities at <200 m typically consist of *Aetoxicon punctatum* growing in association with myrtaceous *Myrceugenia planipes*, *M. ovata* and *Luma apiculata*. At higher elevations (200–400 m) evergreen beech, *Nothofagus dombeyi*, and *Eucryphia cordifolia* are main components of the forest, as well as the species *Laurelia philippiana*, *Caldcluvia paniculata,* and *Lomatia ferruginea*. Valdivian forest contains a diverse assemblage of lianas and ferns. Openings in the forest canopy are frequented commonly by bamboo (*Chusquea quila*) (Heusser et al. 1999).

The *North Patagonian Evergreen Forest* (400–1100 m altitude) lying altitudinally higher than the Valdivian forest in the coastal and Andean cordilleras, constitutes a distinctive zone dominated for the most part by broad-leaved species. At lower altitudes (400–800 m) mountain communities are dominated by *Nothofagus dombeyi* in the Andes and *N. nitida* in the Cordillera de la Costa associated with *Laurelia philippiana* and *Weinmannia trichosperma.* A coniferous element, represented by *Podocarpus nubigena*, *Saxegothaea conspicua*, *Fitzroya cupressoides* and *Pilgerodendron uviferum*, is conspicuous at >600 m. At higher elevations in the Andes (800–1100 m) *Nothofagus dombeyi* forms pure stands with an

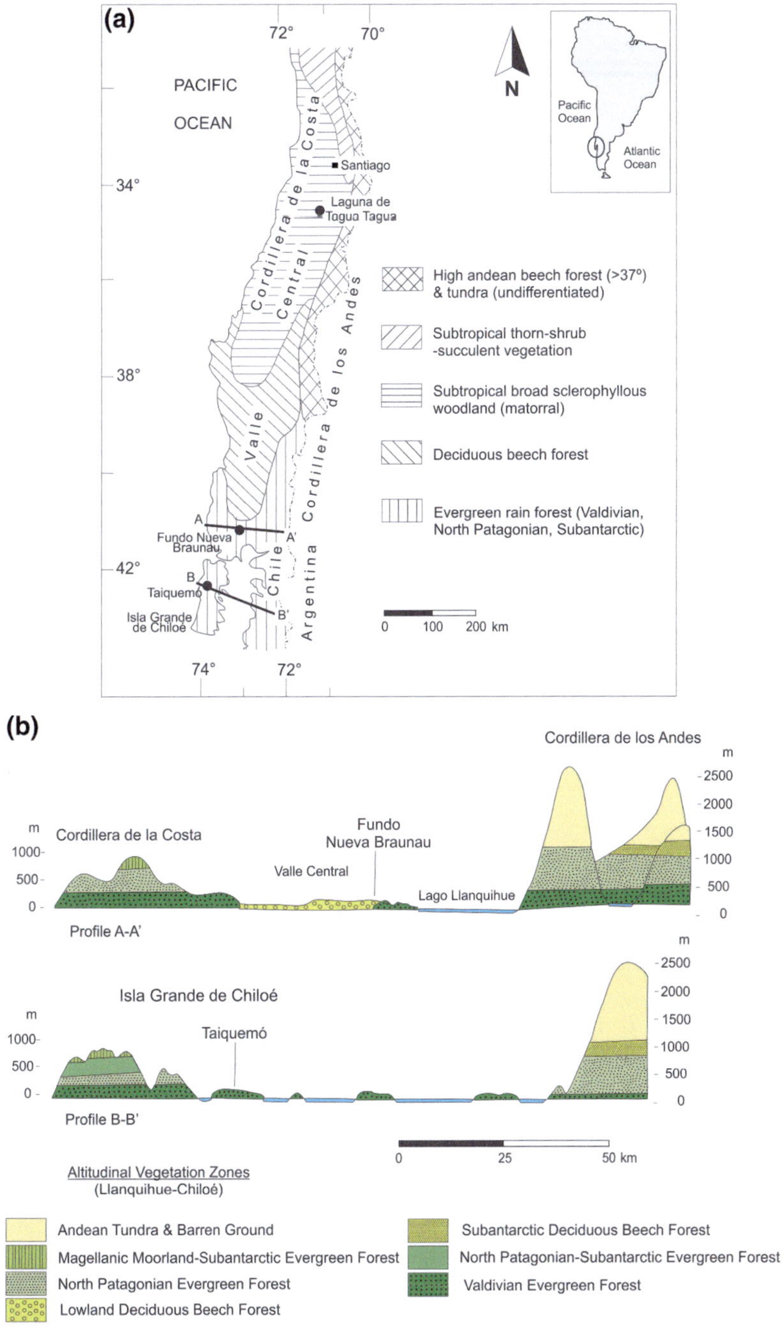

Fig. 1 **a** Map of plant formation of subtropical Chile showing the location of fossil pollen sites (modified from Heusser 1990). **b** Altitudinal distribution of vegetation according to cross-sectional profiles: A–A′ in the Southern Lake District, and B–B′ on Isla Grande de Chiloé (modified from Heusser et al. 1999)

understorey of *Desfontainia spinosa*, *Maytenus magellanica* and *Drimys winteri* var. *andina*. Valdivian structural features and a wealth of lianas, epiphytes and bamboo also characterize the North Patagonian Forest (Heusser et al. 1999).

The *Subantarctic Evergreen Forest* and open *Magellanic Moorland* develop along the crest of the Cordillera de la Costa and on slopes above the North Patagonian Evergreen Forest farther south in the Andes (Heusser et al. 1999). The beech of *Nothofagus betuloides* is found growing with *N. antarctica*, *Drimys winteri* var. *chilensis*, *Lomatia ferruginea*, *Tepualia stipularis* and gymnosperms such as *Podocarpus nubigena*, *Saxegothaea conspicua* and *Fitzroya cupressoides*. Moorland species constitute a mosaic of cushion bogs, mires, and scrub among patches of forest. It includes *Astelia pumila*, *Donatia fascicularis*, *Gaimardia australis*, *Drosera uniflora*, *Oreobolus obtusangulus*, *Gaultheria antarctica* and *Lepidothamus fonkii* (Heusser et al. 2000).

The *Subantarctic Deciduous Beech Forest* (900–1350 m) occupies the subalpine zone to timberline. The beech is characterized by two species, *Nothofagus pumilio* and *N. antarctica*, often with *Chusquea tenuiflora*, making up most high-elevation stands. Shrub cover includes *Drimys winteri* var. *andina*, *Berberis montana*, *Maytenus disticha*, *Escallonia alpina* and *Ribes cucullatum*; representative herbs are *Rubus geoides*, *Gunnera magellanica* and *Valeriana lapathifolia* (Heusser et al. 1999).

2.2 Argentine Sector

Climate over the area south of 40° S in the Patagonian region, is under the influence of prevailing westerlies all year associated with a strong meridional pressure gradient. The region is placed between the subtropical high pressure belt and the subpolar low pressure zone. The Andean Cordillera intersects the westerlies in a perpendicular position, creating a marked climatic contrast between the Pacific and the Atlantic slopes (Coronato et al. 2008). Rainfall drastically changes at both sides of the Andes, with a relationship varying from 5:1 to 10:1, but its seasonality and the patterns of cloudiness and temperature do not behave in such a contrasting way. After crossing the Andean Cordillera, the westerlies create rain shadow conditions in eastern Patagonia and they limit the Atlantic Ocean influence, but expand the Pacific Ocean impact across the region at the same time (Coronato et al. 2008). A sharp change in precipitation regimes is shown between the Andean region and the Patagonian Plateau, where stations east of the 71° W longitude, register less than 350 mm annually. West of the 71° W longitude, isohyets show an abrupt rise in precipitation, with values up to 1500 mm near the Chilean-Argentine border (Prohaska 1976). Almost all of extra-Andean Patagonia gets less than 250 mm per year. South of 52° S, the Andean ranges have a lesser height and loose continuity, shifting to a W–E orientation. Eastern Patagonia has a drier climate with moderate thermal amplitude. The opposite side, western Patagonia, has a markedly oceanic climate, cooler than its counterpart in the continent, particularly due to the absence of summer heat. The extension of Patagonia over more than 2200 km in N–S direction implies significant differences in the incoming solar radiation. The temperature pattern follows an opposite pattern due to the continental narrowing. The mean annual thermal amplitude varies from 16 °C in the north to 8 °C in the south, or even down to 4 °C in the outermost Magellanic Islands. The extreme temperatures follow the same pattern, with maxima of 38 °C recorded at 46° S in eastern Patagonia, but in Tierra del Fuego they do not reach 30 °C, not even 20 °C in the hyperoceanic islands. Minimum temperatures of −30 °C are recorded in the central tablelands at 41° S (Coronato et al. 2008). Along the Canal Beagle, in the Fuegian Archipelago, rainfall decreases eastward, due to the influence of the W-SW winds. South of the Beagle Channel, the lack of mountain obstacles determines that these winds generate an increase in rainfall toward the eastern portion of Tierra del Fuego (Coronato et al. 2008). In the island, mean summer isotherms increase northeastward from 9 to 12 °C and precipitation decreases to the north and east. Mean annual rainfall in Ushuaia is 570 mm, but less than 300 mm in Río Grande in the northern part of the island (Tuhkanen 1992).

Distribution and composition of vegetation in Patagonia (Fig. 2) are subject to strong latitudinal and altitudinal climatic gradients. Controlling parameters are temperature, net precipitation and wind, and to a lesser degree topography and soils (Heusser 2003). South of ~40° S, east of the Andean ranges, the Subantarctic, Patagonian, and Monte phytogeographical provinces are successively juxtaposed (Cabrera 1971).

Subantarctic Province vegetation, forested in part in proximity to the Andes, extends from ~39° S to the Fuegian archipelago, covering much of the southern half of Tierra del Fuego. It comprises four districts: the *Valdivian District* (40°–42° S) to the east slope of the Andes subject to cooler, cloudy climate with heavy precipitation annually exceeding 4000 mm. It is hardly different from the Chilean Valdivian forest. Among arboreal members are *Nothofagus dombeyi*, *Eucryphia cordifolia* and *Aetoxicon punctatum* growing up to 1100–1200 m in altitude. These species and a typical gymnospermous element consisting of *Podocarpus nubigena*, *Saxegothaea conspicua*, *Pilgerodendron uviferum,* and *Fitzroya cupressoides* illustrate the extent of similarity on both slopes of the cordillera. The *Magellanic District*, in parts of the cordillera at around 50° S and south in Tierra del Fuego, is dominated by *Nothofagus betuloides*, *Drimys winteri*, *Maytenus magellanica*, *Tepualia stipularis*, *Pseudopanax laetevirens,* and *Pilgerodendron uviferum*. Its counterpart in Chile is best represented by the Subantarctic Evergreen Forest, in part by North Patagonian Evergreen Forest, and where mires and low scrub prevail, by Magellanic Moorland (Heusser 2003). The *Pehuen District*, occupying the west-central part of the Province of Neuquén (39° S), is dominated by *Araucaria araucana.* It is an endemic species that forms pure stands or commingles with *Nothofagus pumilio* and *N. antarctica* at altitudes above 900 m. The *Deciduous Forest District* occupies the eastern border of the Subantarctic Province. It includes the deciduous beech species of *Nothofagus obliqua*, *N. alpina*, *N. pumilio* and *N. antarctica*. The forest comprises the Lowland Deciduous Beech and Subantarctic Deciduous Beech Forest formations recognized in Chile. In the northern part of the district in proximity to the Patagonian Province are dense stands of *Austrocedrus chilensis*.

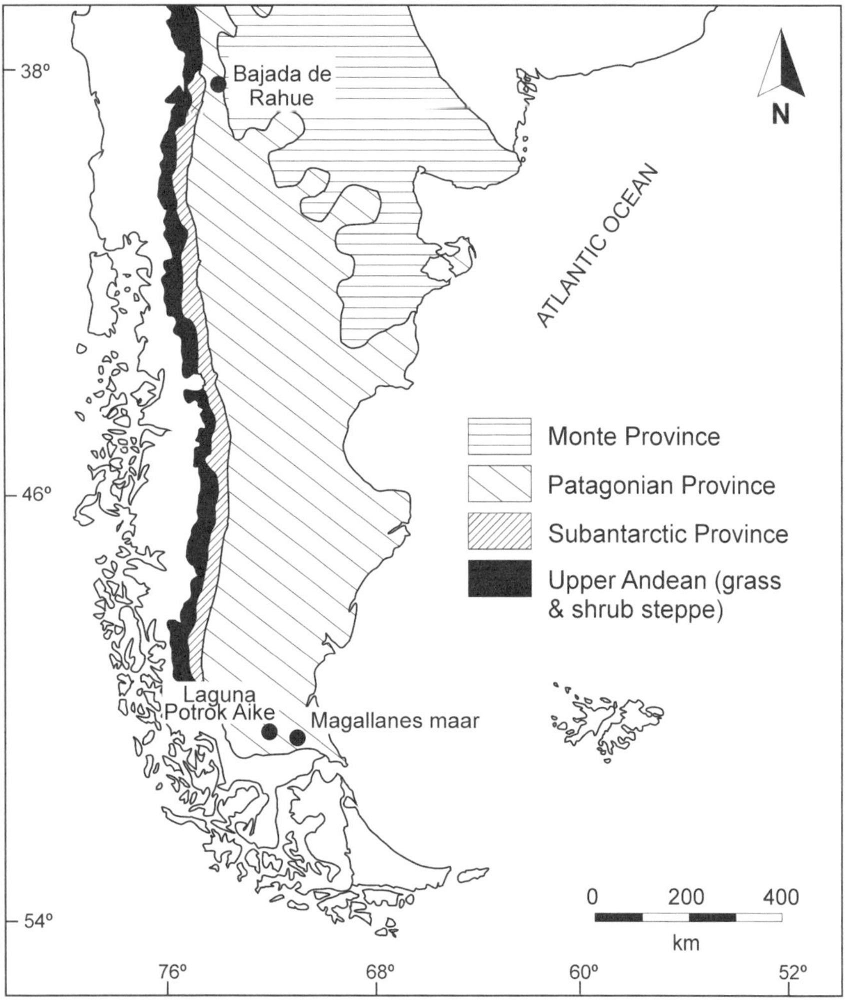

Fig. 2 Map of major Patagonian plant formations (modified from Cabrera 1971) showing the location of fossil pollen sites

The *Patagonian Province* vegetation extends to the Atlantic Ocean coast. Southward in Patagonia, the province narrows as it crosses northern Tierra del Fuego. This province includes the *Subandean District*. Fronting the slopes of the Andes and following the length of the cordillera south of 44° S, a mosaic of grasses dominates the landscape (*Festuca monticola*, *Agrostis pyrogea*, *Deschampsia elongata* and *Poa ligularis*) along with herbs and shrubs (*Mulinum spinosum*, *Nassauvia aculeata,* and *Berberis cuneata*). Over the plateau, the Patagonian steppe comprises the semidesert of *Nassauvia* shrubs, *Chuquiraga aurea,* and *Ephedra frustillata* with dominant *Stipa humilis* among grasses. Communities in the far south (up to 51° S) form the grass steppe (*Festuca gracillima*) owing to an increase in humidity. In the *Western District*, outside the eastern limit of the Subandean District, from 34° to 47° S, vegetation is essentially steppe in which communities are dominated by tussock grasses or "coirones" (*Stipa patagonica*, *S. humilis*, *S. chrysophylla*, *Festuca monticola,* and *F. argentina*) in association with low shrubs (*Mulinum spinosum*, *Adesmia trijunga*, *Senecio filaginoides*, *Lycium tenuispinosum*, *Verbena ligustrina*, *Nassauvia axillaris*, *Berberis cuneata,* and *Ephedra frustillata*).

The *Monte Province* vegetation, occurring 100 km and more distant from the Subantarctic Province, is characterized by a shrub steppe of *Larrea* (*L. divaricata*, *L. cuneifolia*, *L. nitida*) and *Prosopis* (*P. alpataco*, *P. strombulifera*, *P. globosa*) species associated with *Bougainvillea spinosa*, *Cassia aphylla*, *Monttea aphylla* and *Condalia microphylla* among shrubs.

On the northern Tierra del Fuego (Fig. 3), *Steppe* of grassland, scrub and heath occupies the driest areas of the island where mean annual precipitations are less

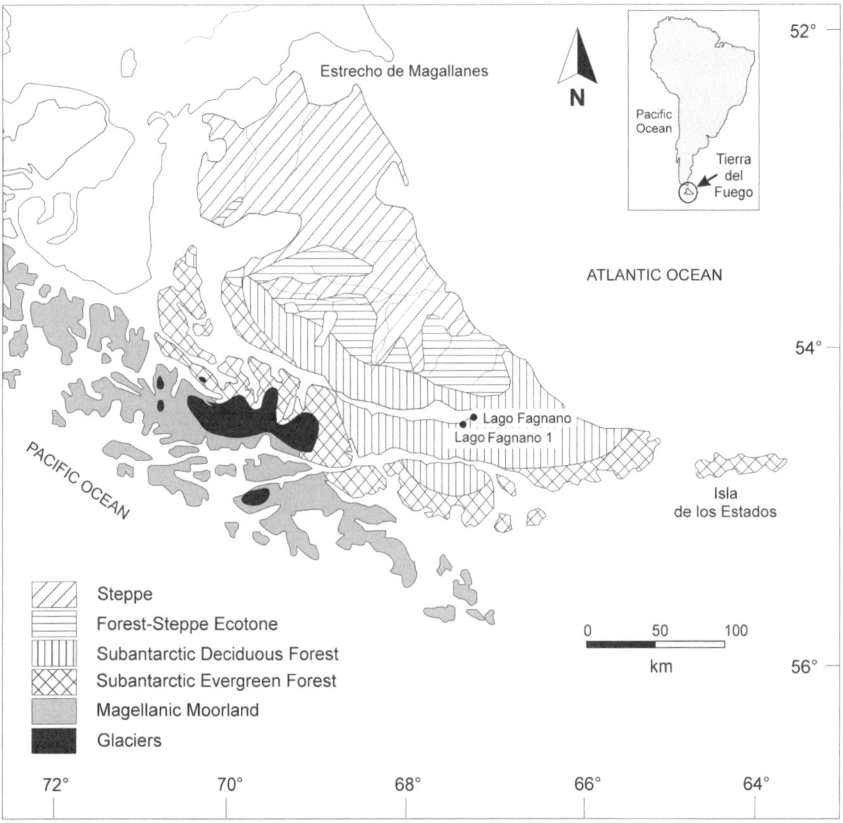

Fig. 3 Map of Isla Grande de Tierra del Fuego vegetation units (modified from Tuhkanen 1992) showing the location of fossil pollen sites

than 400 mm. Grasses include *Festuca gracillima* as the dominant species. Scrub is dominated by shrubby composites, *Lepidophyllum cupressiforme*, in the north, and southward *Chiliotrichum diffusum*. Dwarf shrub heath is typified by *Empetrum rubrum* (Heusser 1989). Contact of steppe with deciduous beech forest occurs through an ecotone under precipitation of 400–500 mm annually. Vegetation at lower altitudes is dominated by the *Subantarctic Deciduous Forest* (Pisano 1977; Moore 1983). It is characterized by two species, *Nothofagus pumilio* and *N. antarctica*, the latter is present at all elevations on poorer and disturbed soils. Both deciduous *Nothofagus* taxa grow from sea level to treeline (550–600 m a.s.l.) and become dominant where precipitation is between 400 and 800 mm/year. With increased precipitation, the *Subantarctic Evergreen Forest* develops and *N. betuloides* assumes dominance westwards and southwards into areas of up to at least 4000 mm annual precipitation. Communities occur either in pure stands or in association with *Drimys winteri*, *Maytenus magellanica* and abundant ferns and mosses (Moore 1983). *Magellanic Moorland* occurs beyond the forest along the exposed outermost coast under conditions of increased precipitation (5000 mm or more), high winds and poor drainage. Mostly treeless and tundra-like, the moorland is distinguished by a profusion of cushion bogs with distinctive species (*Astelia*, *Donatia*, *Gaimardia*, *Phyllachne*). Between treeline and snowline, *Andean Tundra* develops. It is comprised of cushion heath (*Bolax gummifera*), dwarf shrub heath (*Empetrum rubrum*) and meadow communities (Pisano 1977; Heusser 2003).

3 Vegetation Reconstruction and Palaeoclimate

The MIS 3 is characterized by recurring millennial-scale climate oscillations such as the Heinrich (H) events (stadial phases) and the Dansgaard–Oeschger (D-O) events (interstadial phases) that are well-documented in glacial ice, marine and terrestrial cores (Voelker and participants 2002), though the periodic behavior of the well-known millennial-scale variations (H and D-O events) is uncertain (Long and Stoy 2013).

Multiple published pollen records of corresponding MIS 3 interval from Chile have served to reconstruct the overall composition and character of the Late Pleistocene vegetation (Table 1). Particularly, in central Chile the pollen record of lacustrine deposits of **Laguna de Tagua Tagua** (site 1: 200 m a.s.l.; Figs. 1 and 4) is one of the most remarkable ice age sites in southern South America (Heusser 1983, 1990, 1994; Valero-Garcés et al. 2005). It is constrained chronologically by fourteen ^{14}C dates. In spite of the uncertainty of the chronology covered by the infinite dates, the base of the core is estimated to date the MIS 3/4 boundary. During >45,000 years of the last ice age, the vegetation changed dramatically. Thus, according to Heusser (1990), the pollen assemblage shows, throughout the MIS 3, the dominance of trees of *Nothofagus dombeyi* type and *Prumnopitys andina* mixed with considerable amounts of grasses (Poaceae) and shrubs (Asteraceae) spreading on non-glaciated low-lying terrains now occupied by broad

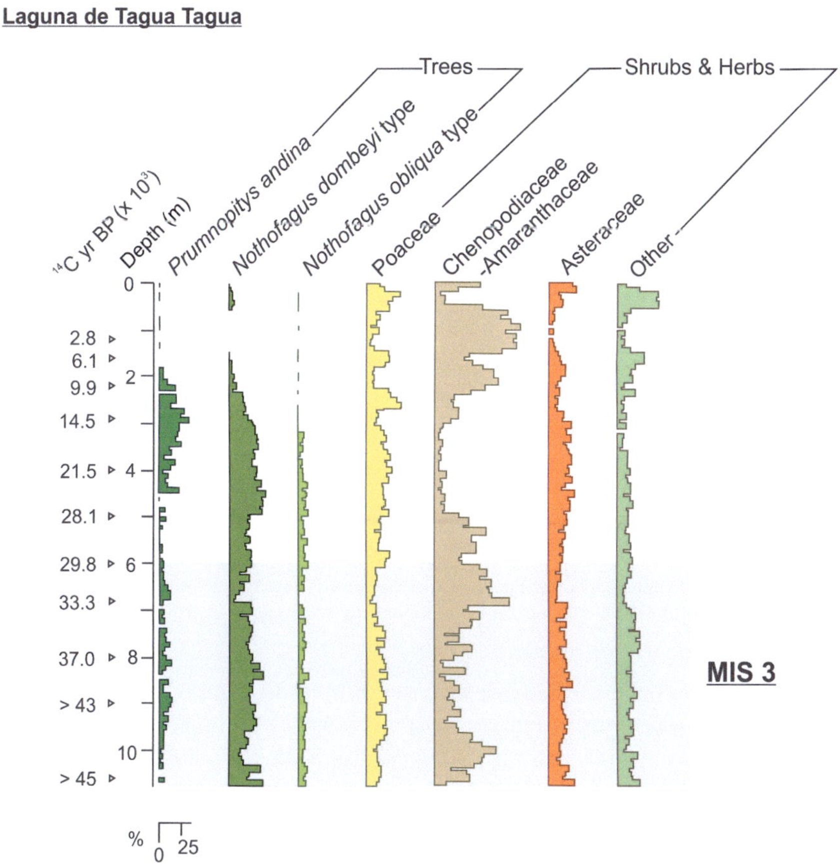

Fig. 4 Pollen diagram (%) of leading trees, shrubs and herbs, with radiocarbon age control, for the section at Laguna de Tagua Tagua (modified from Heusser 1994)

sclerophyllous plant communities. The Chenopodiaceae–Amaranthaceae pollen as xeric indicators, serve as an expression of dryness during times when these plants invaded exposed lake bottom. The episodes of open southern beech–podocarp woodland abundance at 50,000–35,500 ^{14}C year B.P. (MIS 3), and at 28,500–14,500 ^{14}C year B.P. (MIS 3/2) indicate cooler and more humid climate than present. Species of beech are found today in remote mountain areas some 100 km distant or at altitudes of 800–1000 m, and stands of podocarp most proximal are 150 km away at altitudes of more than 1000 m (Heusser 1994). Pollen of chenopods and amaranths, indicative of warmer and drier conditions, reach maximums at >43,000 ^{14}C year B.P., and between *ca.* 37,000 and *ca.* 28,500 ^{14}C year B.P., with a maximum at 33,300 ^{14}C year B.P. when lake level was low. The Chenopodiaceae, such as *Atriplex chilense*, *A. ripandum* and *A. philippi*, are found today in saline basins of the central and northern Chilean provinces (Heusser 1983).

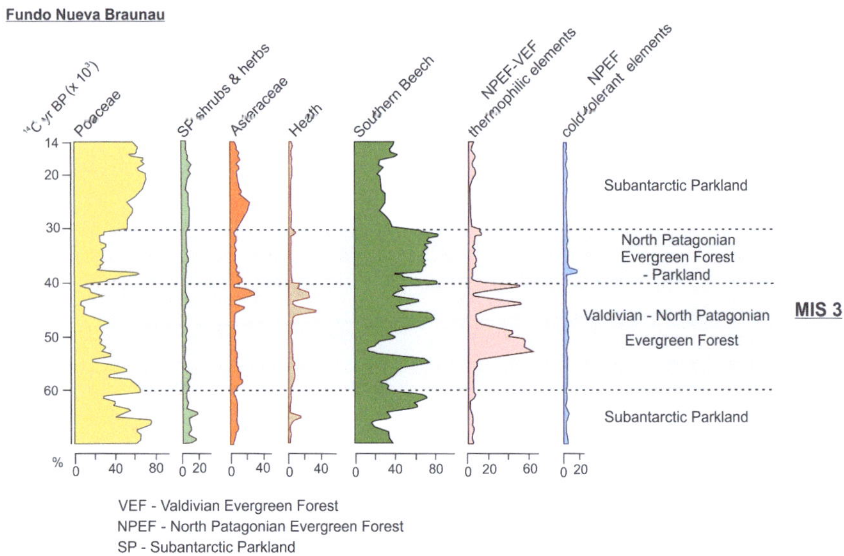

Fig. 5 Age plot of ecologically significant taxa and groups of taxa for the pollen section at Fundo Nueva Braunau until 14,000 ^{14}C year B.P. (modified from Heusser et al. 2000)

In the southern part of the Chilean Lake District, the **Fundo Nueva Braunau** section (site 8: 66 m a.s.l.; Fig. 5) constitutes one of the most valuable record covering the end of MIS 4 to MIS 2 interval (Heusser et al. 2000). The site is a wooded minerotrophic mire located west of Lago Llanquihue (Fig. 1). At present it is in the contact zone between the Lowland Deciduous Beech Forest and the Valdivian Evergreen Forest communities. According to Heusser et al. (2000), the pollen sequence, constrained chronologically by seven ^{14}C dates, shows toward the end of MIS 4 (prior 60,000 ^{14}C year B.P.) a vegetation dominated by Poaceae with colonies of *Nothofagus dombeyi* type conforming Subantarctic Parkland communities, without modern analogs of vegetation, that characterized the outwash plains of the Southern Lake District-Isla Grande de Chiloé. Colder and wetter climatic conditions are suggested by subantarctic shrub and herb indicators such as *Lepidothamnus fonkii*, *Donatia fascicularis*, *Astelia pumila*, *Nanodea muscosa*, *Gaimardia australis*, *Euphrasia* and *Huperzia fuegiana*. At the end of MIS 4, climate moderated and became more temperate during MIS 3, similar to those present conditions at higher altitudes in the cordillera. At this time expansion of *Nothofagus dombeyi* type (up to 70 %) implies more favorable conditions, and contractions of *N. dombeyi* type depict a return of Poaceae-dominated vegetation under colder conditions. Around *ca.* 50,000 ^{14}C year B.P., the increase and richness in thermophilic elements such as trees (Myrtaceae, *Lomatia*, *Podocarpus nubigena*, *Drimys winteri*, *Weinmannia trichosperma* and *Pseudopanax laetevirens*), shrubs and herbs (*Corynabutilon*, *Cissus striata*, *Ovidia* and *Tepualia atipularis*), and ferns suggest the presence of the Valdivian and North Patagonian Evergreen Forest.

These plant communities along with the expansion of *N. dombeyi* type indicate more temperate climate and still humid conditions. Toward *ca.* 40,000 ^{14}C year B.P., peaks of *N. dombeyi* type and Myrtaceae pollen plus the presence of shrubs (*Empetrum*/Ericaceae, Asteraceae) and charcoal particles are indicative of fires. During this time, reduction in Poaceae implies closed canopy of the forest communities under more temperate climate. Shortly after *ca.* 40,000 ^{14}C year B.P., the expansion of the Subantarctic Parkland communities suggesting colder and humid climate is followed by the North Patagonian Evergreen Forest and Subantarctic Parkland vegetation development under still humid but less severe conditions. These vegetal communities include *Podocarpus nubigena*, *Drimys winteri*, *Pilgerodendron* type, and *Embothrium coccineum* from the high-montane near treeline area. Between 30,000 ^{14}C year B.P. and estimated age of 14,000 ^{14}C year B.P. (MIS 3/2), the pollen assemblage shows the Subantarctic Parkland communities as a reflection of colder climate similar to that of the end of MIS 4.

The **Taiquemó** section (site 11: 170 m a.s.l.; Fig. 6) in the northeastern lowland of the Isla Grande de Chiloé and 100 km southwest distant of Fundo Nueva Braunau locality, is the oldest and most complex site at the Isla Grande de Chiloé (Fig. 1). It is a wooded topogenous mire located in the temperate Valdivian Evergreen Forest communities. A total of 27 dates ranging from >49,892 to 10,355

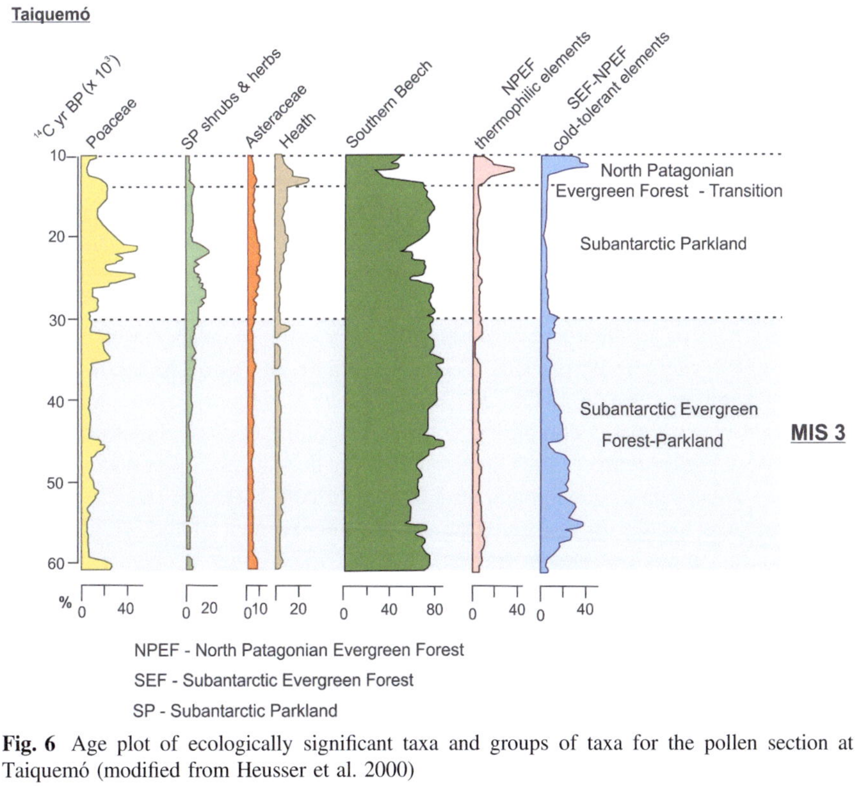

Fig. 6 Age plot of ecologically significant taxa and groups of taxa for the pollen section at Taiquemó (modified from Heusser et al. 2000)

^{14}C year B.P., provide palaeoenvironmental information from MIS 4 to MIS 2 (Heusser et al. 1999; Heusser and Heusser 2006). The pollen record shows the lengthy dominance of *Nothofagus dombeyi* type, coupled with the recurrence of Poaceae maxima. During the very onset of the sequence at infinite radiocarbon age (>49,892 ^{14}C year B.P., MIS 4) the Subantarctic Parkland communities, for which the Magellanic Moorland of Tierra del Fuego is probably its closest modern analog, dominate the landscape under colder climate. Prior to 47,110 ^{14}C year B.P. (early MIS 3), by approximately ten thousand years, the first expansion of *N. dombeyi* type of Subantarctic Evergreen Forest provenance, indicates milder conditions. *N. dombeyi* type occurs in association with cold-resistant species such as *Podocarpus nubigena*, *Pilgerondendron* type and *Pseudopanax laetevirens.* Also thermophilic elements (*Drimys winteri*, *Embothrium coccineum*, *Lomatia*, Myrtaceae, *Desfontainia spinosa* and *Misodendrum*) are recorded. The presence of North Patagonian Evergreen components infiltrating the Subantarctic Evergreen Forest elements implies tracts of interstadial forest under warmer conditions. Similar conditions are also registered between 45,226 and 40,011 ^{14}C year B.P. Expansions of *N. dombeyi* type along with Poaceae maxima of Subantarctic Evergreen Forest and Subantarctic Parkland communities, indicate a return of colder, wetter stadial climate during MIS 3 (46,780–45,226 ^{14}C year B.P., 40,011–32,105 ^{14}C year B.P., and 26,019–24,895 ^{14}C year B.P.) and MIS 2 (23,223–21,430 ^{14}C year B.P., and 16,473–13,148 ^{14}C year B.P.).

In the Argentine sector, the locality **Bajada de Rahue** (site 15: 1000 m a.s.l.) is an exposed sediment section situated in the central Andean cordillera between latitudes 39° and 40° S, near the *Nothofagus* forest-steppe ecotone (Fig. 2). Five radiocarbon dates indicated that the 450-cm section of lacustrine sediments was deposited between 33,500 and 27,900 ^{14}C year B.P. According to Markgraf et al. (1986), the fossil pollen assemblage reflects a local sedge-marsh/shallow pond environment surrounded by steppe-scrub with *Nothofagus* forests nearby. The present pollen assemblage from a surface sample in the steppe/forest zone distant 20 km from the nearest *Austrocedrus* and *N. dombeyi* forests, supplies a modern analog for the vegetation and climate at Bajada de Rahue locality. The similarity of the occurrences of arboreal and non-arboreal components suggests that the floristic setting of the steppe environment 30,000 years ago was also similar to that of today. Therefore, the Bajada de Rahue record supports the notion of interglacial-type climatic conditions with substantially less precipitation and/or considerably higher temperatures than reported for the full-glacial period (MIS 2).

In the southern part of Patagonia, the **Laguna Potrok Aike** (site 16: 113 m a.s.l.) is a maar lake located in the southernmost continental area of the world, in the southeastern Patagonian Steppe (Fig. 2). Climate in the area is characterized by strong westerly winds and the rain shadow effect of the Andes results in less than 300 mm of precipitation per year. The chronology of the two pollen records (Recasens et al. 2012) from the sediment cores 5022-1D and 5022-2C was obtained from three distinct volcanic ash layers from Hudson, Mt Burney and Reclus volcanoes, respectively. The tephra layers were identified in the topmost 18 m of the composite record from site 5022-2CP. These tephra layers were used as isochrones to pin down a chronological framework for the last 14,900 cal year (age of the Reclus tephra at 14.7 m event-corrected sediment depth). To extend this chronology to the base of the record the mean sedimentation rate was used tentatively. Below the Reclus tephra there remain 36 m of chronologically relevant sediments. The authors assumed that the linear sedimentation rate calculated for the Holocene and the Late Glacial parts of the record was applicable to the glacial section, thus a first approximation of its duration was obtained after dividing 36 m by 0.99 mm/year which yielded 36,400 cal year of lacustrine deposition. Adding this value to the age of the Reclus tephra (14,900 cal year B.P.) a preliminary basal age of 51,300 was obtained for the PASADO record (5022-2CP). The basal age of 51,300 estimated in this study was consistent with a first radiocarbon date estimated by Kliem et al. (2011) that revealed a basal age of ca. 51,000 cal year B. P. Recasens et al. (2012) indicated that the area was covered with Patagonian Steppe vegetation during the last 51,300 cal year B.P. The oldest part of both pollen records, corresponding to glacial times, shows a relatively high contribution of *Colobanthus* type, which is nowadays mainly distributed in the dry steppe of Patagonia. Caryophyllaceae represents local vegetation coming from the immediate catchment area of the Laguna Potrok Aike. In contrast, the record of *Nothofagus* pollen is due to its presence in the Andes distant 60 km westward. The pollen signal suggests glacial environments with vegetation dominated by communities of dwarf shrubs under drier conditions than today. In the Andean area *Nothofagus* was present probably as shrubs as it occurs today close to its lower temperature limits. After 80 mcd (meters composite depth), *Nothofagus* and charcoal concentrations decreased and *Nassauvia* increased importantly. The pollen signal indicates glacial conditions with considerable lower temperatures and probably lower humidity than before. Today, *Nassauvia* grows in the driest parts of the Patagonian Steppe and also in the higher altitudes of the southern Andes.

The **Magallanes maar** (site 17), 10 km from the Estrecho de Magallanes (Fig. 2), is a core drilled in soft sediments reaching a maximum depth of 59 m below the surface before stopping at the basaltic bed rock (Corbella et al. 2000). Two ^{14}C radiometric ages were performed on the organic matter fraction of samples: >51,700 ^{14}C year B.P. at 47 m depth and 31,560 ± 480 ^{14}C year B.P. at 37 m depth (Corbella 2002). Only 12 samples, of 22 total samples between 29 and 56 m core depth, contained pollen. The identified pollen assemblages, represented by Poaceae, *Acaena*, Asteraceae subf. Asteroideae, *Nassauvia* and *Empetrum rubrum* as principal contributors, reflect in general terms, grass steppe environments. About >51,700 ^{14}C year B.P., the pollen samples recorded a gramineous steppe vegetation but with elements and indicators that point to a greater aridity as: *Nassauvia*, *Azorella*, a reduced amount of *Empetrum rubrum* and *Acaena*, and an increase in halophytic plants (Chenopodiaceae). Around ca. 31,560 ^{14}C year B.P., the pollen samples recorded a gramineous steppe environment accompanied by dwarf shrub heaths (*Empetrum rubrum*) under conditions of increased relative humidity. *Nassauvia* and *Ephedra*, plants that characterize the xeric steppes, were not important.

In Tierra del Fuego, a glaciolacustrine sequence (site 18) presently located into the Subantarctic Deciduous Forest communities, is exposed at the southeastern end of **Lago Fagnano** (Fig. 3). The organic mud contained in these glaciolacustrine deltaic sediments dates to >58,000 and 39,560 ^{14}C year B.P. (Bujalesky et al. 1997). Pollen in the fossil peat beds is mainly of Poaceae, *Empetrum rubrum* and Asteraceae, besides significant amounts of *Acaena*, *Gunnera* and Caryophyllaceae. The pollen assemblages indicate regional steppe-like vegetation of grasses, dwarf shrub heaths and scrubs. Records of Cyperaceae and *Myriophyllum* suggest the existence of shallow ponds, and *Litorella*, from mud areas near-shore. A very low content of *Nothofagus* pollen supports the interpretation of steppe/tundra environments instead of the present closed-canopy *Nothofagus* forest. The climatic implication is of colder and drier conditions than present semi-oceanic ones.

The locality **Lago Fagnano 1** (site 19) is situated 2.3 km westward from the Lago Fagnano record (site 18) (Fig. 3). The pollen analysis was performed on four fossil peat beds interbedded with till cropping out in a cliff, which were radiocarbon dated at 48,200 ± 3300, >44,800, 44,800 ± 2300, and 31,080 ± 510 ^{14}C year B.P. (Ponce et al. 2014). The pollen assemblages indicate the development of herbs (Poaceae), dwarf shrub heaths (*Empetrum rubrum*), scrubs (Asteraceae subf. Asteroideae) and cushion plants (*Azorella*, *Myrteola nummularia*), with low amounts of *Nothofagus* pollen. The pollen record also suggests the presence of minerotrophic environments (Cyperaceae). These plant communities indicate colder and drier climate conditions than present ones, and similar to those of the Lago Fagnano locality (Bujalesky et al. 1997).

4 Conclusions

The reconstruction of the vegetation pattern from the existing pollen data allows only a broad climatic comparison between Chilean and Argentine sectors because of the uncertain chronology, since conventional radiocarbon dates yielding ages older than 35 ka are rarely reproducible. By comparison, the persistent hyperhumid climate in the region of Southern Lake District–northern Isla Grande de Chiloé (40–42° S) during the MIS 3 period contrasts with the central part of Chile (32–38° S) that exhibits periods of lower humidity broken by intervals of wetter climate. According to the published pollen records, in the mid-latitude of Chile (40–42° S), the climate toward the end of MIS 4 (prior 60,000 ^{14}C year B.P.) was cold, Subantarctic type. Vegetation consisted of Subantarctic Parkland (park tundra) communities marked essentially by the interplay between southern beech (*Nothofagus dombeyi* type) and grasses (Poaceae). During MIS 3 (between 60,000 and 30,000 ^{14}C year B.P.) climate moderated and became more temperate and humid. At this time, interstadial episodes alternated with stadial events. The first were characterized by the development of mixed forests of *Nothofagus* and thermophilic/cold-tolerant elements indicative of Valdivian-North Patagonian Evergreen Forest in Fundo Nueva Braunau site, and by the development of a Subantarctic Evergreen Forest/Parkland in Taiquemó site (Heusser et al. 2000). During this forest expansion, the glaciers were in a state of recession, while the stadial episodes showed a Subantarctic Parkland vegetation suggesting colder climate. Toward MIS 3/2 boundary (between 30,000 and 14,000 ^{14}C year B.P.), the grass-dominated Subantarctic Parkland (~Magellanic Moorland of Tierra del Fuego) vegetation expanded under colder climatic conditions and the piedmont glacial lobes reached their maxima. Calibrated from *N. dombeyi* type-Gramineae ratios, mean summer temperatures at 11–12 °C in MIS 3 fell to 6–8 °C in MIS 2 (LGM), a depression of as much as −6 and −8 °C compared with the present conditions in the Isla Grande de Chiloé (Heusser 2003).

MIS 3 climate in semiarid–arid subtropical Chile (32–38° S) was characterized by temperate and humid conditions (between 50,000–35,000 and 28,500–14,500 ^{14}C year B.P.), that favored the development of a woodland of southern beech (*Nothofagus dombeyi* type) and podocarp (*Prumnopitys andina*). Climate conditions to support these vegetal palaeocommunities indicate an average of summer temperature of at least 7 °C lower and annual precipitation at least 1200 mm greater than present. These episodes of more humid conditions alternated with episodes of aridity with rise of temperature (centered about >43,000 and 33,000 ^{14}C year B.P.) when Chenopodiaceae/Amaranthaceae expanded and lake levels were comparatively lower as can be seen in the pollen record from Laguna de Tagua Tagua (Heusser 1990). Aridity with a rise of temperature, which is indicated by a chenopod/amaranth peak dated at 37,000 and 33,300 ^{14}C year B.P., coincides with millennia of glacier minima in the Southern Lake District. Meanwhile, the southern beech expansion at Tagua Tagua between 28,500 and 14,500 ^{14}C year B.P., corresponds closely with dated glacier maxima (Heusser et al. 2000).

In the Argentine sector, low resolution of the pollen records and the usually finite radiocarbon dates make the palaeoenvironmental interpretation speculative. In northwestern Patagonia at 39° S, the pollen record from Bajada de Rahue (Markgraf et al. 1986) suggests a steppe-scrub vegetation between 33,000 and 27,000 ^{14}C year B.P., indicative of warmer and drier interglacial conditions than the following full-glacial period (MIS 2). It compares well with increases of chenopods and amaranths in the Laguna de Tagua Tagua pollen record, but contrasts to the hyperhumid vegetation at Taiquemó and Fundo Nueva Braunau sites.

In southern Patagonia (51–52° S), according to the radiocarbon dates, drier and colder conditions were inferred by pollen data from the Laguna Potrok Aike section after 51,300 cal year B.P. (Recasens et al. 2012). At the Magallanes maar, the pollen assemblage reflects lower relative humidity by about >51,700 ^{14}C year B. P. and relatively more humid conditions about 31,560 ^{14}C year B.P. (Corbella et al. 2000). In Tierra del Fuego (54° S) the two studied pollen sections located in Lago Fagnano revealed similar pollen assemblages indicative of drier and colder conditions earlier than ~40,000 ^{14}C year B.P. (Bujalesky et al. 1997; Ponce et al. 2014).

According to Heusser (1990) and Heusser et al. (1999), it becomes axiomatic that the vegetal changes observed in the pollen records and glacier stadial/interstadial behavior are related to fluctuations in intensity of the Southern Westerlies, marked by oscillations of the Polar Front. The influence of Southern Westerlies may have been greater at the time of MIS 3, and the effect of the subtropical high-pressure cells was apparently weakened (Heusser 1983; Villagrán 1990). Opinions differ about the direction of wind movement and sources of moisture. Heusser (1983, 1990, 1994) established that during the Last Glacial Maximum and the Late Glacial period, the westerly wind system was strengthened equatorward with precipitation coming from storm systems advancing farther northward than today, and thus the mesic forests expanded northwards. Meanwhile, Markgraf (1987, 1989) and Markgraf et al. (1992) postulated that during Full- and Late-Glacial times, the westerlies intensified and shifted poleward, and a temperature decrease of about 6 °C would explain the greater effective moisture availability in the lowlands at central Chile. Thus, the full-glacial vegetation represents a montane open forest that expanded into the lowlands due to cooler conditions. In general, according to Villagrán et al. (1995) the Chilean pollen records suggest altitudinal and latitudinal forest displacements during glaciation and show a decrease in temperature and an increase in precipitation in these regions.

Acknowledgments Support was provided by Consejo Nacional de Investigaciones Científicas y Técnicas, Argentina (PIP 2011–2013). The authors are very grateful to the Editors for the invitation to participate in this volume.

References

Bujalesky G, Heusser C, Coronato A, Roig C, Rabassa J (1997) Pleistocene glaciolacustrine sedimentation at Lago Fagnano, Andes of Tierra del Fuego, Southernmost South America. Quat Sci Rev 16(1–2):767–777

Cabrera AL (1971) Fitogeografía de la República Argentina. Boletín de la Sociedad Argentina de Botánica XIV 1–2:1–50

Corbella H (2002) El campo volcánico-tectónico de Pali-Aike. In: Haller MJ (ed) Geología y recursos naturales de Santa Cruz. Relatorio del XV Cong Geol Arg, Buenos Aires, pp 287–303

Corbella H, Borromei AM, Quattrocchio M (2000) Quaternary climate changes in Southernmost South America inferred from lacustrine sediments preserved in volcanic maars. In: Smolka PP, Volkheimer W (eds) Southern hemisphere paleo- and neoclimates. Springer, Germany, pp 263–273

Coronato AMJ, Coronato F, Mazzoni E, Vázquez M (2008) The physical geography of Patagonia and Tierra del Fuego. In: Rabassa J (ed) The Late Cenozoic of Patagonia and Tierra del Fuego, Developments in Quaternary Science, Elsevier, Amsterdam, vol. 11, chapter 3, 13–56 pp

Heusser CJ (1981) Palynology of the last interglacial—glacial cycle in midlatitudes of southern Chile. Quat Res 16:293–321

Heusser CJ (1983) Quaternary pollen record from Laguna de Tagua Tagua, Chile. Science 219:1429–1432

Heusser CJ (1989) Late Quaternary Vegetation and Climate of Southern Tierra del Fuego. Quat Res 31:396–406

Heusser CJ (1990) Ice age vegetation and climate of subtropical Chile. Palaeogeogr Palaeoclimatol Palaeoecol 80:107–127

Heusser CJ (1994) Pattern of glacial-interglacial vegetation in subtropical Chile. Hist Biol 9:35–45

Heusser CJ (2003) Ice Age Southern Andes—a chronicle of paleoecological events. Developments in Quaternary Science 3. Elsevier, Amsterdam

Heusser CJ, Heusser LE (2006) Submillennial palynology and palaeoecology of the last glaciation at Taiquemó (50,000 cal year, MIS 2–4) in southern Chile. Quat Sci Rev 25:446–454. doi:10.1016/j.quascirev.2005.04.008

Heusser CJ, Heusser LE, Lowell TV (1999) Paleoecology of the Southern Chilean Lake District—Isla Grande de Chiloé during middle—late Llanquihue glaciation and deglaciation. Geogr Ann 81 A (2): 231–284

Heusser CJ, Lowell TV, Heusser LE, Moreira A, Moreira S (2000) Pollen sequence from the Chilean Lake District during the Llanquihue glaciation in marine oxygen isotope stages 4–2. J Quat Sci 15:115–125

Heusser L, Heusser C, Pisias N (2006) Vegetation and climate dynamics of southern Chile during the past 50,000 years: results of ODP Site 1233 pollen analysis. Quat Sci Rev 25:474–485

Kaplan MR, Moreno PI, Rojas M (2008) Glacial dynamics in southernmost South America during Marine Isotope Stage 5e to the Younger Dryas chron: a brief review with a focus on cosmogenic nuclide measurements. J Quat Sci 23:649–658

Kliem P, Buylaert JP, Hahn A et al (2011) Dating, age-depth modeling and hydrological interpretation of the 51 cal. ka BP composite profile from Laguna Potrok Aike in Southern Patagonia, Argentina. In: 3rd international PASADO workshop. Montreal, Canada, 21–23 Mar, pp 31–32. Available at http://www.pasado.uni-bremen.de/Files/3rd_Pasado_WSMtl_2011.pdf

Long J, Stoy PC (2013) Quantifying the periodicity of Heinrich and Dansgaard-Oeschger events during marine oxygen isotope stage 3. Quat Res 79:413–423

Markgraf V, Bradbury JP, Fernandez J (1986) Bajada de Rahue, Province of Neuquén, Argentina: an interstadial deposit in northern Patagonia. Palaeogeogr Palaeoclimatol Palaeoecol 56:251–258

Markgraf V (1987) Paleoenvironmental changes at the northern limit of the Subantarctic *Nothofagus* forest, lat 37° S, Argentina. Quat Res 28:119–129

Markgraf V (1989) Reply to C. J. Heusser's 'Southern Westerlies' during the Last Glacial Maximum. Quat Res 31:426–432

Markgraf V, Dodson JR, Kershaw AP, McGlone MS, Nicholls N (1992) Evolution of late Pleistocene and Holocene climates in the circum-South Pacific land areas. Clim Dynam 6:193–211

Miller A (1976) The climate of Chile. In: Schwerdtfeger W (ed) World survey of climatology, climate of Central and South America, Elsevier, Amsterdam, vol 12, 113–145 pp

Moore DM (1983) Flora of Tierra del Fuego. Nelson, Oswestry

Pisano E (1977) Fitogeografía de Fuego-Patagonia Chilena. I.- Comunidades vegetales entre las latitudes 52° y 56° S. Anales Inst Patag 8:121–250

Ponce JF, Borromei AM, Coronato A, Rabassa J, Onorato R (2014) Evidencias sedimentológicas y palinológicas de fluctuaciones glaciales durante el EI3 en Lago Fagnano, Tierra del Fuego. XIX Cong Geol Arg. Resúmenes, Córdoba

Prohaska F (1976) The climate of Argentina, Paraguay and Uruguay. In: Schwertfeger W (ed) Climates of Central and South America. World survey, climatology. Elsevier, Amsterdam

Recasens C, Ariztegui D, Gebhardt C, Gogorza C, Haberzettl T, Hahn A, Kliem P, Lisé-Pronovost A, Lücke A, Maidana N, Mayr C, Ohlendorf C, Schäbitz F, St-Onge G, Wille M, Zolitschka B and The PASADO Science Team (2012) New insights into paleoenvironmental changes in Laguna Potrok Aike, southern Patagonia, since the Late Pleistocene: The PASADO multiproxy record. Holocene 22(11):1323–1335

Schmithüsen J (1956) Die räumliche Ordnung der chilenischen Vegetation. Bonner Geographische Abhandlungen 17:1–86

Tuhkanen S (1992) The climate of Tierra del Fuego from a vegetation geographical point of view and its ecoclimatic counterparts elsewhere. Acta Bot Fenn 125:4–17

Valero-Garcés BL, Jenny B, Rondanelli M, Delgado-Huertas A, Burns SJ, Veit H, Moreno A (2005) Palaeohydrology of Laguna de Tagua Tagua (34° 30′ S) and moisture fluctuations in Central Chile for the last 46,000 year. J Quat Sci 20:625–641

Villagrán C (1985) Análisis palinológico de los cambios vegetacionales durante Tardiglacial y Postglacial en Chiloé, Chile. Rev Chil Hist Nat 58:57–69

Villagrán C (1990) Glacial climates and their effects on the history of the vegetation of Chile: a synthesis based on palynological evidence from Isla de Chiloé. Rev Palaeobot Palynol 65:17–24

Villagrán C, Moreno P, Villa R (1995) Antecedentes palinológicos acerca de la historia cuaternaria de los bosques chilenos. In: Armesto JJ, Villagrán C, Arroyo MK (eds) Ecología de los bosques nativos de Chile. Editorial Universitaria, Santiago de Chile, pp 51–69

Voelker AHL, workshop participants (2002) Global distribution of centennial-scale records for Marine Isotope Stage (MIS) 3: a database. Quat Sci Rev 21:1185–1212

Response of Diatoms to Late Quaternary Climate Changes

Marcela Alcira Espinosa

Abstract Diatoms are very useful proxy indicators to reconstruct past climate changes. Studies are based on qualitative and quantitative analyses that allow to infer variables related directly to climate as temperature, or indirectly as salinity, depth, productivity, and pH. Reconstructions based on these methods rely on the general assumption that past environmental requirements of the fossil diatom taxa have remained similar to those of their closest living representatives. In this way, the environmental information obtained from living organisms can be used as analogs and extrapolated to the fossil record, particularly in Late Quaternary studies. Diatom records from lacustrine deposits from Argentina, ancient lakes from South America, and marine cores from Southeastern Atlantic Ocean and Eastern Equatorial Pacific were reinterpreted with the aim to correlate them to climatic changes during Marine Isotope Stage 3 (MIS 3) in the Southern Hemisphere. Marine records allowed paleo-reconstructions of productivity and upwelling conditions; at the same time continental records were used to interpret the lake-level histories. The high temporal resolution of diatom assemblages in both environments makes it possible to identify abrupt climate changes between *ca*. 60 and 30 cal. ka B.P. The future integration of diatom datasets constructed from different environments will solve the analogy problems between fossil and modern assemblages and increase the potential for reliable quantitative reconstructions of Late Quaternary climate in southern South America.

M.A. Espinosa (✉)
Instituto de Geología de Costas y del Cuaternario, Universidad Nacional de Mar del Plata, CC 722 7600 Mar del Plata, Argentina
e-mail: maespin@mdp.edu.ar

M.A. Espinosa
Instituto de Investigaciones Marinas y Costeras, CONICET-UNMDP, CC 722 7600 Mar del Plata, Argentina

G.M. Gasparini et al. (eds.), *Marine Isotope Stage 3 in Southern South America, 60 ka B.P.–30 ka B.P.*, Springer Earth System Sciences,
DOI 10.1007/978-3-319-40000-6_16

Keywords Diatoms · Paleoenvironments · Climate change · Argentina · South America

1 Introduction

Diatoms are unicellular algae with a siliceous exoskeleton (frustule) resistant to decay. They are ubiquitous, occurring in almost all aquatic habitats, where they may be planktonic, benthic, periphytic (growing on plant or seaweed surfaces), epizoic (on animals) or endozoic (within animals). Because of their siliceous composition, they are often very well preserved. Once incorporated in sediments, diatom frustules remain forming a record of the populations from the benthos or the water column.

The estimation of environmental variables from diatom data sets has become an important tool for paleoenvironmental reconstructions. Diatoms are used extensively in these studies because they are excellent indicators of past conditions and are particularly useful because they can be identified to the species level using light and scanning electron microscopy. Thus by inspection of assemblages in sedimentary records, we can make direct and indirect inferences about past environmental conditions (Mackay et al. 2003). Diatoms have been used as proxy indicators to reconstruct Quaternary climate variability in every continent, being on continental aquatic ecosystems much more common than marine or coastal studies (Mackay et al. 2003). However, despite the high potential of diatoms records for Late Quaternary environmental reconstructions, a limited number of detailed studies have been conducted in South America.

During the Late Pleistocene, the MIS 3 was an interstadial stage, a relatively warmer climatic period which developed roughly between 60–50 and 30 cal. ka B. P. (ca. 56 to 25 ^{14}C ka B.P.). The climatic conditions fluctuated over a broad range of millennial time scales (Rabassa and Ponce 2013). Ever since these millennial-scale climate oscillations were identified in ice cores from Greenland (Grootes et al. 1993) and Antarctica (Brandefelt et al. 2011), one major aim in climate research has been to study these oscillations also in marine and terrestrial records of different latitudes and, thereby, to gain insights into their driving factors.

The studies to reconstruct past climatic variables with diatoms may be either directly, such as surface water temperature and air temperature, or indirectly by inferring for example, salinity, conductivity and pH. Qualitative and quantitative information are provided by assessing changes in diatom species themselves in both freshwater and marine environments. As high-resolution terrestrial records are sparse, there are many gaps on the Southern South America map (Fig. 1) especially

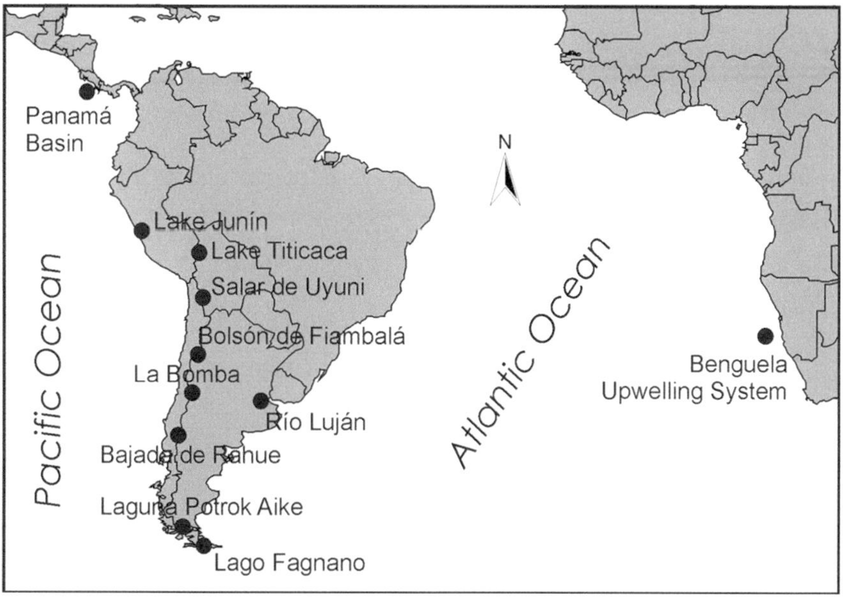

Fig. 1 Location map of MIS 3 diatom records

in Brazil, Colombia, Venezuela, and Ecuador. A few low-resolution records, which might reveal millennial-scale climate oscillations at a higher temporal resolution are available from some lacustrine deposits from Argentina. For some ancient lakes of Peru (Lake Junín), Bolivia (Lake Titicaca and Salar de Uyuni) and Argentina (Laguna Potrok Aike), various high-resolution diatom records of low latitudes were published recently (Tapia et al. 2006; Fritz et al. 2004, 2007; Recasens et al. 2015) as well as for the paleoceanographic community (Cermeño et al. 2013; Romero 2010).

This article presents a brief overview of the results concerning the use of diatom analysis to reconstruct Quaternary climate variability in South America. The outline and reinterpretation of studies that make use of diatom assemblages (including both qualitative and quantitative approaches) during MIS 3 aims to improve the climatic reconstructions and highlight the importance of site selection and sampling resolution, and the need for further studies in the region.

To provide an overview on the spatial distribution of diatom records of MIS 3 in South America, some terrestrial and marine sites have been compiled according to their location and temporal resolution (Table 1). For a total of 11 sites, detailed information on their geographical setting are presented being terrestrial records the most abundant.

Table 1 List of ages (^{14}C, OSL, U/Th), location and references of diatom records of MIS 3

Location	Age years BP	Cal. years BP	Latitude	Longitude	References	Localities
Argentina, Catamarca	32,000 ± 520 (^{14}C) 29,380 ± 410 (^{14}C)		28° 08′ S	68° 10′ W	Garleff et al. (1993)	Bolsón de Fiambalá
Argentina, Buenos Aires	56,400 ± 6500 (^{14}C) 50,400 ± 10,200 (^{14}C) 44,000 ± 6500 (^{14}C) 37,710 ± 840 (^{14}C) 32,500 ± 4700 (^{14}C)		34° 34′ 08″ S 34° 34′ 54″ S 34° 34′ 08″ S 34° 34′ 08″ S 34° 34′ 55″ S	59° 07′ 29″ W 59° 10′ 20″ W 59° 07′ 29″ W 59° 07′ 29″ W 59° 10′ 20″ W	Blasi et al. (2010)	Río Luján
Argentina, Mendoza	35,460 ± 740 (^{14}C) 35,170 ± 670 (^{14}C) 31,570 ± 440(^{14}C) 31,520 ± 520(^{14}C)		33° 28′ S	69° 03′ W	Hassan et al. (2013)	La Bomba
Argentina, Neuquén	33,500 ± 1500 (^{14}C) 32,600 ± 1500 (^{14}C) 29,000 ± 1400 (^{14}C) 28,800 ± 1100 (^{14}C) 27,900 ± 1200 (^{14}C)		39° 22′ S	70° 56′ W	Markgraf et al. (1986)	Bajada de Rahue
Argentina, Santa Cruz	44,800 ± 2000 (OSL)	48,500 53,000	51° 57′–59′ S	70° 24′–21′ W	Haberzettl et al. (2008)	Laguna Potrok Aike
Argentina, Tierra del Fuego	39,560 ± 3980 (^{14}C)		54° 33′ S	67° 19′–68° 48′ W	Bujalesky et al. (1992, 1997)	Lago Fagnano
Bolivia, Bolivian Altiplano	32,600 ± 410 (^{14}C) 31,500 ± 480 (^{14}C) 36,300 ± 590 (^{14}C) 36,200 ± 800 (U/Th) 46,200 ± 2800 (U/Th) 50,500 ± 2700 (U/Th)	37,830 36,630 41,840	20° 14.97′ S	67° 30.03′ W	Fritz et al. (2004)	Salar de Uyuni

(continued)

Table 1 (continued)

Location	Age years BP	Cal. years BP	Latitude	Longitude	References	Localities
Bolivia/Perú Tropical Andes	28,390 ± 180 (^{14}C) 33,370 ± 200 (^{14}C) 31,230 ± 660 (^{14}C) 36,680 ± 270 (^{14}C) 37,900 ± 1900 (^{14}C) >43,000 >52,000	32,444 37,925 35,569 41,142 41,416	14° 09′–17° 08′ S	68° 03′–71° 04′ W	Fritz et al. (2007)	Lake Titicaca
Perú, Ondores	25,700 ± 330 (^{14}C) 39,020 ± 1045 (^{14}C)	30,018 44,728	11° 03.52′ S	76° 07.27′ W	Tapia et al. (2006)	Lake Junín
Southwestern Africa	40230 +1480/− 1250 42770 +2120/− 1680	44,174 ± 1057 46,535 ± 1899	25° 28.0′ S	13° 05.0′ E	Romero (2010)	Benguela Upwelling System
Panamá Basin	Between ~ 98,000 to ~ 10,000	Last 40,000	08° 12.33′ N	84° 07.32′ W	Romero et al. (2011)	Costa Rica margin

2 Lacustrine Records from Argentina

2.1 Bolsón de Fiambalá, Catamarca, Argentina

Bolsón de Fiambalá is located at 27°–28° S, 67°–68° W in the eastern side of the so-called Arid Diagonal Zone of South America (Fig. 1). Late Quaternary climatic changes are of special interest because this is a transitional zone between the tropical westerlies and the tropical monsoonal system.

West of the mountains of the Bolsón there are lake deposits between ca. 3830 and 3820 m a.s.l., 40–30 m above the basin of Las Lozas (28° 08′ S, 68° 10′W). These lacustrine limestones were dated between 32,000 ± 520 and 29,380 ± 410 years ^{14}C B.P. (Garleff et al. 1993).

The stratigraphic section is mainly composed of coarse gravels and fanglomerates, interbedded by fine sands layers with diatoms. Six samples were analyzed but only four contained diatom frustules. A total of 25 taxa were identified (Fig. 2). The assemblages are dominated by freshwater/brackish species, benthos, aerophilous, and epiphytes. *Navicula cryptotenella*, *Nitzschia amphibia* fa. *umbrosa*, *Pseudostaurosira brevistriata,* and *Planothidium lanceolatum* are the most important taxa characterizing a shallow environment with the presence of salt. Pleistocene diatom taxa from the Las Lozas basin were found in modern assemblages from high-altitude aquatic environments of Catamarca province studied by Maidana and Seeligmann (2006). Then, the limnic sedimentation studied in Bolsón de Fiambalá sequence represents humid phases with milder climate during MIS 3.

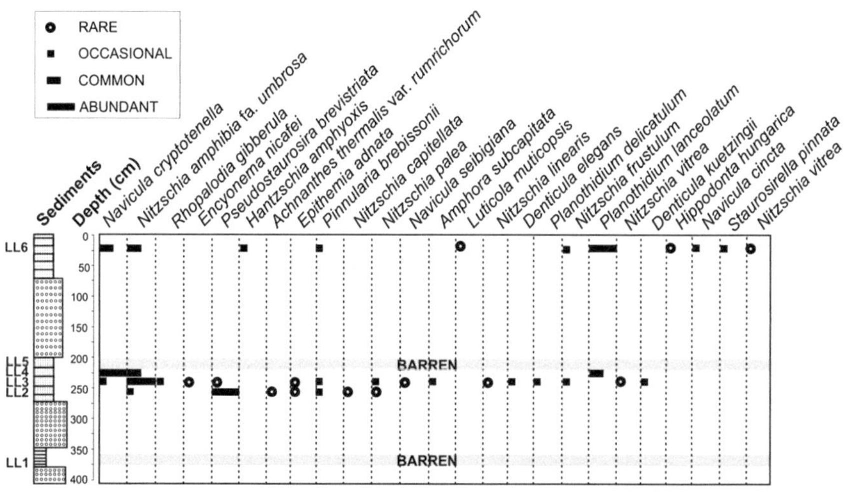

Fig. 2 Diatom relative abundances from the Bolsón de Fiambalá section

2.2 *Río Luján, Buenos Aires, Argentina*

Sedimentary sequences located on the middle course of the Río Luján River, Buenos Aires (Fig. 1) were analyzed by Blasi et al. (2010) with the aim to define depositional units and infer the paleoenvironmental conditions from the analysis of mollusks, phytoliths, and diatoms. The sequences dated between *ca.* 70,000 and 11,000 year B.P. (MIS 4–2) preserve paleoenvironmental information of the Late Quaternary climatic evolution from the northeastern Pampean region. Blasi et al. (2010) recognized five facies in the Late Pleistocene sedimentary record (Fig. 3). Facies 1 represents ephemeral lacustrine–fluvial deposits dominated by epiphytes and benthic diatoms. Facies 2 (~56,000 to ~50,000 year B.P.) are aeolian deposits where diatom content is very low. The minimal depositional age was inferred as being *ca.* 60,000 year B.P. Facies 3 (between ~44,000 and ~37,700 years B.P.) are lacustrine deposits with abundant diatoms. The planktonic and brackish *Cyclotella meneghiniana* dominate accompanied by the epiphytes *Epithemia adnata*, *Nitzschia amphibia*, *Cocconeis placentula,* and *Amphora copulata*. Benthos epipelon and aerophilous taxa are well represented too. This assemblage indicates increased salinity and depth and allows inferring higher temperature. Facies 4 and Facies 5 have brackish diatoms with different degree of preservation. According to Blasi et al. (2010), the accumulation started with settling in permanent lakes or ponds with variable inputs of aeolian sand and dust, due to wind storms, under temperate and sub-humid climatic conditions (Facies 4). Subsequently, these lentic water bodies were degraded by dystrophy (Facies 5).

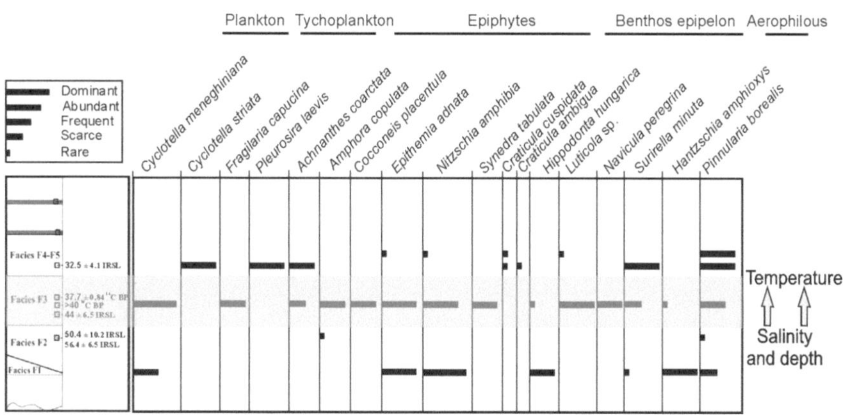

Fig. 3 Diatom relative abundances from the Río Luján section (data taken from Blasi et al. 2010)

2.3 La Bomba, Mendoza, Argentina

Diatom content of an alluvial sequence cropping out on the right margin of the Arroyo La Estacada, Mendoza (33° 28′ S, 69° 03′ W; Fig. 1) was analyzed by Hassan et al. (2013). The 3 m thick section comprised the interval between ~35,000 ^{14} C year B.P. and ~31,000 ^{14} C year B.P. The base of the sequence (~35,000 ^{14}C year B.P.) was characterized by the dominance of the fresh/brackish tychoplankton *Pseudostaurosira brevistriata* and *Staurosira venter*, accompanied by the brackish epiphyte *Planothidium delicatulum* (Fig. 4). The brackish planktonic species *Cyclotella meneghiniana* increased to the top dominating the lapse comprised between ~35,000 and ~31,000 ^{14}C year B.P. According to these results, it can be stated that the La Bomba sequence represents the evolution of a freshwater/brackish and cooler shallow lake or pond towards more saline and warmer environment. The diatom record of La Bomba suggests a relatively humid and mild interval in agreement with regional patterns during the MIS 3 (Rabassa and Ponce 2013) a period characterized by a general tendency to global warming. Salinity change would likely have been driven by an increase in temperature increasing the evaporation rates leading to more saline conditions (Hassan et al. 2013).

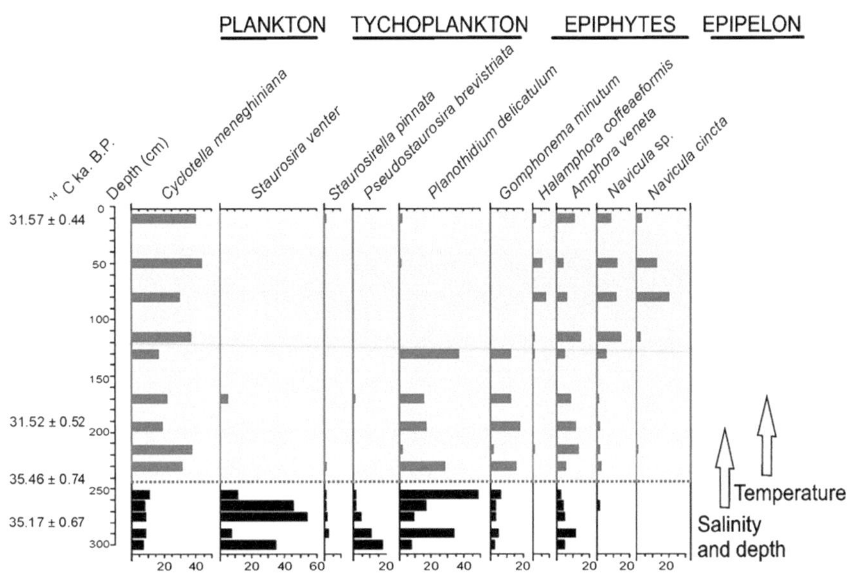

Fig. 4 Diatom diagram of the La Bomba section (modified from Hassan et al. 2013)

2.4 *Bajada de Rahue, Neuquén, Argentina*

A sequence deposited between 33,500 and 27,900 ^{14}C year B.P. in Bajada de Rahue, Neuquén (39° 22′ S, 70° 56′ W, 1000 m a.s.l., Fig. 1) was analyzed by Markgraf et al. (1986). A semiquantitative evaluation of diatom concentration on a section of 450 cm was performed. The tychoplanktonic *Staurosirella pinnata* is the dominant taxon accompanied by the epiphytes *Epithemia adnata* and *Cymbella cymbiformis* indicating shallow waters, probably not exceeding 1–2 m in depth (Fig. 5). This is particularly valid for the top and base of the section. The diatoms from the Rahue section are all species that live in shallow, freshwater environments. Diatom concentration varied considerably throughout the section; the upper and lower units are entirely composed of diatoms (diatomites) and the intermediate units range from barren to highly diatomaceous levels. Less diatom concentration and the presence of *Aulacoseira distans* and *Eunotia* spp. in the middle part of the profile indicate that probably during this period the environment was receiving an abundant supply of terrestrial organic material (Markgraf et al. 1986). Lacustrine levels would have been deposited during a relative short time interval under an environmental regime resembling the modern, local environments. According to Markgraf et al. (1986) the influx of non-diatomaceous materials probably reflects stream activity and water depth. The units with higher diatom concentration might have presumably formed in somewhat deeper water in areas of a marsh or pond surrounded by extensive emergent and submerged vegetation that traps incoming clastic sediment.

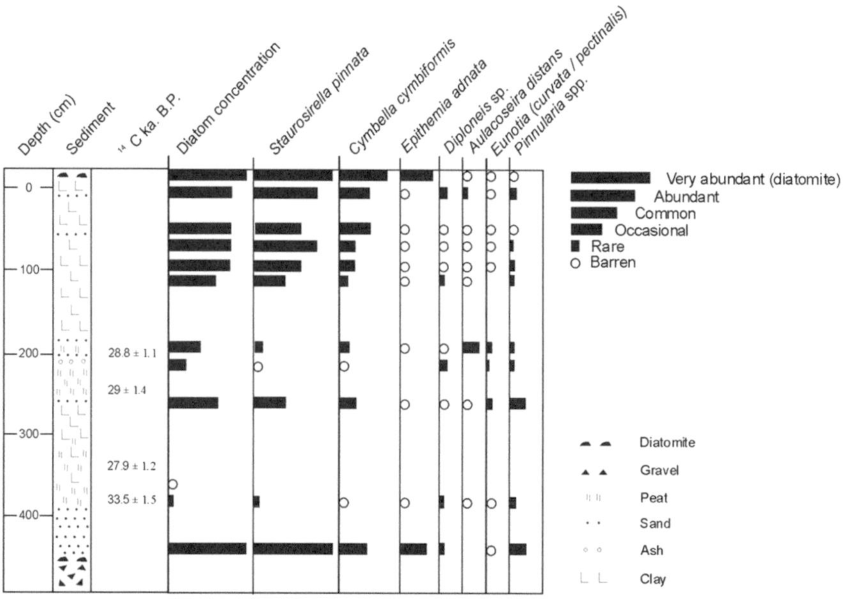

Fig. 5 Diatom relative abundances from the Rahue section (modified from Markgraf et al. 1986)

2.5 Lago Fagnano, Andes of Tierra del Fuego

Bujalesky et al. (1992, 1997) described a Pleistocene glaciolacustrine sequence exposed at the southeastern end of Lago Fagnano, Tierra del Fuego (54° 33′ S, 67° 19′–68° 48′ W; 80 m a.s.l; Fig. 1). One level at the middle of the sequence was dated at 39,560 ± 3,980 ^{14}C years B.P. This radiocarbon date should be considered as a minimum age, since another layer in the sequence provided an infinite age (>58,000 ^{14}C years B.P.; Bujalesky et al. 1997, p. 773). The diatom analysis from the dated layer was interpreted as a freshwater, cool, temperate, low energy environment, with alkaline pH and relative stable trophic conditions (Bujalesky et al. 1992). Sediments with diatoms contain an assemblage dominated by *Epithemia adnata* (freshwater epiphyte; 70 %) accompanied by different species of Fragilariaceae (freshwater/brackish tychoplankton) and the freshwater/brackish epiphytes *Cocconeis placentula*, *Cymbella cymbiformis*, *Gomphonema olivaceum*, and *Planothidium hauckianum* (Fig. 6). This assemblage could be related to climatic warming during the end of a glacial stage.

2.6 Laguna Potrok Aike

A diatom record of a core covering over the last 50,000 cal. year B.P. was studied in Laguna Potrok Aike, Santa Cruz, Patagonia (Fig. 1) by Recasens et al. (2015). Between ~51,000 and ~49,000 cal. year B.P., the assemblages were dominated by the planktonic and freshwater/brackish taxa *Discostella stelligera* and *Discostella stelligera* morph 1, with high diatom concentration. The tychoplanktonic *Staurosirella pinnata* was always present, accompanied by isolated peaks of several benthic and epiphytic diatoms. These alternations would indicate little changes in the depth of the lake and probably temperature changes. The assemblages showed a sudden drop in the relative abundance of *Discostella stelligera* M1 at ~48,000 cal. year B.P. and the dominance of *Cyclostephanos patagonicus*, a big planktonic diatom that was not previously found in the record and appeared as a

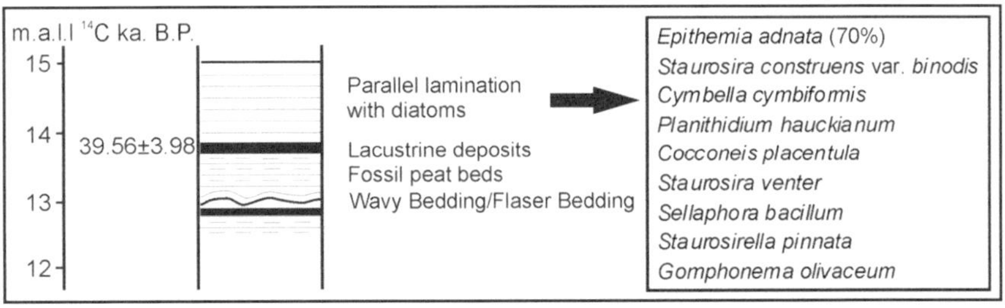

Fig. 6 Stratigraphic section of Lago Fagnano at *ca.* 40 ka B.P. and the most common diatom taxa (modified from Bujalesky et al. 1997)

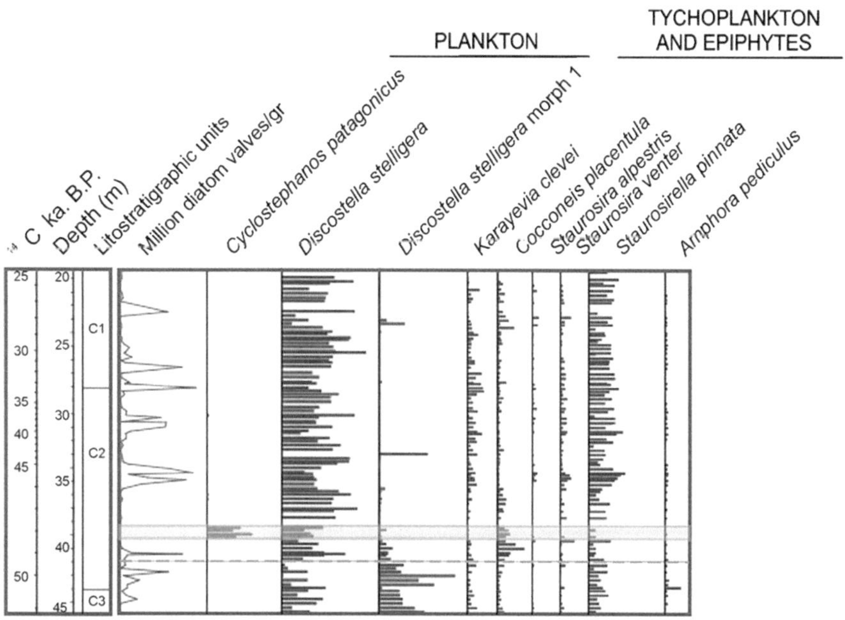

Fig. 7 Diatom diagram of Laguna Potrok Aike core. *Gray shaded zone* shows the peak of *Cyclostephanos patagonicus* and the lowest diatom concentration (modified from Recasens et al. 2015)

pronounced peak. This species has been found only in freshwater and oligotrophic ecosystems in southern Patagonia and indicates a decrease of the temperature. After this zone, *C. patagonicus* drastically disappeared until more recent times. Then, the relative abundances of the different taxa are quite constant and show no big variations. The diatom assemblages are mainly dominated by *Discostella stelligera* and *Staurosirella pinnata*. Also, *Karayevia clevei* occurred in relatively high abundances (Fig. 7). The MIS 3 record in Laguna Potrok Aike shows fluctuations in diatom abundances that coincide with Antarctic warmer events described for ice cores at around 44,500 and 38,500 cal. year B.P., respectively (Blunier and Brook 2001). Low diatom concentration and high proportion of *C. patagonicus* could indicate colder conditions between 47,500 and 48,500 cal. year B.P.

3 Ancient Lakes from South America

A limited number of long continental records allow evaluation of whether the tropics have experienced the same high frequency of climate variations evident in ice cores from northern high latitudes and some marine sediment cores. The only long drilled cores with studied diatom records of tropical South America are Salar de Uyuni, farther south on the Bolivian Altiplano (Fritz et al. 2004), Lake Titicaca in the tropical Andes (Fritz et al. 2012) and Lake Junín, Peru (Tapia et al. 2006).

3.1 Salar de Uyuni, Bolivia

Salar de Uyuni (20° S 68° W, 3653 m a.s.l.) is the largest salt flat of the world. It is located on the Bolivian Altiplano, a high-elevation plateau between the eastern and western cordilleras of the central Andes (Fig. 1). Water inputs to the modern salt-lake are from local precipitation and river runoff.

A ~170,000 year B.P. sequence hydrologic variation from the Salar de Uyuni in tropical South America was analyzed (Fritz et al. 2004). One sediment core of 220.6 m in the central portion of the Salar de Uyuni (20° 14.97′ S, 67° 30.03′ W) was drilled. Alternating mud and salt units in the core reflect alternations between wet and dry periods. The most striking feature of the sequence is that the duration of paleolakes increased in the late Quaternary.

Diatom assemblages are diverse and preservation is highly variable. The uppermost section analyzed (~60,000 to ~20,000 ^{14}C year B.P.) corresponding to MIS 3, is composed by alternating levels containing plankton and benthos diatoms (Fig. 8). Between ~60,000 and ~55,000 year B.P. (46.5–41.6 m) is distinct, because of the high percentages of *Denticula subtilis, Navicula salinicola*, and *Navicula* sp., which together suggest shallow moderately saline waters. A mixture of freshwater/brackish tychoplankton *Pseudostaurosira, Staurosira,* and *Staurosirella*, dominate the diatom-bearing mud units below 50 m. The planktonic taxa *Cyclotella meneghiniana* and *Discostella stelligera*, which tolerate a wide range of salinity, dominate between ~55,000 to ~42,000 year B.P. indicating a deeper environment. Then, the dominance of *Denticula seriata* between ~42,000 to ~35,000 year B.P. suggests increased salinity and decreased depth just prior to precipitation of the overlying salts. The planktonic and saline species *Cyclotella choctawatcheeana* is dominant in the uppermost major mud unit (9.75–18.42 m) between 35,000 and 20,000 year B.P. The diatom composition suggests that lake depths varied from moderate, with high proportions of the plankton, to shallower, with a diverse flora of tychoplankton, epipelon, and epiphytes species (Fig. 8). Many of these species are common in freshwater springs at the margin of the present saline depression (Sylvestre et al. 2001). These changes may reflect increased/decreased precipitation, geomorphic, or tectonic processes that affected the basin hydrology, or some combination of both. Fritz et al. (2004) postulated that the relative influence of insolation forcing on regional moisture budgets may have been stronger during the past 50,000 years than in earlier times.

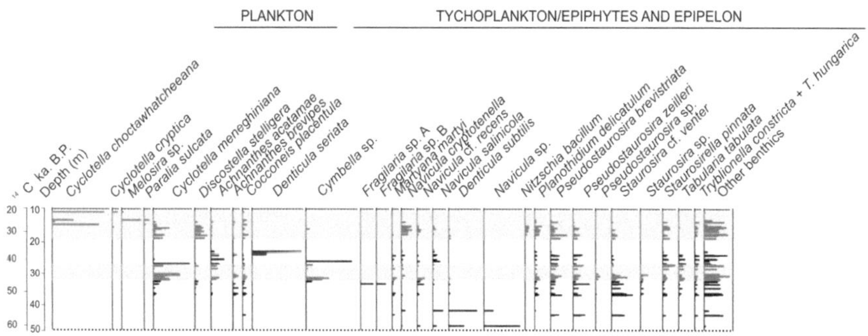

Fig. 8 Diatom percentages of the common taxa in the upper 50 m of the Salar de Uyuni core. *Gray shaded zones* indicate the levels dominate by plankton (modified from Fritz et al. 2004)

3.2 Lake Titicaca

Lake Titicaca (14° 09′–17° 08′ S, 68° 03′–71° 04′ W) is a large, high altitude (3812 m a.s.l.) lake in the tropical Andes (Fig. 1). Fritz et al. (2007, 2010 and 2012) drilled the sediments of Lake Titicaca to obtain a paleoclimatic record that extended to periods prior to the Last Glacial Maximum (LGM). The detailed diatom stratigraphy of one of those cores was studied to interpret the lake-level history in relation to climatic changes (Fritz et al. 2012). The core record that span for the last ~370,000 years replicate both the influence of climate and long-term evolution of the lake basin, and its diatom biota. The first 40 m of the core (~60,000 to ~20,000 ^{14}C year B.P.) includes plankton, tychoplankton, and epiphytes diatoms that are characteristic of freshwater to saline environments (Fig. 9). Cold intervals were deep and dominated by freshwater planktonic taxa as *Cyclostephanos andinus* and *Discostella stelligera*, and peak milder intervals were shallow and dominated

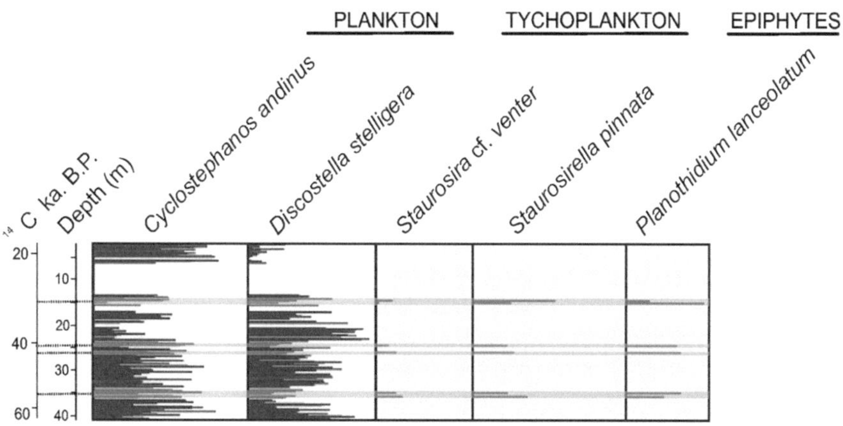

Fig. 9 Diatom percentages of the common taxa in the upper 40 m of Lake Titicaca core. *Gray shaded zones* indicate levels dominate by tychoplankton and epiphytes (modified from Fritz et al. 2012)

by the freshwater/brackish *Staurosira* cf. *venter* and *Staurosirella pinnata* (tychoplankton) and *Planothidium lanceolatum* (epiphyte).

3.3 Lake Junín, Perú

A sediment core (JU96-A) was taken near the town of Ondores on the shore of Lake Junín, Perú (11° 03.52′ S, 76° 07.27′ W; Fig. 1) in 2-m water depth at the edge of a floating sedge mat (Tapia et al. 2006). It spanned between 44,728 cal year B.P. to present days. Diatom analysis of the oldest layers (~45,000 to ~30,000 cal year B.P.; Fig. 10) is dominated by *Staurosira* cf. *venter*, which is indicative of a shallow freshwater lake. The assemblage contains freshwater tychoplankton and epiphytes diatoms during part of MIS 3 (Fig. 10). In the modern Lake Junín samples, *Staurosira* cf. *venter* comprises 55 % of the total assemblage, and it is also found in stream and spring areas. Based on its diatom composition, this environment resembles present day conditions (Tapia et al. 2006). Geomorphological evidence suggests that the Junín Plains were surrounded by glacial ice during the Río Blanco glaciation phase (Wright 1983) favoring the development of Lake Junín in the middle of the high plateau before 40,000 ^{14}C year B.P. Palynological evidence from Lake Junín suggests the development of cold Puna vegetation from ~42,000 to 39,000 ^{14}C year B.P. (Hansen et al. 1984). Plant remains and gastropod shells found in the same levels analyzed for diatoms suggest the presence of aquatic vegetation in shallow waters. The dominance of the epiphytes diatoms *Cocconeis placentula* and *Gomphonema pumilum* at ~43,000 cal. year B.P. and *Cocconeis placentula* between ~38,000 and ~36,000 cal. year B.P. is indicating that the shallow lake environment was punctuated by short-term decreases in water

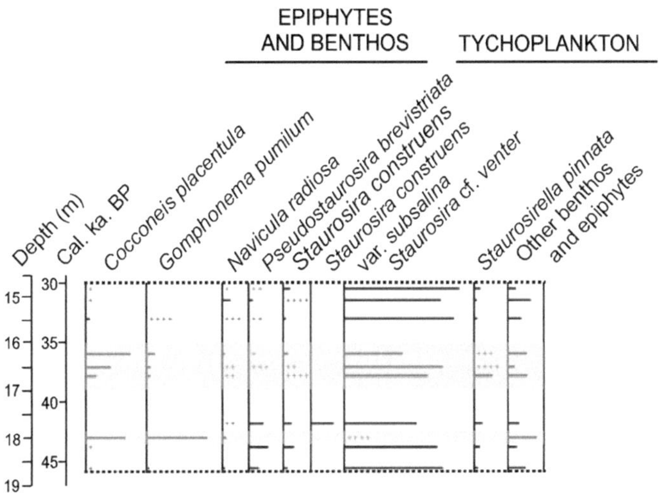

Fig. 10 Diatom percentages of the common taxa in the base of Lake Junín core (between 45 and 30 cal. ka B.P.). *Shaded gray zones* show levels with epiphytes (modified from Tapia et al. 2006)

level with aquatic vegetation. An extensive zone barren of diatom frustules began after around 30,000 cal year B.P. (Fig. 10).

4 Marine Sediment Cores

Paleotemperature records from high latitudes indicate that rapid climatic changes occurred on a millennial timescale during MIS 3 (Heinrich-Dansgaard/Oeschger—H-DO—variability). These features are also observed in tropical areas sensitive to summer monsoon fluctuations, linked to latitudinal shifts of the Intertropical Convergence Zone (ITCZ). To explain the existence of H-DO features at different latitudes, it is necessary to understand how climatic connections are set up between higher and lower latitudes and/or different ocean basins (Leduc et al. 2007). Several very cold periods, known as the Heinrich (H) events, developed during MIS 3 as a result of partial collapse of the North American ice sheet margins and are thought to have impacted global climate; however, their effects on upper ocean dynamics and plankton ecology varied substantially across different regions (Cermeño et al. 2013).

4.1 Benguela Upwelling System

Romero (2010) described rapid paleoceanographic changes that occurred during the last 70,000 years B.P. in the southeastern Atlantic Ocean. The study is based on diatom assemblage analysis, the concentration and the bulk biogenic components of three gravity cores recovered from the Benguela Upwelling System (BUS) between 19° and 25° S (Fig. 1). Diatoms are the main contributors to the biogenic siliceous fraction within the BUS. The dominance of species of upwelling-associated *Chaetoceros* spp. spores, *Delphineis karstenii,* and the neritic *Cyclotella litoralis* shows that the most favorable upwelling conditions occurred. According to Pokras (1991), when *D. karstenii* associated with *Chaetoceros* spores occurred, they are reflecting intense upwelling in a neritic environment. Independent of the core sites, the highest diatom concentration and accumulation rate, as well as the strongest fluctuations in magnitude, occurred during MIS 3 (Fig. 11). At 25° S, the more intense upwelling location was placed, due to the combination of strong seaward extending upwelling areas, shoaling of the upwelled water, and the influence of silicate-rich waters of Antarctic origin. Abrupt diatom accumulation rate decreases were observed in different moments during MIS 3. These are related to Heinrich (H) events: H3, H4, H5, and H6, approximately ~31, ~39, ~46, and ~58 ka, respectively (Fig. 11). According to Romero (2010), Late Quaternary variations of productivity and upwelling intensity are linked to the variability in wind stress, but there were several oceanographic processes that influenced the temporal variation pattern of diatom productivity along the continental margin off southwestern Africa during MIS 3.

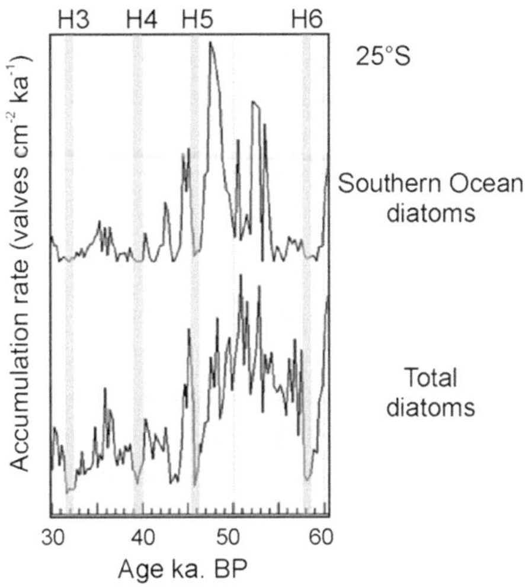

Fig. 11 Accumulation rate (valves cm^{-2} ka^{-1}) of Southern Ocean diatoms and total diatoms at the core GeoB3606-1 (25° S) in the Benguela Upwelling System for the MIS 3 (modified from Romero 2010)

4.2 Panamá Basin

Tropical communities of marine diatoms responded through changes in total abundance and species richness to events of rapid changes in climate initiated at higher latitudes (Cermeño et al. 2013). Total diatom concentrations analysis from MD02-2529 piston core (08° 12.33′ N; 84° 07.32′ W; Fig. 1) provided clear evidence for millennial variability in diatom productivity at the Panamá Basin during the last glacial period (Romero et al. 2011). Diatom production is expected to respond to changes in surface water nutrient concentration as triggered by changes in wind conditions. During the Heinrich events, diatoms strongly reduced their production, probably because of enhanced stratification in the upper water column (Fig. 12). The broad dispersal of marine microbial plankton species (Cermeño and Falkowski 2009) and their rapid response to environmental variability confer on microbial plankton community low resistance to change but enormous potential for recovery. Marine planktonic diatoms from low-latitude coastal ecosystems, responded to climate perturbations through drastic changes in population

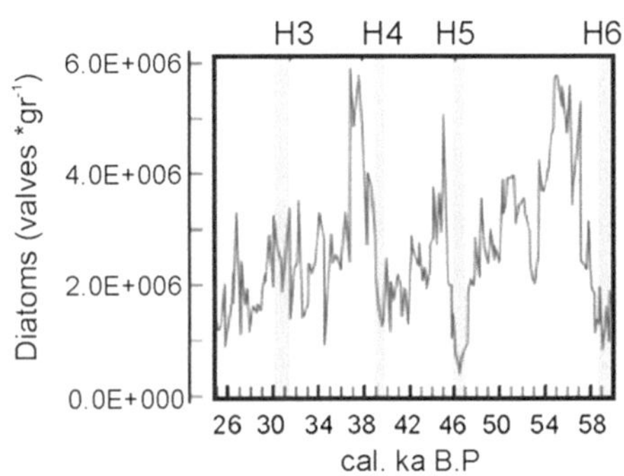

Fig. 12 Variations of total diatom concentration (valves g^{-1}) for the time period 60–25 ka B.P. at site MD02-2529 from the Panamá basin (modified from Romero et al. 2011)

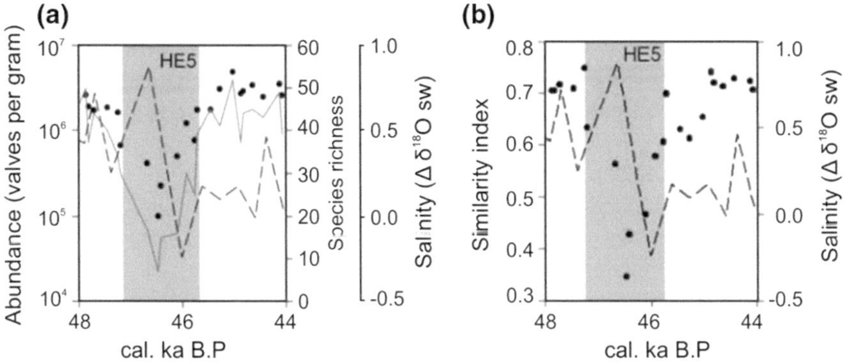

Fig. 13 **a** Temporal evolution of diatom abundance (*dots*), species richness (*gray line*) and sea surface salinity (*black dashed line*) in the Panamá Basin during HE5. The *shaded areas* denote the time interval of the corresponding climatic episode. **b** Community similarity through time (*dots*) and sea surface salinity (*dashed line*) in the Panamá Basin during HE5 (modified from Cermeño et al. 2013)

abundance, species richness, and taxonomic composition but they rapidly reestablished once climatic forcing and oceanographic conditions returned to previous states (Cermeño et al. 2013). An example is what is occurring in the Panamá Basin where the abundance (Fig. 13a) and the similarity index (Fig. 13b) changed sharply during the perturbation after H5 (about 46,000 cal year B.P.), showing the potential of these microbial communities to recover once climatic or oceanographic conditions are fully reestablished. Then, in the marine microplankton realm, analogous climate conditions produce similar communities. Cermeño et al. (2013) found that patterns of community change in the Panamá Basin coincide with periods of higher sea surface salinity (Fig. 13b), which is indicative of enhanced subtropical influence in this region. These authors concluded that these diatom communities have been relatively stable (resilient) over the Late Pleistocene despite abrupt changes in climate and hydrographic conditions.

5 Discussion

Past glacial periods were punctuated by pseudo-periodical abrupt temperature oscillations (the D–O cycles and the Heinrich events), which are particularly marked in ancient records from the middle to higher latitudes of the Northern Hemisphere (Greenland ice cores, ocean, and lacustrine sediments, and cave stalagmites). Associated events (oscillations of temperatures and/or hydrological cycle) are observed in the intertropical zone and in Southern Hemisphere surface records, as well as in the intermediate and deep oceans.

In diatom studies, knowledge of the autoecology of individual diatom species is implicit. Both direct and indirect climate reconstructions have furthered our understanding between diatoms and their environment, although the results are often contentious; whereas quantitative studies are increasingly being improved, interpretations from more qualitative work should not be ignored (Mackay et al. 2003). There are differences in the environmental requirements (e.g., light and nutrients) among freshwater and marine diatoms, and these diatom taxon-specific differences can be used to interpret climate-induced changes in aquatic systems over time (Rühland et al. 2015). In recent years, the number of diatom-based Quaternary climate reconstructions has increased. The studies are either from terrestrial ecosystems as marine and coastal environments.

Diatom assemblages can be especially sensitive to warming-induced changes in lakes that favor the growth of small and planktonic species (Rühland et al. 2008) or of more complex and diverse periphytic diatom assemblages (Douglas and Smol 2010). Shifts in the relative abundances between planktonic and periphytic diatoms (fragilarioid taxa) may have been in response to changes in the duration of ice cover (Smol and Douglas 2007). The high-resolution diatom stratigraphy of a paleolake at Les Echets, eastern France, has revealed distinct species turnover events and large fluctuations in stable oxygen isotope values in diatom frustules, as a response to DO climate variability (Ampel et al. 2010). Transitions from stadial to interstadial conditions show Fragilaria–Cyclotella succession, which resembles an example given by Rühland et al. (2008) for Arctic and Subarctic freshwater ecosystems in the Northern Hemisphere, where limnological changes with distinct diatom assemblages turnovers as a result of the temperature increase during the last 150–200 years. Same assemblages became reestablished during each stadial and interstadial phases. The relative abundance of the most dominant species within these assemblages varies and might indicate differences in climatic conditions. This type of succession was observed in ancient lakes of South America. For example, in the stratigraphic record of the Salar de Uyuni, Bolivia (Fritz et al. 2004) *Cyclotella* spp. and *Discostella stelligera* dominate the levels between *ca.* 25 to 33 ka B.P., and *ca.* 41 to 46 ka B.P. (as shown in Fig. 8). These taxa would be indicating warmer conditions. In the Lake Titicaca core (Fritz et al. 2012) important peaks of fragilariod diatoms (colder conditions) alternating with *Discostella stelligera* peaks, were observed at ca. 31, 40, 43 and 56 ka B.P. (see Fig. 9). The Lake Titicaca drill-core sequence represents the longest continuous record of both glaciation and hydrologic variation from the Southern Hemisphere tropics of South America and documents four regional glacial–interglacial cycles in the tropical Andes. The general correspondence of regional glacial periods with intervals of higher lake level and of periods of reduced glaciations with times of lower lake level indicates that colder–wetter and warmer–drier conditions are the regional models found (Fritz et al. 2007).

Some lacustrine deposits from Argentina, as the La Bomba sequence in Mendoza province, presented the succession Fragilariod-Cyclotella between *ca.* 35 and 31 ka B.P. (Fig. 4) showing a probably milder climate conditions. Before and after this interval, the dominance of Fragilarioid taxa was recorded at the Rahue section, Neuquén (Fig. 5) and Lago Fagnano, Tierra del Fuego (Fig. 6).

Sediment cores from marine environment of lower and middle latitudes revealed that the maxima in total diatom concentration occurred during the MIS 3 (Romero 2010; Romero et al. 2011 and Cermeño et al. 2013). The enhanced primary productivity is the result of intense upwelling. Rapid changes in diatom concentrations were recorded. Such patterns indicate a linkage between diatom production in the coastal Eastern Equatorial Pacific and Benguella Upwelling System with rapid climate changes in the high-latitude North Atlantic Ocean. As shown at Figs. 11, 12 and 13 diatoms strongly reduced their production during Heinrich events (colder), probably because of enhanced stratification in the upper water column. The dominance of upwelling-associated *Chaetoceros* spores shows that the most favorable upwelling conditions occurred in this interval.

More studies are needed in order to analyze the evolution of water bodies and coasts of South America related to climatic change and correlate them with other records of high latitudes. The importance of obtaining paleoclimatic archives of continuous, high-quality records spanning the Quaternary should be emphasized, especially in the MIS 3 period, in addition to reliable dates and model ages.

Many challenges to the discipline remain, including the understanding of the ecological requirements of diatom species and diatom life cycle strategies. At the same time databases should be used more widely to explore the relationships between diatom biogeography and climate, habitat, and environmental chemistry, thereby generating new hypotheses for future study (Mackay et al. 2003).

References

Ampel L, Wohlfarth B, Risberg J, Veres D, Leng MJ, Tillman PK (2010) Diatom assemblage dynamics during abrupt climate change: the response of lacustrine diatoms to Dansgaard-Oeschger cycles during the last glacial period. J Paleolimnol 44:397–404

Blasi A, Castiñeira Latorre C, Del Puerto L, Prieto A, Fucks E, De Francesco C, Hanson P, García Rodriguez F, Huarte R, Carbonari J, Young A (2010) Paleoambientes de la cuenca media del Río Luján (Buenos Aires, Argentina) durante el último período glacial (EIO 4-2). Lat Am J Sedimentol Basin Anal 17(2):85–111

Blunier T, Brook EJ (2001) Timing of millennial-scale climate changes in Antarctica and Greenland during the last glacial period. Science 291:109–112

Brandefelt J, Kjellstrom E, Naslund JO, Strandberg G, Voelker AHL, Wohlfarth B (2011) A coupled climate model simulation of marine isotope stage 3 stadial climate. Clim Past 7:649–670

Bujalesky GG, Coronato AM, Roig CE, Rabassa JR, Espinosa MA (1992) Facies deltaicas proglaciales pleistocenas del Lago Fagnano, Tierra del Fuego, Argentina. IV Reunión Argentina de Sedimentología Actas, La Plata, pp 235–242

Bujalesky GG, Heusser CJ, Coronato AM, Roig C, Rabassa JR (1997) Pleistocene glaciolacustrine sedimentation at Lago Fagnano, Andes of Tierra del Fuego, southernmost South America. Quat Sci Rev 16:767–778

Cermeño P, Falkowski PG (2009) Controls on diatom biogeography in the ocean. Science 325:1539–1541

Cermeño P, Marañón E, Romero O (2013) Response of marine diatom communities to Late Quaternary abrupt climate changes. J Plankton Res 35(1):12–21

Douglas MSV, Smol JP (2010) Freshwater diatoms as indicators of environmental change in the high Arctic. In: Smol JP, Stoermer EF (eds) The diatoms: applications for the environmental and earth sciences, 2nd edn. Cambridge University Press, Cambridge, pp 249–266

Fritz SC, Baker PA, Lowenstein TK, Seltzer GO, Rigsby CA, Dwyer GS, Tapia PM, Arnold KK, Ku TL, Luo S (2004) Hydrologic variation during the last 170,000 years in the southern hemisphere tropics of South America. Quat Res 61:95–104

Fritz SC, Baker PA, Seltzer GO, Ballantyne A, Tapia PM, Cheng H, Edwards RL (2007) Quaternary glaciation and hydrologic variation in the South American tropics as reconstructed from the Lake Titicaca drilling project. Quat Res 68:410–420

Fritz SC, Baker PA, Ekdahl E, Seltzer GO, Stevens LR (2010) Millennial-scale climate variability during the last glacial period in the tropical Andes. Quat Sci Rev 29:1017–1024

Fritz SC, Baker PA, Tapia PM, Spanbauer T, Westover K (2012) Evolution of the Lake Titicaca basin and its diatom flora over the last ~370,000 years. Palaeogeogr Palaeoclimatol Palaeoecol 317–318:93–103

Garleff K, Helmut Stingl B, Heinz Veit B (1993) New dates on the late quaternary history of landscape and climate in the Bolsón of Fiambalá/NW Argentina (Province Catamarca). Zbl Geol Paläont Teil I 1–2:333–341

Grootes PM, Stuiver M, White JWC, Johnsen SJ, Jouzel J (1993) Comparison of oxygen isotope records from the GISP2 and GRIP Greenland ice cores. Nature 366:552–554

Haberzettl T, Kuck B, Wulf S, Anselmetti F, Ariztegui D, Corbella H, Fey M, Janssen S, Lücke A, Mayr C, Ohlendorf C, Schäbitz F, Schleser GH, Wille M, Zolitschka B (2008) Hydrological variability in southeastern Patagonia and explosive volcanic activity in the southern Andean Cordillera during oxygen isotope stage 3 and the Holocene inferred from lake sediments of Laguna Potrok Aike, Argentina. Palaeogeogr Palaeoclimatol Palaeoecol 259:213–229

Hansen BCS, Wright HE Jr, Bradbury JP (1984) Pollen studies in the Junín area, central Peruvian Andes. Geol Soc Am Bull 95:1454–1465

Hassan GS, De Francesco CG, Dieguez S (2013) The significance of modern diatoms as paleoenvironmental indicators along an altitudinal gradient in the of central Argentina. Palaeogeogr Palaeoclimatol Palaeoecol 369:249–360

Leduc G, Vidal L, Tachikawa K, Rostek F, Sonzogni C, Beaufort L, Bard E (2007) Moisture transport across Central America as a positive feedback on abrupt climatic changes. Nature 445:908–911

Mackay AW, Jones VJ, Battarbee RW (2003) Approaches to Holocene climate reconstructions using diatoms. In: Mackay AW, Battarbee AW, Birks RW, Oldfield F (eds) Global change in the Holocene, pp 294–309

Maidana NI, Seeligmann C (2006) Diatomeas (Bacillariophyceae) de Ambientes Acuáticos de Altura de la Provincia de Catamarca, Argentina. Bol Soc Arg Bot 41(1–2):1–13

Markgraf V, Bradbury JP, Fernández J (1986) Bajada de Rahue, Province of Neuquén, Argentina: an interstadial deposit in northern Patagonia. Palaeogeogr Palaeoclimatol Palaeoecol 56:251–258

Pokras EM (1991) A displaced neritic diatom (Delphineis karstenii) in pelagic sediments of the southeast Atlantic. Mar Micropaleontol 17:311–317

Rabassa J, Ponce JF (2013) The Heinrich and Dansgaard-Oeschger climatic events during marine isotopic stage 3: searching for appropriate times for human colonization of the Americas. Quat Internat 299:94–105

Recasens C, Ariztegui D, Maidana N, Zolitschka B (2015) Diatoms as indicators of hydrological and climatic changes in Laguna Potrok Aike (Patagonia) since the Late Pleistocene. Palaeogeogr Palaeoclimatol Palaeoecol 417:309–319

Romero OE (2010) Changes in style and intensity of production in the Southeastern Atlantic over the last 70,000 yr. Mar Micropal 74:15–28

Romero OE, Leduc G, Vidal L, Fischer G (2011) Millennial variability and long-term changes of the diatom production in the eastern equatorial Pacific during the last glacial cycle. Paleoceanography 26:2212. doi:10.1029/2010PA002099

Rühland K, Paterson AM, Smol JP (2008) Hemispheric-scale patterns of climate-related shifts in planktonic diatoms from North American and European lakes. Global Change Biol 14:2740–2754
Rühland K, Paterson A, Smol JP (2015) Lake diatom responses to warming: reviewing the evidence. J Paleolimnol 54:1–35
Smol JP, Douglas MSV (2007) From controversy to consensus: making the case for recent climate change in the Arctic using lake sediments. Frontiers Ecol Environ 5:466–474
Sylvestre F, Servant-Vildary S, Roux M (2001) Diatom-based ionic concentration and salinity models from the south Bolivian Altiplano (15–23 degrees South). J Paleolimnol 25:279–295
Tapia PM, Fritz SC, Seltzer GO, Rodbell DT, Metivier SP (2006) Contemporary distribution and late-quaternary stratigraphy of diatoms in the Junín Plain, Central Andes, Peru. Bol Soc Geol Perú 101:19–42
Wright HE Jr (1983) Late Pleistocene glaciation and climate around the Junín Plain, central Peruvian highlands. Geogr Ann 65A(1–2):35–43

Silicophytolith Studies in South America and Argentina: Scope and Limitations for Paleoenvironmental Reconstruction of the Marine Isotope Stage 3 (MIS3)

Margarita Osterrieth, María Fernanda Alvarez, Mariana Fernández Honaine and Georgina Erra

Abstract Silicophytoliths are amorphous silica biomineralizations deposited in intracellular or extracellular spaces of plant tissues. Due to their taxonomic value and their high preservation in a variety of soils and sediments, they are widely used as indicators of past plant communities. Numerous phytolith studies show the presence of past grass-dominated ecosystems in the Late Cenozoic, including changes between glacial and interglacial periods. Studies in South America are scarce, particularly those associated to the temporal interval corresponding to the Marine Isotopic Stage 3 (MIS3). A synthesis of silicophytolith studies on pedosedimentary sequences of MIS3 age in South America is herein presented and, particularly, our own work carried out in Argentina. Integrated profiles' representatives of typical pedostratigraphic sequences from two regional geomorphological units (Mesopotamia and the Pampean Plain) were analyzed. Samples from pedostratigraphic sequences were subjected to routine analysis. Silicophytoliths were extracted after the elimination of carbonates, organic matter, and clay; and their morphologies were described under optical and scanning electron microscopes (SEMs). Profiles from both regions show the presence of conspicuous paleopedological levels, developed in the MIS3 interval. C_3 grasses (Pooideae and/or

M. Osterrieth (✉) · M.F. Alvarez · M. Fernández Honaine
Instituto de Geología de Costas y del Cuaternario, Facultad de Ciencias Exactas y Naturales, Universidad Nacional de Mar del Plata-CIC, CC 722, Correo Central, 7600 Mar del Plata, Argentina
e-mail: mosterri@yahoo.com.ar

M. Osterrieth · M.F. Alvarez · M. Fernández Honaine
Instituto de Investigaciones Marinas y Costeras, Universidad Nacional de Mar del Plata-CONICET, Peña 4046, 7600 Mar del Plata, Argentina

G. Erra
Facultad de Ciencias Naturales y Museo de La Plata, Universidad Nacional de La Plata, Paseo del Bosque s/n, 1900 La Plata, Argentina

M.F. Alvarez · M. Fernández Honaine · G. Erra
CONICET, Buenos Aires, Argentina

G.M. Gasparini et al. (eds.), *Marine Isotope Stage 3 in Southern South America, 60 ka B.P.–30 ka B.P.*, Springer Earth System Sciences,
DOI 10.1007/978-3-319-40000-6_17

Panicoideae subfamilies) and, in a lesser proportion, C_4 grasses (Chloridoideae and/or Panicoideae subfamilies) were present in both areas. This indicates the development of mesothermal grass-dominated ecosystems, which nowadays grow mainly in warm-temperate regions. Within the MIS3, frequent climatic environmental variations during the Late Pleistocene may have led to a fluctuation in biogeographic connections between the Mesopotamian region and other parts of South America, closely linked to the Chaco-Pampean plain and, at other times, to inter-tropical regions.

Keywords Pedostratigraphic · Phytolith · Grasses · Pampean plain · Mesopotamia

1 Introduction

Biomineralizations are present in all levels of the biosphere. They are biogenic inorganic–organic composites, crystalline or amorphous, deposited in intracellular or extracellular spaces as a consequence of the metabolic activity of organisms (Lowenstam 1981; Osterrieth 2004). Phytoliths are biomineralizations present in cell walls and/or extra or intracellular spaces of plant tissues and can be compounds of different elements, such as calcium or silicon (Parry and Smithson 1964; Bertoldi de Pomar 1975; Piperno 2006; Gomes Coe et al. 2014). The most common phytoliths in vascular plants are calcium carbonates and oxalates (calciphytoliths), and amorphous silica hydrated deposits or silicophytoliths (also known as silica phytoliths, opal phytoliths, silica plant bodies, amorphous silica biomineralizations, etc.) (Gomes Coe et al. 2014) (Fig. 1). Both types of phytoliths are widespread throughout the plant kingdom; however, in general, those taxa that commonly produce calcium phytoliths do not produce silica phytoliths in abundance, and vice versa, except for some families such as Ulmaceae, Urticaceae, and Cannabinaceae, which produce both calciphytoliths and silica phytoliths (Fig. 2) (Fernández Honaine et al. 2005; Borrelli et al. 2011). Although the analysis of phytoliths composed by calcium (calciphytoliths) and amorphous silica is very useful for anatomical, taxonomical, and physiological studies, the silica phytoliths are widely applied in paleoenvironmental, paleontological, and archaeological studies due to their high preservation in soils and sediments and their taxonomic relevance. Silicophytoliths are incorporated in the soil through pre-, sin- and post-pedogenetic processes, whereby they are good indicators of past plant communities and the associated pedological development (Osterrieth 2008a, b). In contrast, calciphytoliths are rapidly dissolved in slightly acidic environments and they are seldom found in soils and sediments. Fungi also produce the same morphologies as plants, and their low preservation in soils makes them unsuitable as indicators of past plant communities (Osterrieth et al. 2014a).

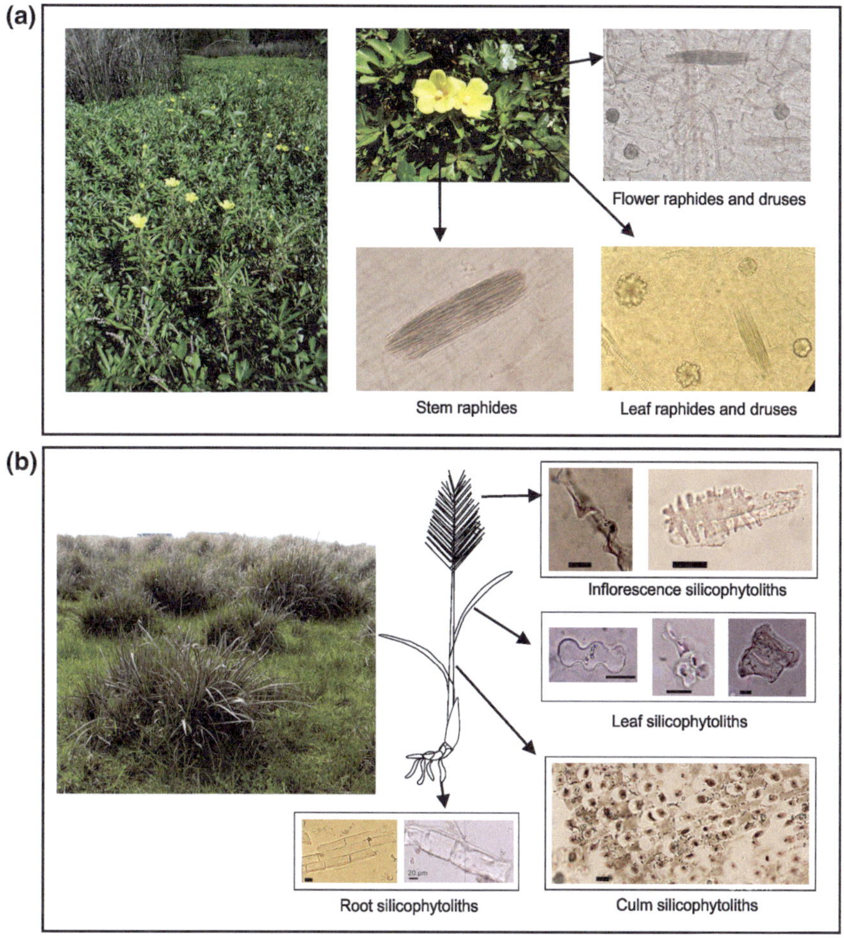

Fig. 1 Production of calciphytoliths and silicophytoliths in two different taxa. **a** Raphides and druses (calciphytoliths) in flower, leaf and stem of *Ludwigia peploides* (Onagraceae) (Borrelli et al. 2011, Altamirano et al. unpublished), **b** Different morphotypes of silicophytoliths in inflorescence, leaf, culm and root of *Paspalum quadrifarium* (Poaceae) (Modified from Gomes Coe et al. 2014)

1.1 How and Where the Silicophytoliths Are Produced?

The formation of silicophytoliths in plant tissues is the result of the Si polymerization. Silicon (Si) is present in soils as monosilicic acid and plants absorb it through their roots (Ma and Takahashi 2002; Piperno 2006). In the plant root, the uptake of Si involves protein-mediated active transport and/or passive transport through diffusion; the relative importance of each type of transportation present in a

Fig. 2 Production of calciphytoliths and silicophytoliths in *Celtis ehrenbergiana* ("tala"), Cannabinaceae family. **a** Plant aspect, **b** Phytoliths in leaves, **c** Phytoliths in fruit. (Fernández Honaine et al. 2005; Borrelli 2008)

plant depend on the taxon involved (Ma and Takahashi 2002; Ma and Yamaji 2006; Cornelis et al. 2010). Once silicic acid is in xylem, it is translocated to the stem by the transpiration stream, and transported to parenchymatic cells (Yamaji et al. 2008). Silicic acid is further concentrated through loss of water and polymerized when concentration exceeds 2 mM. The process of Si polymerization converts silicic acid to colloidal silicic acid and finally to silica gel (Ma and Takahashi 2002).

The distribution of silica deposits is mainly controlled by the transpiration process (Ma and Takahashi 2002). However, there are other factors that influence the silicification process, such as the systematic position of a taxon, the phenological stage of the individuals, the age of the organ and the mechanism of Si uptake (Hodson et al. 2005; Motomura et al. 2002, 2004; Ma et al. 2011; Fernández Honaine and Osterrieth 2012; Fernández Honaine et al. 2013). Also, the availability and content of silicic acid in soils (which also depends on the presence of Fe and Al oxides and pH), the temperature (which affects transpiration rate) and the incidence of herbivore animals, may also influence on silicophytolith content (Jones and Handreck 1967; McNaughton et al. 1985; Epstein 1994; Massey et al. 2007). The presence of silica biomineralization in plant species has numerous benefits such as improvement of biomass production, anti-herbivore defense, and the amelioration of heavy metal toxicity, among others (e.g., Jones and Handreck 1967; Epstein1994; Ma and Takahashi 2002).

The deposits of amorphous silica (silicophytoliths) within the lumen cells can adopt different forms. Some silicophytolith morphologies may reflect the shape of the cell, enabling the association between the morphology and a type of cell or tissue, and hence the taxon that produced it (e.g., Metcalfe 1960; Twiss 1992; Piperno 2006). In the particular case of Poaceae, where it is possible to assign a silicophytolith morphotype to a subfamily level, it is also possible to assign them to a photosynthetic pathway (C_3 or C_4) (Twiss 1992; Gallego and Distel 2004; Fernández Honaine et al. 2006; Piperno 2006). In other cases, such as in Cyperaceae and Arecaceae families, the resulting morphologies do not resemble cellular forms; however, the silicophytoliths that are formed are unique to each family and allow the assignation to specific taxa (Metcalfe 1960; Tomlinson 1961; Ollendorf 1992). For these reason, silicophytoliths are considered as an important taxonomic tool for plant differentiation (e.g., Metcalfe 1960; Piperno 2006; Fernández Honaine et al. 2006) (Table 1). The main silicophytolith producers are

Table 1 Main silicophytolith morphologies present in soils, paleosoils and sediments and their systematic affinity

Morphotype	Taxon	Photo
Simple lobate[a]	Poaceae family: mainly in Panicoideae, Stipoideae, Arundinoideae; in some species of Pooideae	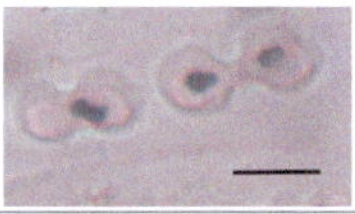
Panicoid bilobate[a]	Poaceae family: mainly in Panicoideae, Arundinoideae, Bambusoideae subfamilies	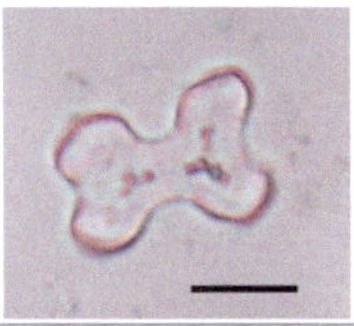
Stipa type bilobate[a]	Poaceae family: mainly in Pooideae, Stipoideae subfamilies	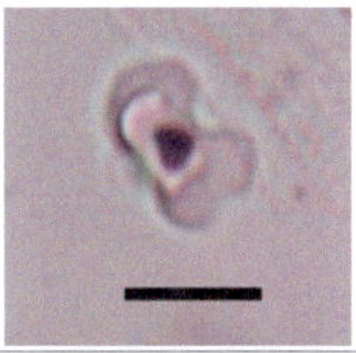
Saddle[b]	Poaceae family: mainly in Chloridoideae and in some species of Bambusoideae and Arundinoideae subfamilies	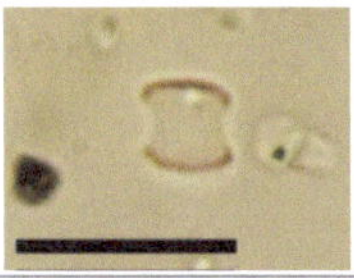
Cross[c]	Poaceae family: mainly Panicoideae, Bambusoideae subfamilies	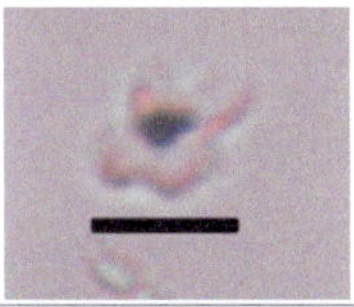
Rondel[b]	Poaceae family: mainly Pooideae and Stipoideae, and in some species of Bambusoideae subfamilies	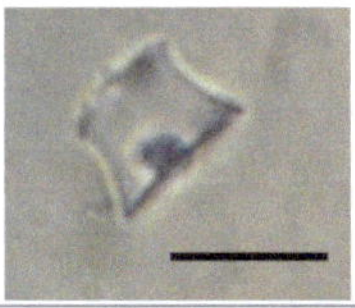

(continued)

Table 1 (continued)

Morphotype	Taxon	Photo
Polylobate[b]	Poaceae family: mainly Panicoideae	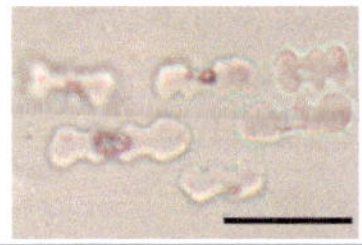
Trapeziform/trapeziform crenate[b]	Poaceae family: mainly Pooideae	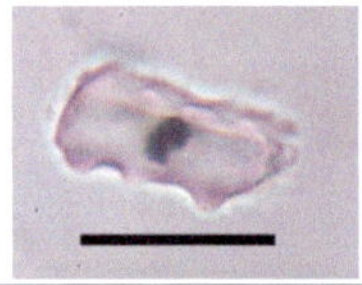
Elongate psilate, crenate or echinate[b]	Monocots and dicots	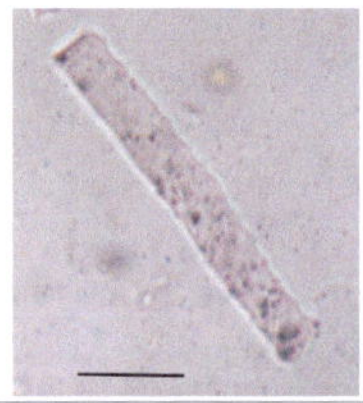
Point-shaped[c]	Poaceae and dicots	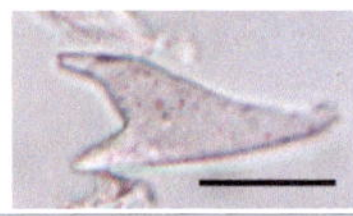
Bulliform cell[b]	Mainly Poaceae family	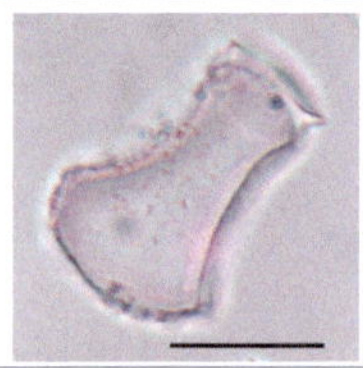
Conical[d]	Cyperaceae	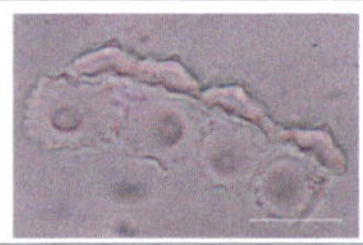
Globular psilate[b]	Monocots, woody dicots, Poaceae (roots)	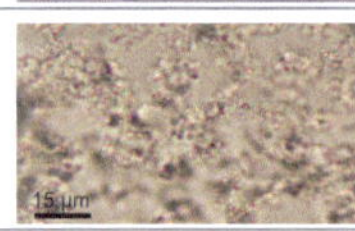
Globular granulate[b]	Woody dicots, Cannaceae	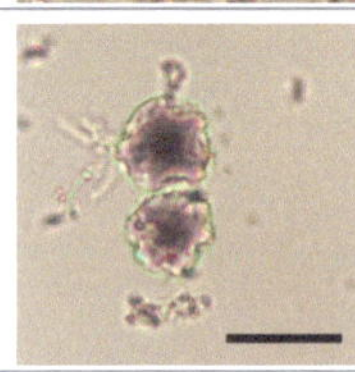

(continued)

Table 1 (continued)

Morphotype	Taxon	Photo
Globular echinate[b]	Arecaceae, Bromeliaceae	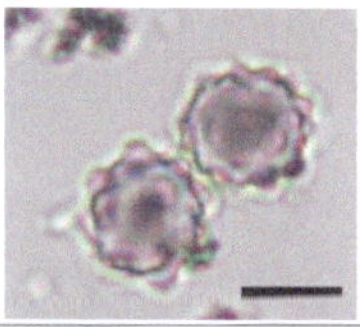
Tabular polihedrical[b]	Dicots	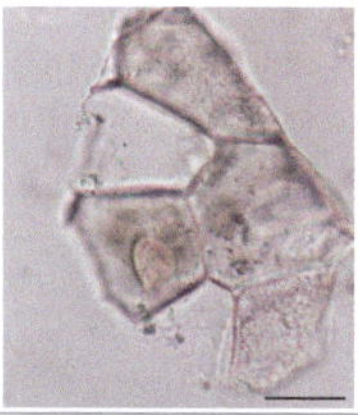
Cylindrical sulcated tracheids[b]	Monocots and dicots	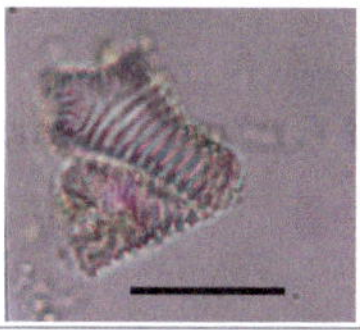

Morphotypes names according to: *a* Fredlund and Tieszen 1994; *b* Madella et al. 2005; c Twiss 1992; d Ollendorf 1992. Scale bar: 25 μm, except in Simple lobate, Panicoid bilobate, *Stipa* type bilobate, Cross, Rondel, Globular echinate (10 μm), and Globular psilate (15 μm)

the Poaceae, Cyperaceae, Arecaceae, and Equisetaceae families, among others (Metcalfe 1960; Piperno 2006).

The taxonomic-systematic problem is well known in silicophytolith studies; fortunately, work is being done to develop an integrated nomenclature through the International Code for Phytolith Nomenclature (Madella et al. 2005). The multiplicity and redundancy problem, which affects the paleoenvironmental, paleobotanical, and archaeological interpretations, has also been discussed (Rovner 1971; Piperno 1988, 2006).

1.2 *Silicophytoliths in Soils, Paleosols, and Sediments*

Once the plant or the organ that contains the phytoliths falls to the ground and decomposes, silicophytoliths are released and incorporated to soils. Due to their siliceous composition, they can be preserved in many types of environments, and comprise an important fraction of soil and sediment particles. Depending on the environmental and pedological conditions, silicophytoliths are affected by diverse taphonomical processes, both in natural and anthropic environments during the Quaternary. They can be preserved, dissolved, or fragmented and also be transported by different agents (wind, water, animals and people). In other words, the silicophytolith record represents a fragmentary sample of the phytolith production of plants, therefore the study of the taphonomical processes in relation to their presence/absence in pedosedimentary sequences is essential but it has been scarcely addressed (Osterrieth et al. 2009; Fernández Honaine et al. 2009; Borrelli 2008; Osterrieth et al. 2014a, b; 2015). Other taphonomical aspects also important to be evaluated are the methodologies used in silicophytolith studies, from soil sampling to studies at a submicroscopic level. Currently, there are several methodologies and the work is essentially done at a very detailed resolution level, which could lead to interpretation errors if the environmental or paleoenvironmental context of the study material is unknown or not clearly stated (Osterrieth et al. 2009, 2014a).

Grasslands and savannas occupy ¼ of the total surface of the South America continent and can be grouped in tropical and subtropical savannas, temperate grasslands, and cold-temperate grasslands (Burkart 1975; Soriano 1979). Tropical and subtropical grasslands include the "llanos" associated with the Río Orinoco system in Venezuela and Colombia, highlands and basins of Guiana between the Orinoco and Amazons rivers, and the "campos cerrados" of central Brazil. The temperate grasslands include the Río de la Plata grasslands, which encompass the "pampas" from Argentina and "campos" from Brazil and Uruguay. Finally, cold-temperate grasslands are located in the Patagonian region (Coupland 1992). Different proxies, such as pollen, vertebrate fossils and silicophytoliths, allowed the interpretation of the evolution of the grasslands in South America. Data from pollen assemblages and endemic ungulates (notoungulates) in the early and late Paleocene showed the presence of grass species in these times (Jacobs et al. 1999). Paleobotanical and paleontological evidence suggests that C_3 grasses may have been a major component in some South American Late Oligocene ecosystems. Evidence of C_4 grasses as a dietary component is first apparent in the enamel of 10 Ma herbivores from Bolivia. Overall, the vertebrate and isotope data suggest that the initial spread of grass-dominated ecosystems and the coevolution of grazing herbivores involved C_3 grasses (Jacobs et al. 1999). After their Late Miocene expansion, C_4 grasses were more important in herbivore diets at lower latitudes, indicating a latitudinal gradient in C_4 grass abundance (Strömberg 2005, 2011).

All these communities (grasslands) are represented by species belonging to Poaceae, the main silicophytolith producer among plant families. Due to the taxonomic relevance that silicophytoliths have in grass families, and their good preservation in a wide range of soils, paleosols and sediments, they have been largely applied as good indicators of paleograsslands in paleoenvironmental studies. Numerous global and regional studies have proven their usefulness in the interpretation of past grass-dominated ecosystems in the Cenozoic, including changes between glacial and interglacial periods (e.g., Fredlund and Tieszen 1994, 1997; Alexandre et al. 1997; Strömberg 2005; Strömberg et al. 2007; Osterrieth et al. 2014b). However, studies in South America are scarce, particularly those associated to the temporal interval corresponding to the Marine Isotopic Stage 3 (MIS3).

The first silicophytolith studies were those carried out by Ehrenberg, with samples brought by Charles Darwin from his voyage of the "Beagle" (Ehrenberg 1854). Then, many studies described the presence of silica deposits in tissues of many plant families, but all of them have an anatomical and taxonomical purpose (e.g., Metcalfe 1960; Prychid et al. 2004). In the 1970s, Rovner and other researchers applied phytolith studies in order to solve different archaeological, archaeobotanical, and in a lesser proportion, paleobotanical and paleoenvironmental questions. During the last 25–30 years, phytolith studies have been applied to understand pedoarchaeological, paleoecological and paleoenvironmental changes during the Late Quaternary–Holocene, both in global and South American scenarios (Piperno 2006; Gomes Coe and Osterrieth 2014).

Since more than 100 years, there has been an international background in studies of soil silicophytoliths on the plains. In Central and South America there are a few but important contributions from Colombia (Florez and Parra 2008) and Uruguay (Campos et al. 2001; Iriarte 2001; Iriarte et al. 2012; Del Puerto and Inda 2008). In Argentina, research was scarce at first, but established strong foundations and fostered later continuity for the phytolith studies of soils (e.g., Frenguelli 1930; Teruggi 1955, 1957; Andreis 1972; Bertoldi de Pomar 1970, 1971, 1975). Since the 1990s, phytolith studies have been ongoing, and even though the number of researchers is still low, the number of studies has increased over the last 10 years (Osterrieth et al. 2014a).

Joaquín Frenguelli was the first author to mention silicophytoliths in Quaternary soils and sediments in the Pampean region in 1930. Later on, Teruggi (1955, 1957) pointed out the presence of grass vegetation during the formation of the loess deposits in the area. In addition, Bertoldi de Pomar (1967–1980) analyzed several aspects of the phytoliths found in continental soils and sediments on the plains of the province of Santa Fe (Argentina). The silicophytolith studies in pedosedimentary sequences of loess-paleosols in the Quaternary of the Pampean Plains have been based on the fact that the mineral studies were not very efficient due to the origin of the transported and re-transported sediments under the influence of wind and water action (Osterrieth and Martínez 1993, among others). Furthermore, in the mineralogy of these soils and sediments of the Late Quaternary it is evident that light minerals are more abundant and, among them, those composed of amorphous silica of organic and inorganic origin reach percentages of over 20 % (Osterrieth et al. 2014a). Finally, different studies have determined the contents of silicophytoliths in soils, paleosols, loess sediments, and typical loess of the different environments in southeastern Buenos Aires province, as well as their variations through the Late Cenozoic (Osterrieth 1998, 2000, 2004, 2006; Osterrieth et al. 2004–2015).

Silicophytoliths studies in Central and South America are scarce, particularly those associated with the Marine Isotopic Stage 3 (MIS3) (Fig. 3). In Mexico, silicophytoliths studies in the pedostratigraphic sequences associated with the Volcanic Transmexican Range show that during this period, marked fluctuations of shorter–cooler with longer–warmer periods occurred. This was demonstrated by the presence of paleosols buried by volcanic deposits, which in turn reflect higher precipitations at the beginning of MIS3 (Sedov et al. 2003, 2009). At the end of the MIS3, the conditions were drier with a marked seasonality (Tovar et al. 2013). In Southern Brazil, silicophytolith studies carried on paleosols dated in 60–25 ka cal

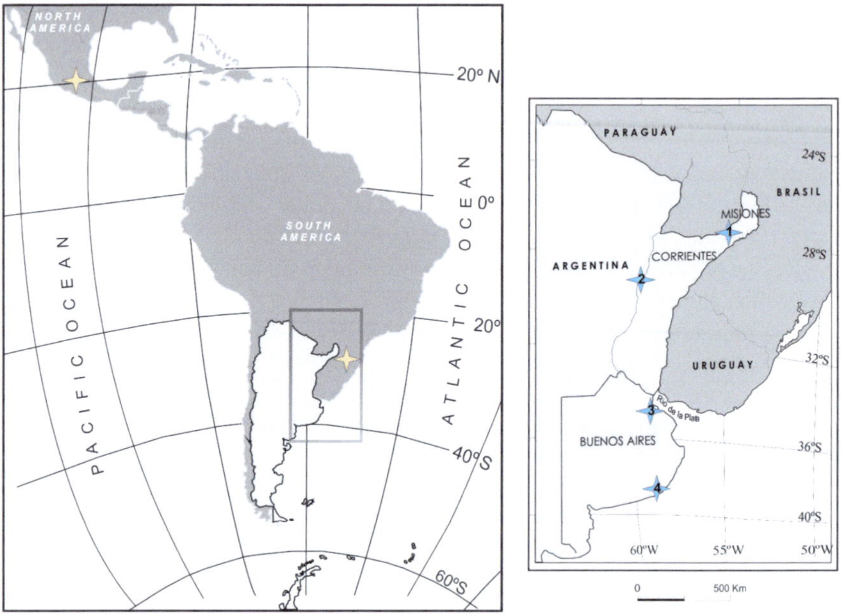

Fig. 3 Location of profiles of silicophytolith studies of sequences including MIS3 in Central and South America (*brown crosses*) and in Argentina (*light blue crosses*). *1* Oberá profile, *2* Bella Vista profile, *3* Gaona profile, *4* Mar del Plata profile

B.P. from pedostratigraphic sequences of the Araucaria Plateau showed that the water regime was sufficiently humid to develop hydromorphic horizons. These paleosols developed from a drier hydric regime occurred in the MIS2, when the erosion of the paleosols predominated (Paisani et al. 2014).

In this special volume, we present a synthesis of the scarce works about silicophytoliths analyses in pedosedimentary sequences of the Marine Isotopic Stage 3 (MIS3) in South America, and particularly, our own work carried on pedosedimentary sequences from Argentina.

2 Materials and Methods

The study area is located in two regional geomorphological units of Argentina, the Mesopotamia and the Pampean Plains. In each region, two integrated profiles, representative of typical pedoestratigraphic sequences were analyzed (Fig. 3). They are located in: Obera region, Misiones province (site D4); Bella Vista region, Corrientes province (site S1); Gaona region (site GAO) and Mar del Plata region (site MDP), Buenos Aires province.

Samples taken for the analyses were first air-dried. The morphological and physico-chemical properties of soils, paleosols, and sediments were made according to the standards established by the Soil Survey Staff (2010). The organic matter content and the particle size distribution were determined for each sample (Walkley and Black 1965; Ingram 1971; Galehouse 1971).

In order to determine the percentage of silicophytoliths in relation to the total sum of mineralogical components, 5 g soil samples were taken from each level. Organic matter was oxidized with 30 % hydrogen peroxide at 70 °C and the clay minerals were extracted by repeated centrifugation at 1000 rpm for 3 min (Alvarez et al. 2008). Once the sample was cleaned it was mounted on immersion oil and 500 grains were counted under an optical microscope (OM) Olimpus (450× magnification) and under SEM at Universidad Nacional de Mar del Plata, Argentina. The qualitative analysis of the silicophytolith morphologies was made according to Twiss (1992), Fredlund and Tieszen (1994) and the ICPN (Madella et al. 2005).

The data were subjected to cluster analyses through Paleontological Statistics (PAST, Hammeer et al. 2001). Principal component analyses of data from Mar del Plata profile were performed on the basis of variance—covariance matrix.

3 Results and Discussion

3.1 Mesopotamia Region

The Mesopotamia region of Argentina covers approximately 200,000 km^2 extending between the current Paraná and Uruguay rivers, and includes the territories of the provinces of Misiones, Corrientes and Entre Ríos (Aceñolaza 2007). Scarce silicophytolith studies have been conducted on pedostratigraphic sequences, of the Marine Isotope Stage 3 (MIS3), in different sites of the Mesopotamia region (Osterrieth et al. 2005; Erra et al. 2006; Osterrieth et al. 2008b; Montti et al. 2009; Erra 2011; Erra et al. 2011, 2013).

3.1.1 Oberá Pedoestratigraphic Sequences

The Province of Misiones is located in northeastern Argentina. Its climate is subtropical humid; its mean annual temperature is 20 °C and its mean annual rainfall 1850 mm. The area is included in the Neotropical phytogeographical region known as the Paranaense province (Cabrera 1976). The present vegetation is a subtropical forest, composed by different strata. In the south-western part of the province, the vegetation is characterized by different grass communities (Cabrera 1976). The geomorphological unit known as the "Preserved Central Plateau" has an undulated topography in its central part, developed on faulted basalt rocks. Deep red soils, mainly Ultisols, with a solum thickness of about 3–7 m above the weathered basalt,

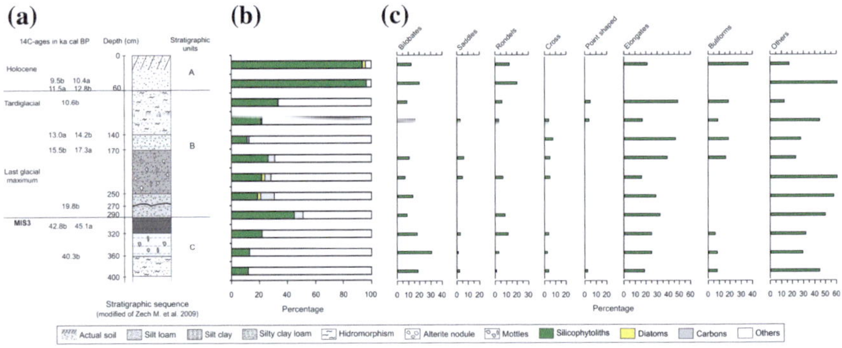

Fig. 4 Oberá pedoestratigraphic sequence, D4 profile **a** Pedoestratigraphic sequence (modified of Zech et al. 2009a, b). **b** Percentages of bioliths and other particles, **c** Percentage of silicophytolith morphologies

cover the unit (Morrás et al. 2009; Moretti and Morrás 2013). The soils have sedimentary and hydromorphic features, and from a taxonomic point of view they are classified as Mollisols, Alfisols, and Inceptisols.

Several recent studies on red soils and hydromorphic soils indicate paleoecological fluctuations in the province of Misiones related to climatic changes during the Early Quaternary (Morrás et al. 2009; Zech et al. 2009a, b). Particularly, multi-proxy studies carried out by these authors on a soil-sediment sequence sampled in a weakly flooded small basin located northeast of the city of Oberá (site D4) indicate several climatic fluctuations from MIS3 to the Holocene. In this pedostratigraphic sequence, three stratigraphic units have been defined as A, B, and C from the top to the base (Morrás et al. 2009; Zech et al. 2009a, b) (Fig. 4). According to the grain size data provided in the aforementioned studies, the parent material in unit A is silty fine sand. In unit B, the texture changes abruptly to very dense, light gray mottled clay; at the base of this unit, at a depth between 2.5 and 2.9 m, there is a silty layer with a thin intercalated layer of humus. The top of unit C is characterized by silty clay dark sediment, followed by a coarser grained and paler material. The total organic carbon content (TOC) is very high at the top of unit A, reaching around 15 %; it decreases sharply below a depth of 30 cm. In units B and C, organic carbon values are mostly below 2 %, though an increase reaching a maximum of 5.1 % is observed in the middle part of unit B, between a depth of 1.5 and 2.5 m (Fig. 4) (Morrás et al. 2009; Zech et al. 2009a, b).

In the D4 soil-sediment sequence, the silicophytolith content for the samples shows contrasting values (Osterrieth et al. 2008). In the present soil (unit A) more than 90 % of the sample consists of silicophytoliths, which would imply the development of histic horizons during most of the Holocene. The silicophytolith content in relation to the total mineralogy decreases progressively in the upper part of section B, up to a minimum level of 12 %. Then, in the lower part of unit B, a substantial increase in phytolith content is observed. At the top of unit C, the proportion of silicophytoliths reaches a new maximum of about 45 %. In addition,

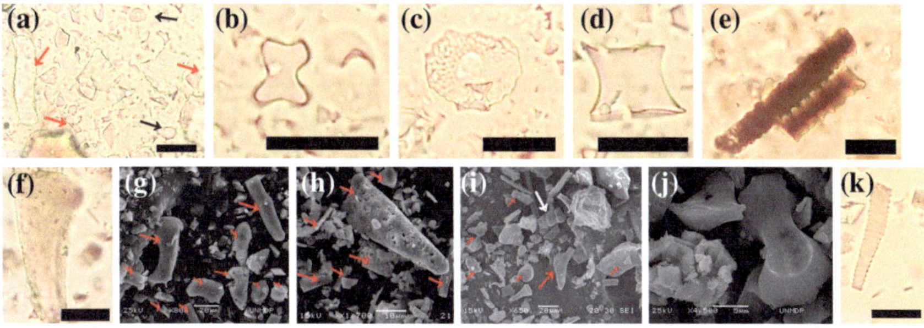

Fig. 5 Microphotographs of silicophytolith morphologies observed in the Oberá D4 profile, and taken under optical microscope (**a–e**, **f**, **k**) and MEB (**g–j**) (*red arrow*). **a** Elongate and broken silicophytoliths, **b** panicoid bilobate, **c** conical, **d** rondel, **e** elongate crenate articulated, **f** bulliform, **g** bulliform and elongate, **h** point-shaped degraded, **i** point-shaped and cylindrical sulcate tracheid (*white arrow*), **j** bilobate and broken bilobate, **k** cylindrical sulcate tracheid. Scale bar = 25 μm in **a–e**, **f**, **k**

from the middle part of unit B to the point of contact with unit C, the proportion of diatoms clearly increases (Fig. 4).

The observed increase of silicophytoliths and diatoms in the lower half part of unit B, between 1.7 and 2.5 m depth, in coincidence with the reported organic carbon increase, and sedge phytoliths (Fig. 5c), suggests the presence of a paleosol that would have evolved under conditions of saturation and high environmental humidity. At the top of unit C, the content of phytoliths increases markedly again; this would indicate the presence of another paleosol developed under conditions of less saturation and better pedological conditions, which could sustain a thicker vegetation cover. At the base of the sequence, silicophytolith content is low, scarcely reaching 10 % of the total mineralogy.

The highest diversity of morphotypes appears in the present soil (Figs. 4 and 5), with a predominance of bilobate short cells, among which panicoid bilobates (mainly derived from C_4 grasses) are very abundant (Figs. 4, 5b, i, j). Within the same affinity, the saddle and cross types (mainly derived from Chloridoid and Panicoid grasses), are abundant (Fig. 5). Elongate smooth cells, bulliform cells, point-shaped cells (Fig. 5a, e, f, g, h) and numerous cylindrical sulcate tracheids of different sizes (Fig. 5i, k), (especially big size) are also common. Several of these forms could be attributed not only to grasses, but also to dicotyledons. Articulated silicophytoliths (Fig. 5e) are present and are commonly used as indicators of environmental stability. The percentage of altered silicophytoliths is also high in several horizons, which could be related to pH variations, specific saturation conditions and/or the presence of iron oxides (Fig. 5g, h). There is a close relation between silicophytolith diversity and pedogenesis, since there is less diversity associated to low pedological development.

To sum up, the silicophytoliths and other amorphous silica biomorphs indicate a continuous growth of vegetation, and silicophytoliths predominate in the inorganic fraction of the first 50 cm of the profile. The two levels rich in silicophytoliths described above would correspond to paleoedaphic levels developed under different conditions of saturation and environmental humidity from the Late Pleistocene until the transition with the Holocene. The vegetation cover would be represented by different species of grasses (belonging to Panicoideae, Pooideae, and Choridoideae subfamilies), as well as dicotyledons.

Specifically for the MIS3 interval, an increase in the content of silicophytoliths in relation to total mineralogy was observed. The values were twice the contents observed in the over- and underlying corresponding levels. These values are evidence of an intense pedogenesis related to an abundant plant cover dominated mainly by grasses. The abundance of bilobates, saddles, crosses, rondels, bulliforms, and dicotyledon phytolith morphotypes was also highest in this interval. The presence of carbons in unit B and at the top of unit C is possibly related to fires or persistent reducing conditions.

In general, the quantitative–qualitative data obtained from silicophytolith analyses show the presence of conspicuous paleopedological levels. For the MIS3 interval, it can be inferred that C_3 grasses (mainly Pooid and/or Panicoid types) and, to a lesser proportion, C_4 grasses (Chloridoid and/or Panicoid types) were present in the area. These data indicate semiarid-subhumid, cooler climatic conditions, with little soil moisture and marked seasonality. However, in the middle levels of MIS3, warmer and wetter conditions may have been developed. These conditions allowed the development of a conspicuous paleosol, which had high amounts of silicophytoliths derived from C_3 grasses (40 %), and showed an increase in C_4 grasses, some trees and diatoms (10 %). This paleosol may have evolved from silt clay parent material, which could also be associated to wetter conditions.

3.1.2 Bella Vista Pedoestratigraphic Sequences

The province of Corrientes is located in northeastern Argentina and covers an area of 88,800 km^2. The climate of the area is described as wet subtropical, characterized by relatively equable temperatures and regular rains throughout the year. Phytogeographically, the area studied is included in the Chaco Eastern District, province of Chaco (Cabrera 1976). The vegetation is characterized by xerophytic forests, palm forests and savannas (Cabrera 1976).

Geomorphologically, the province belongs to a vast sedimentary basin that constitutes a part of ancient shelf relief (Erra et al. 2013). Two Pleistocene sedimentary units are recognized in the province of Corrientes: the Toropí and Yupoí formations, both representing floodplain sedimentation. Recent dating analyses using Optically Stimulated Luminescence (OSL) of the deposits have resulted in ages between 52 ka (for the Toropí Formation) and 36 ka (for the Yupoí Formation) (Francia et al. 2012a, b), sometime within the "Lujanense" Stage *sensu strict* (Late Pleistocene–Early Holocene). Lithologically, the Toropí Formation is composed of clayey sand, sandy limestone and sandy clay. The Yupoí Formation consists of muddy sandstone with variable portions of sandy limestone and sandy clay. These units represent floodplain deposits, and are broadly distributed, covering a large part of the western and eastern sections of the Corrientes province along the Paraná and Uruguay rivers, respectively. The Toropí and Yupoí formations are both highly fossiliferous and preserve an important and varied fauna of mega- and micro-mammals (Herbst 1971; Iriondo 1973; Herbst and Álvarez 1975; Tonni et al. 2005). The Late Pleistocene vertebrate faunal record in the province of Corrientes shows clear compositional changes through time, linked to fluctuations in climate. Specifically, the changes in diversity are consistent with pulses of colder and arid to semiarid climate (presence of *Dolichotis patagonum*, *Pampatherium typum*, *Chaetophracus villosus*, *Neosclerocalyptus paskoensis*, *Glyptodon reticulatus*., etc.), alternating with wetter and warmer climate (presence of *Holmesina paulacoutoi*, *Tapirus* sp., *Euphracthus* aff. *E. sexcintus*, *Boa constrictor*, etc.) (Tonni et al. 2005; Francia 2014; Francia et al. 2015).

One profile (S1 profile) was selected, where the remains of one Quaternary mammal, *Lestodon* (Xenarthra, Phyllophaga), have been previously found (Scillato-Yané et al. 1998; Carlini et al. 2008). The Toropí Formation (2.50–7.50 m) in this profile S1 (containing specimens of *Lestodon*) is relatively clay-rich, as is typical of this unit, and comparatively indurated and with high color chromaticity and value. The Yupoí Formation (above 2.50 m) is characterized by typical features of the parental materials: argillic endopedons and mollic epipedons. These features of well-developed Mollisol and Alfisol paleosols are predominant in these sequences (Fig. 6).

All samples studied contained silica biomorphs (silicophytoliths, diatoms, sponge spicules, etc.). The amount of silicophytoliths as a weight percentage of the total inorganic component varies between 4.03 and 21.85 % through the profile S1 pedosequence. The highest yields were from the modern soil, and the lowest yields from samples from the Yupoí Formation. Apart from silicophytoliths, other silica biomorphs were found, such as sponge spicules, diatoms, and chrysophyte cysts (Fig. 8a, b, c). The spicules were found in every sample in low frequencies (0.2–1.5 %), as well as diatoms (0–2 %), which reached their maximum amount in

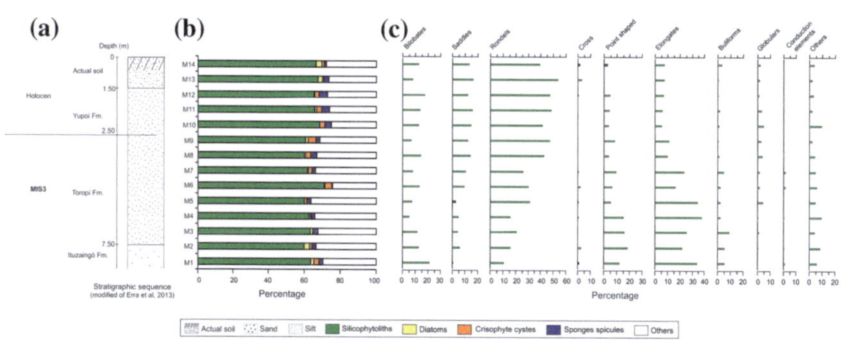

Fig. 6 Bella Vista pedoestratigraphic sequence, profile (S1). **a** Pedoestratigraphic sequence (modified of Erra et al. 2013). **b** Percentages of bioliths and other particles, **c** Percentage of silicophytolith morphologies

the Toropí Fm. (M1) and modern soil samples (M12, M13 and M14). Chrysophyte cysts were found exclusively in the modern soil samples in lower abundance (0–4 %) (Fig. 6).

The cluster analysis performed based on silicophytolith morphotypes from profile S1 ("Lestodon") resulted in three main groups (Fig. 7). The oldest samples in the section (M1–M7) form a group that is characterized by higher abundance of rondels and saddles (Fig. 8d, f), and the presence of globular smooth and equinated phytoliths (Fig. 8e). The next group (M8–M10) has high abundance of rondels, elongate sinuate (Fig. 8g), and presence of crosses (Fig. 8h), and silicophytoliths derived from conduction elements. A group containing the youngest samples (M11–M14) is characterized by a lower abundance of rondels and saddles and higher abundance of elongate psilate, bulliform cells and pointed shaped (Fig. 8).

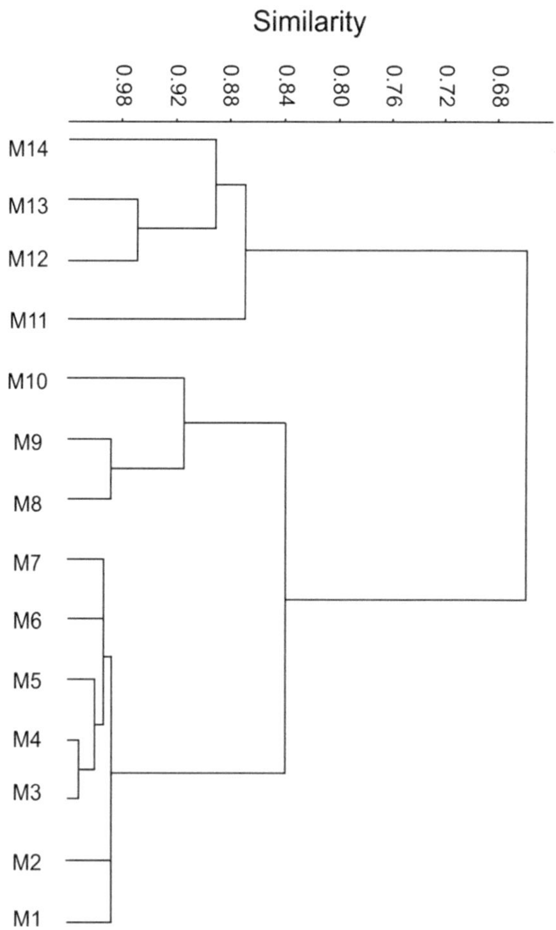

Fig. 7 Dendrogram showing profile samples grouping based on their silicophytolith assemblages

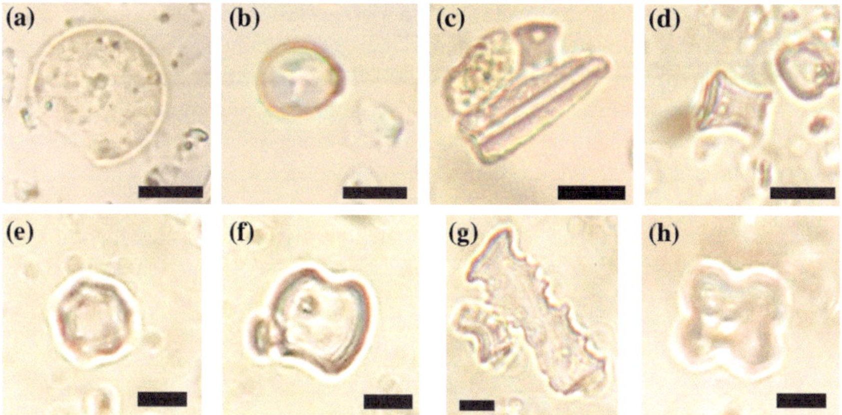

Fig. 8 Microphotographs of bioliths and silicophytolith morphologies observed under optical microscope in Bella Vista (S1) samples. **a** diatom, **b** chrysophyte cyst, **c** sponges spicules, **d** rondel, **e** globular echinate, **f** saddle, **g** elongate, **h** cross. Scale bar = 25 μm

Silicophytolith records from the environments in Argentine Mesopotamia by the Middle Pleistocene. Changes in composition in profile S1 indicate that these grass-dominated habitats were initially dominated by Pooid grasses (producing rondels) with a lower but consistent contribution from Chloridoid C_4 grasses (producing saddles). Further up in this section, in particular in the modern soil, these grass lineages appear to have become slightly less abundant, resulting in a relatively higher contribution from potential Panicoid grasses (C_3 or C_4)—or other grasses within the PACMAD (Panicoideae, Arundinoideae, Chloridoideae, Micrairoideae, Aristidoideae, and Danthonioideae; Aliscioni et al. 2012) clade producing bilobates and crosses—, which remain moderately abundant through the section.

Because particular grass sub-clades tend to be associated with different climatic conditions, the silicophytolith assemblages can also shed light on climate during the Pleistocene of the Argentine Mesopotamia. While the Pooideae subfamily C_3 grasses are associated to cool-temperate or high-elevation grasslands and the Chloridoideae subfamily contains grasses with C_4 photosynthetic pathway found in warmer and more arid climates, the Panicoideae subfamily, containing either C_3 or C_4 species, is generally found in tropical or subtropical regions (Twiss 1992; Kellogg 2001; Edwards and Smith 2010). The mix of Pooids, Chloridoids, and potential Panicoids found in the phytolith record analyzed in the present study should point to a relatively warm, semiarid climate.

However, the relatively consistent presence of palm silicophytoliths and other biosilica wetland indicators (diatoms, sponges, and chrysophyte cysts) point to a riparian element on the landscape. These reconstructed climate and vegetation (grasslands with groves of palms and other trees of shrubs) are consistent with paleoenvironmental interpretations based on faunas recovered from the same sediments (Carlini et al. 2004; Zurita and Lutz 2002; Zurita and Ferrero 2009; Zurita et al. 2014). They also match studies dating the sediments to between 58 and 36 ka (within MIS3) (Francia et al. 2012a, b). Based on this evidence, frequent climatic environmental variations during the Late Pleistocene led to a fluctuation in biogeographic connections between the Mesopotamia region and other parts of South America. At times, Mesopotamia would have been more closely linked to the Chaco-Pampean plain and, at other times, to inter-tropical regions. This may have influenced the development of mesothermal grass-dominated ecosystems (*sensu* Burkart 1975), formed by a mix of C_3 and C_4 grasses, which inhabit mainly warm-temperate (Erra et al. 2013).

3.2 Pampean Plain

The Pampean region covers an area of approximately 1,200,000 km^2 and is located in the central area of eastern Argentina. It includes the provinces of Buenos Aires and Entre Ríos, and part of Santa Fe, Córdoba and La Pampa (Rolleri 1975). The climate is temperate-humid, with abundant rainfalls. From a phytogeographic point of view, this region is included in the Chaco Domain, with Espinal and part of the Pampean Plain (Cabrera 1976). In general, the relief ranges from undulated to plain. The evolution of the Pampean Plain involves the combination of marine erosion, vegetal cover, eolian action, pedological development and anthropic activity (Schnack et al. 1982). The eolian-loessic, fluvial and fluvio-eolian sediments deposited in the Pampean Plain during the Quaternary are composed mainly of lighter minerals. Among them, those composed of amorphous silica, which has an organic origin, represent more than 20 % (Osterrieth and Martínez 1993; Osterrieth 2000, 2008a, b). The Pampean Plain is one of the most fertile regions of the world. Intense agricultural activities are carried out there and this, in turn, has strongly modified the native plant communities, especially grasslands. The soils are generally deep, developed from well-drained loessic parental material and characterized by a silty-loam texture. In topographically low and plain sectors, hydromorphic soils are developed. Several silicophytolith studies have been conducted on Neogene deposits in different sectors of the Pampean Plain; most of them have been mentioned in previous sections.

3.2.1 Gaona Pedoestratigraphic Sequences

The Undulating Pampa is an area including parts of the following provinces: northern Buenos Aires, southern Santa Fe, and southeast Córdoba. The area has a temperate climate, with average temperatures between 24 and 10 °C, in summer and winter, respectively, and an average precipitation of 1000 mm/year (Grill and Morrás 2010). From a phytogeographic point of view, the study area belongs to the Pampean Province (Oriental District) (Cabrera 1976). It is characterized by temperate-humid grasslands, dominated by Poaceae species, scarce shrubs and a halophyte steppe. In some eastern sectors, xerophytic vegetation, characterized by the presence of *Celtis ehrenbergiana*, is also observed (Leon 1992; Grill and Morrás 2010).

The area is located in the southern sector of the Chaco-Paraná basin, in the geological province of the Chaco-Pampean Plain. The Undulating Pampa is a NW-SE fringe generated by fluvio-eolian actions and a slight elevation of the Crystalline Basement, where different pedological processes acted throughout the Late Cenozoic. In general, the relief is slightly undulated and it is drained by several streams, with well-defined lakes and shallow water bodies. In the watersheds there are loess sediments of Pleistocene age, belonging to the Buenos Aires Formation, other older sediments from the Ensenada Formation (Early Pleistocene), and in the lowlands, fluvial sediments from the Luján Formation (Late Pleistocene). All these deposits are covered by recent sediments, of fluvial, marshy, and eolian origin. Towards the east, the Quaternary sediments are interfingered with marine sediments, corresponding to the different marine ingressions of the Late Cenozoic (Nabel et al. 1999, 2005). The soils are mainly Molisols and Vertisols, some of which have saturation and salinity problems (Argiaquols, Argialbols, among others) (Morrás 2004; Morrás et al. 1998a, b).

The studied sequence is located along a 5 km SE-NW cut across section and situated at the highest topographic position at 22 ma.s.l (Fig. 9). The GAO profile was described and sampled in a building excavation at a depth of 6 m. The base of the sequence has loess deposits from the Brunhes Chron, whose lower limit is at 780 ka (Nabel et al. 1999, 2005). In the upper sector of the profile, organic matter associated to a loess level was dated by AMS at 24,098 years B.P. Two morphologically different calcrete levels were present. Integrated analysis of sedimentology, magnetic susceptibility, palynofacies, total organic carbon and silicophytolith studies made it possible to detect paleoenvironmental and paleoclimatic fluctuations during the period under analysis: from the Middle Pleistocene to the Holocene

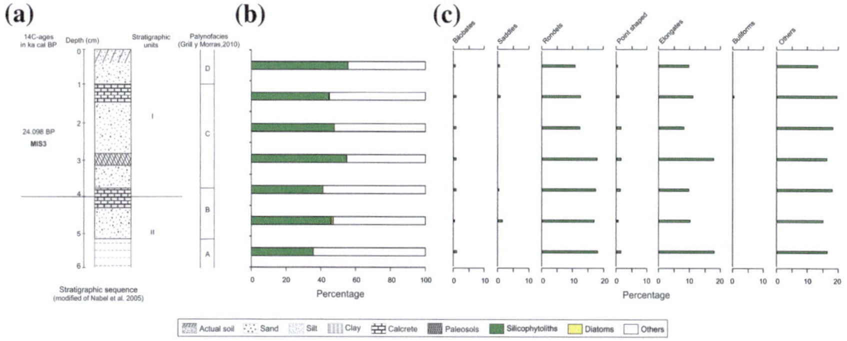

Fig. 9 Gaona pedoestratigraphic sequence, GAO profile. **a** Pedoestratigraphic sequence (modified of Nabel et al. 2005), **b** Percentages of bioliths and other particles, **c** Percentage of silicophytolith morphologies

(Morrás et al. 1998a, b; Nabel et al. 1999; Grill and Morrás 2010; Morrás 1997, 2003, 2004).

Two silicophytolith zones can be identified along the profile, at a depth of 4 m, which are coincident with Units I and II and include palynofacies D–C and B–A, respectively, previously defined in other works (Grill and Morrás 2010) (Fig. 9). Unit I (U.I) has a silt clay texture, with high contents of organic matter in its upper sector; Unit II (U.II) has a sand texture, with low content of organic matter. The results showed that biomorphs made of amorphous silica comprise up to 50 % of the total components of the samples studied. Diatoms are observed only in the base of the sequence (Unit II), whereas in Unit I only fragments of them are present. Within the amorphous silica fraction, silicophytoliths are the predominant component. The total silicophytolith content ranged between 40 and 60 %, and reached the highest values in the epipedon of the present soil and in the level at a depth of 3 m of the profile, defined as a buried soil, in the medium-lower sector of Unit I.

Within Unit I, palynological data from the Late Holocene (2582 cal year B.P.) modern soils (palynofacies type D) are coincident with the silicophytolith content, and show no differences with contents found in other soils from the Pampean Plain (Grill and Morrás 2010, Osterrieth et al. 2008a, 2014a). Grass silicophytoliths (elongates and rondels) dominate in Unit I samples; point-shaped and globular silicophytoliths and tracheid elements are scarce (Figs. 9 and 10). Within grass silicophytoliths, bilobates are present in the sequence with values under 5 %. Broken and eroded bilobates are also abundant (Figs. 9, 10a, b, d, e). Saddles (assigned to Chloridoid grasses) are found in the present soil, the calcrete level and the paleosol, and also in the middle sector of Unit II. These morphologies are related to C_4 grasses and suggest drier conditions, with limitations of soil moisture, higher radiation and/or presence of salts. The abundance of weathered, broken and darkened silicophytoliths, also noticeable, makes their taxonomic assignment difficult; therefore, they have been classified under the categories of "undefined" (Fig. 10a–e).

Unit II presents low silicophytolith content in the upper and lower levels, with a maximum content in the middle sector (Fig. 9). This level may correspond to a higher pedological development, higher moisture and hydromorphism. Silicophytoliths produced by C_3 grasses, commonly developed in tall grasslands, in a cool-humid to

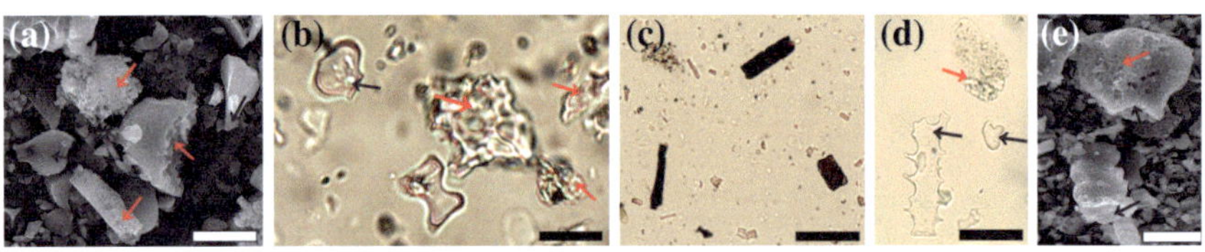

Fig. 10 Microphotographs of silicophytolith morphologies observed in the GAO profile and taken under optical microscope (**b**–**d**) and MEB (**a**, **e**). **a** bilobates and elongates broken and weathered, **b** bilobates, broken bilobate and undefined silicophytoliths, **c** carbonized silicophytoliths, **d**–**e** elongated and broken and weathered bilobate (broken silicophytoliths, *black arrow*; weathered silicophytoliths, *red arrow*). Scale bar = in **a**, **b**, **e** 10 μm; **c**, **d**: 25 μm

temperate climate, were dominant in this sector. Saddles, associated to C_4 grasses, are also common in these paleosols. This is concordant with the palynofacies (A/B) defined by Grill and Morrás (2010), and related to semiarid-warmer regional conditions and wetter local pulses. Sedge silicophytoliths, commonly associated to wetter conditions, are scarcely found, probably due to their lower preservation in soils (Benvenuto et al. 2013).

In summary, a high content of silicophytoliths is found in all the levels studied in the representative pedostratigraphic sequence of the Undulating Plain. In general, the abundance and recurrence of silicophytoliths is a common feature of pedosedimentary sequences of the Late Quaternary in the Pampas (Osterrieth et al. 2005). The silicophytolith analyses of the sequence do not show evidence of changes related to clay-sand texture variations, which in turn may indicate noticeable aquatic or fluvial sedimentary processes in the interface from Unit I to Unit II. The characteristics of the palynofacies defined by Grill and Morrás (2010) indicate eolian loess facies and poor edaphic levels for the MIS3, with a lower vegetation cover. However, silicophytoliths show abundant and homogeneous contents along the profile. The highest content of these silica biomorphs is found in the paleosol developed during the MIS3 interval, which may indicate a higher plant cover. The main silicophytolith morphologies observed were elongate, rondels, point-shaped, bilobates and saddles, representing more than 50 % of the total morphologies. They are mainly assigned to C_3 grasses, and in the case of bilobates and saddles, to C_4 species. Specific wetter conditions were also detected by the presence of sedge silicophytoliths and some anaerobic conditions by the abundance of carbonized silicophytoliths. Finally, some drier periods could have generated conditions of salinity and alkalinity, which in turn may have allowed the development of C_4 grasses (mainly represented in the fossil record by saddle morphotypes).

3.2.2 Mar del Plata Pedoestratigraphic Sequences

This area involves the SE sector of Buenos Aires Province. Its climate is mesothermic and subhumid, with little or no water deficiency (Burgos and Vidal 1951). The mean annual precipitation is 809 mm and the annual average temperature is 13.7 °C (data from the National Weather Service of Argentina, according to the 1920–1980 record). It is included in the Austral district of the Pampean phytogeographical province, where grass steppes and some shrub communities in rocky areas dominate (Cabrera 1976).

Three geological units are distinguished in the study area. There is a crystalline basement of metamorphic and pegmatite rocks of Precambrian age, a sedimentary layer of quartzites from the Paleozoic and a Cenozoic loess and reworked-loess layer, associated with eolian and fluvio-eolian environments. In this region, four geomorphological units are also developed: the Ranges, the Perirange eolian hills, the Fluvio-eolian Plain and the Coastal Plain (Schnack et al. 1982; Osterrieth et al. 1988; Martínez 2001). The Ranges unit belongs to the Tandilia system and is composed of a group of table-like hills with a flat top. The Perirange eolian hills are a relief of morphologically complex hills, with relative heights of up to 30 m and concave–convex outline with intermediate straight patches and slopes between 6 and 8 %. They originated from processes of primary eolian accumulation, later modified by superficial wash (Osterrieth and Martínez 1993). The Fluvio-eolian Plain presents a very gentle slope relief. The Coastal Plain comprises a system of beaches and dunes and is characterized by active morphodynamics and different soil types buried by diverse cycles of coastal dunes and estuarine sediments (Osterrieth 1998; Osterrieth et al. 1988, 2004, 2014b). The main soils developed in this study area are Argiudols, Hapludols, Alfisols and Entisols (Osterrieth et al. 2002).

Typical pedostratigraphic sequences of loess-paleosols from the fluvio-eolian plain and the perirange areas in the NE sector of the Tandilia hills were studied (Bidegain et al. 2005; Osterrieth and Martínez 1993; Osterrieth 2000; Osterrieth et al. 2009, 2014a). The ten levels analyzed included present soil, eolian and fluvio-eolian levels, as well as local and regional paleosols (N1–N10).

The pedostratigraphic sequence is over 10 m long. It is predominantly composed of silty sands, with scarce clays and a low content of organic matter, except for horizons A and B of the present soil (Fig. 11a). At the top of the sequence, there is a typical Argiudol of 100 cm of depth, developed above typical eolian sediments, the loess. Under this sediment, there is a calcrete level associated to a paleosol (P1) widely present in the area, dated at 9040 ± 80 year B.P. The sequence continues with fluvio-eolian and lacustrine sediments, which determine the development of shallow soils and comprises irregular paleopedological facies (p1/p2). Some paleoponds developed hydromorphic soils 10 m wide and 50 cm thick, with an irregular undulated upper limit affected by fluvial erosion and a plain lower limit. In this level, a volcanic ash layer dated at 21,100 ± 7000 year B.P. is present. Another calcrete level dated at 21,190 ± 270 year B.P. and 23,090 ± 330 year B. P. overlies a paleopedological level, widely present in the area, 30 cm thick,

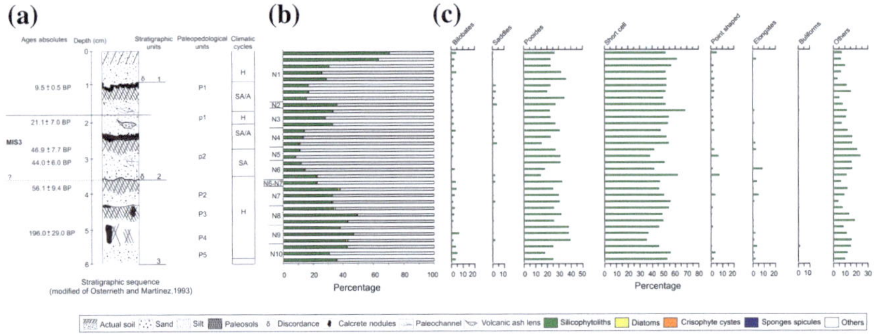

Fig. 11 Mar del Plata pedostratigraphic sequence, profile MDP. **a** Pedostratigraphic sequence (modified of Osterrieth and Martínez 1993), **b** Percentages of bioliths and other particles, **c** Percentage of silicophytolith morphologies

reddish-brown in color and of a moderate prismatic structure (P2). Under this paleosol, eolian and fluvio-eolian sediments dated at 44,000 ± 6000 year, 46,900 ± 7.700 year and 56,100 ± 9400 year from top to base are present. Underlying them, there is another well-developed paleopedological level (P3) with different calcrete sectors that favor their preservation. It is part of a complex polygenetic paleosol, partially truncated, which may have developed in wetter and warmer conditions at the end of MIS4 or beginning of MIS3. Finally, the sequence has two well developed complex paleopedological levels (P4 and P5), widely present in the region, with overlapping horizons, generated in more than one warmer and wetter cycles, during the Middle Pleistocene (168,000 ± 25,000 year and 196,000 ± 29000 year) (Osterrieth and Martínez 1993; Bidegain et al. 2005; Osterrieth 2008a, b; Alvarez 2003).

The percentage of amorphous silica biomineralizations gradually decreases in the present soil (80–16 %), from the soil surface to its base (horizons B and C). In the paleopedological P1 level, the percentage increases, reaching over 30 %. The lower percentages were detected in loess and fluvio-eolian parental material. The P3 (N6–N7) paleosol has a silicophytolith content of 30 %. Silicophytoliths increase in the regional P4 and P5 (N7, N8, N9, and N10) paleosols, with a mean content of 40 %. In general, the silicophytoliths were the dominant fraction within silica biomorphs. Diatoms, chrysophyceae cysts and sponge spicules were present in very low percentages (Fig. 11b, c).

The predominant morphotypes correspond to Pooid grasses (35–70 %); the elongated showed values under 30 %; the chloridoids, <10 %; the panicoids, <7 %; the bulliforms, <8 %; and the point-shaped ones, <10 % (Figs. 11b, 12, 13). In the surface horizons of level 1 (horizon A of the present soil), a great variety of morphologies was detected, whereas in the other levels, the diversity of forms

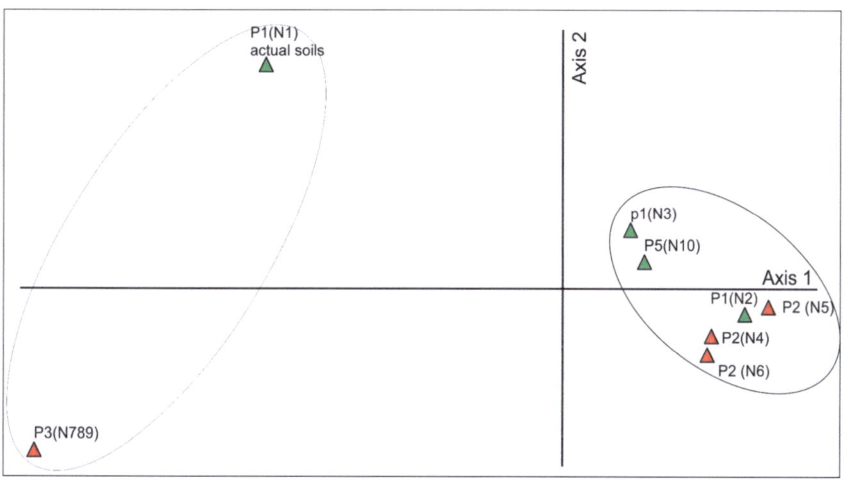

Fig. 12 Principal component analysis of MDP samples profile based on their bioliths and silicophytolith assemblages. *Red triangle* paleosols developed during the MIS3

decreased (Fig. 11c). Very low percentages of C_4 metabolic pathway chloridoid and panicoid silicophytoliths were found in the P2, P3 and P5 paleosols. They are indicators of drier and warmer environmental conditions; the maximum (10 %) was found in the P2 paleosols, developed from the middle to the end of MIS3. High values of silicophytoliths included in the category of undefined forms were found in the whole sequence. They have been affected by several kinds of both physical and chemical alteration and/or by taphonomic processes of different types and degrees of intensity (Osterrieth et al. 2009, 2014b).

Principal component analysis (PCA) shows the separation of a group formed by the samples of the present soil P1 and P3; thus, axis 1(79 %) explains the variation. Such grouping is given by the presence in these soils of the rondel and bilobate morphologies, typical of Pooid grasses. In the second group, the remaining levels of P2, P4, P5 paleosols and local paleosols p1/p2, are associated, characterized by the presence of cysts of crysophyceae and elongated silicophytoliths, one of them developed during the MIS3. Within this large group, P5 and P2 paleosols are associated by the presence of long-cell phytoliths and panicoids, whereas the rest of the paleosols are characterized by the presence of chrysophyte cysts and diatoms (Fig. 12, 13).

In summary, all the analyzed samples show predominance of silicophytoliths over diatoms, chrysophyceae cysts and spicules. The presence of these three microfossils in different levels, together with the iron–manganese flecked ones, may indicate episodic hydromorphic conditions of considerable intensity at the beginning on MIS3. The phytolith percentages indicate variations in the pedological development, associated with the variable vegetal covers, with a predominance of type C_3 Poaceae and Panicoid grasses that accompany the development of the paleosols (Twiss 1992). The drier and warmer stages, marked by the presence of silicophytoliths of the chloridoid type (C_4), are scarce; the highest value appeared in the middle and the end of the MIS3, during the P2 paleosol development.

Statistical analysis allowed establishing different groups among the analyzed soils. The paleosols developed during the MIS3 were included in one group, associated by the number and morphologies of silicophytoliths and the presence of diatoms, cysts and spicules found in each one.

The results permitted to infer temperate–humidity conditions for the present soil and the local hydromorphic (P2) and P3 paleosols, alternating with periods of drought and/or lower water availability during the development of the P1 and P2 soils. From 196 $\pm$ 29 to 56 $\pm$ 9 ka, a humid cycle developed, corresponding to the last interglacial period, with a predominance of C_3 grasses (P4 and P5 paleosols). At the beginning of MIS3, the presence of a well-developed paleosol (P3) and wide covers of tall grasses (C_3 Pooideae) indicated a temperate and humid climate. Within the middle MIS3, conditions of lower humidity and drier periods which favor the development of a moderate paleosol (P2) and poor vegetal cover. Toward the end of MIS3 (21 ka ago) drier conditions with a hydric deficit are characterized by the presence of calcrete soils (P1) and shallow water bodies that end up as hydromorphic paleosols (p1) with vertic features.

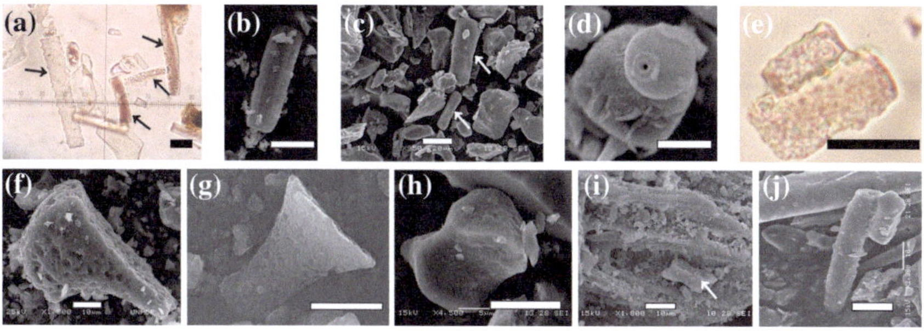

Fig. 13 Microphotographs of bioliths and silicophytolith morphologies observed under optical microscope (**a**, **e**) and MEB (**b**–**d**, **f**–**j**) in the Mar del Plata samples. **a** Elongates (*arrow*), **b** diatom, **c** elongates (*arrow*), **d** chrysophyte cyst, **e** elongate articulated, **f** bulliform degraded, **g** rondel, **h** saddle, **i** bilobate (*arrow*), **j** elongated articulate. Scale bar = in **a**, **e**: 25 μm; in **b**, **d**, **f**–**j**:10 μm; in **c**: 20 μ

4 Final Remarks

In spite of the scarcity of silicophytolith studies in pedostratigraphic sequences involving the MIS3, some interesting remarks can be made. Silicophytolith studies enabled the analysis of the intensity of pedological processes and the extent of erosion and morphodynamic processes affecting different areas of the Mesopotamia and Pampean Plain regions of Argentina, from the Late Pleistocene to the present, especially during the MIS3. These results, together with other available evidence, contribute to the interpretation of the different environmental and paleoenvironmental processes that have taken place, especially those involved in the plant–soil–environment relationships.

In all the pedostratigraphic sequences analyzed, the presence of silicophytoliths was significant, with average values of 40 % of the total mineralogical component. Both in the Mesopotamia and in the Pampean Plain region, values ranged from 10 % for the eolian and fluvio-eolian sedimentary levels, weakly pedogenized to a maximum of 60 % in the pedological and paleopedological levels.

In general, at the beginning of the MIS3 period, the presence of a well-developed paleosol and wide covers of tall grasses has been indicated. The morphologies found show a predominance of C_3 grasses (belonging to Pooideae and/or Panicoideae subfamilies), indicating colder to temperate–humidity conditions. In a lower proportion, silicophytolith from the so-called C_4 grasses (belonging to Chloridoideae and/or Panicoideae subfamilies) were also detected, indicating semiarid to sub-humid climatic conditions with lower soil moisture, a noticeable seasonality, and/or the presence of salts. This indicates the development of mesothermal grass-dominated ecosystems, which nowadays grow mainly in warm-temperate regions.

Within the MIS3, frequent climatic environmental variations during the Late Pleistocene may have led to a fluctuation in biogeographical connections between the Mesopotamia region and other parts of South America, closely linked to the Chaco Pampean plain and, at other times, to inter-tropical regions.

Acknowledgments This study was financially supported by the Universidad Nacional de Mar del Plata (EXA 741/15), the Agencia Nacional de Promoción Científica y Tecnológica (ANPCYT) of Argentina (ANPCyT BID PICT N°1583), and the CONICET-PIP 112-20130100145CO. The authors are especially grateful to Dr. Héctor Morrás for providing us the Oberá and Gaona samples and their valuable comments and contributions to this work, and to Ing. José Vila for their assistance with the SEM analysis.

References

Aliscioni S, Bell HL, Besnard G, Christin PA, Columbus JT, Duvall MR, Edwards EJ, Giussani L, Hasenstab-Lehman K, Hilu KW, Hodkinson TR, Ingram AL, Kellogg EA, Mashayekhi S, Morrone O, Osborne CP, Salamin N, Schaefer H, Spriggs E, Smith SA, Zuloaga F (2012) New grass phylogeny resolves deep evolutionary relationships and discovers C4 origins. New Phytologist 193:304–312

Aceñolaza F (2007) Geología y recursos geológicos de la mesopotamia argentina. Serie de Correlación Geológica 22:149

Alexandre A, Meunier JD, Lezine AM, Vincens A, Schwartz D (1997) Phytoliths: indicators of grassland dynamics during the late Holocene in intertropical Africa. Palaeogeogr Palaeoclimatol Palaeoecol 136:213–229

Alvarez MF (2003) Biominerales de sílice en suelos y paleosuelos presentes en secuencias del Cuaternario tardío en el sudeste bonaerense. Unpublished graduation thesis, Universidad Nacional de Mar del Plata, Mar del Plata, 42 p

Alvarez MF, Borrelli N, Osterrieth M (2008) Extracción de silicobiolitos en distintos sedimentos utilizando dos técnicas básicas. In: Korstanje MA, M del P Babot (eds) Matices Interdisciplinarios en Estudios Fitolíticos y de otros Microfósiles. British Archaeological Reports (BAR) Internacional Series S 1870, pp 31–38

Andreis RR (1972) Paleosuelos de la Formación Musters (Eoceno Medio), Laguna del Mate, Provincia de Chubut, República Argentina. Revista Asociación Argentina de Mineralogía, Petrología y Sedimentología 3:91–97

Benvenuto L, Fernández Honaine M, Osterrieth M, Coronato A, Rabassa J (2013) Silicophytoliths in Holocene peatlands and fossil peat layers from Tierra del Fuego, Argentina, Southernmost South America. Quat Int 287:20–33

Bertoldi de Pomar H (1970) Fitolitos y Zoolitos. Su significado geológico en sedimentos continentales. Boletín Asociación Geológica de Córdoba, Córdoba, Argentina 1(1):21–31

Bertoldi de Pomar H (1971) Ensayo de clasificación morfológica de los silicofitolitos. Ameghiniana 8:317–327

Bertoldi de Pomar H (1975) Los silicofitolitos. Sinopsis de su conocimiento. Darwiniana 19:173–206

Bidegain JC, Osterrieth M, Van Velzen A, Rico Y (2005) Geología y registros magnéticos entre arroyo La Tapera y Santa Clara del Mar, Mar del Plata. Revista Asociación Geológica Argentina, Buenos Aires 60(3): 599–604

Borrelli N (2008) Biominerales de sílice y de calcio. Su rol en la biogeoquímica de Argiudoles típicos del sudeste bonaerense. Unpublished graduation thesis, Universidad Nacional de Mar del Plata, Mar del Plata, 355 p

Borrelli N, Fernández Honaine M, Altamirano S, Osterrieth M (2011) Calcium and silica biomineralizations in species associated to aquatic environments of the Pampean Plain, Argentina. Aquat Bot 94:29–36
Burgos JJ, Vidal A (1951) Los climas de la República Argentina según la nueva clasificación de Thornthwaite. Meteoros, Buenos Aires 1(1):3–32
Burkart A (1975) Evolution of grasses and grasslands in South America. Taxon 24:53–66
Cabrera AL (1976) Regiones Fitogeográficas Argentinas. Enciclopedia Argentina de Agricultura y Jardinería, Segunda edición, Tomo II, Buenos Aires 85 p
Campos S, del Puerto L, Inda H, Castiñeira C (2001) Opal phytoliths analysis: its application to the archaeobotanical record in the East of Uruguay. In: Meunier JD, Colin F (eds) Phytoliths: applications in earth sciences and human history, A.A.Balkema Publishers, pp 129–142
Carlini AA, Zurita AE, Gasparini GM, Noriega JI (2004) Los mamíferos del Pleistoceno de la Mesopotamia argentina y su relación con aquéllos del Centro-Norte de la Argentina, Paraguay y Sur de Bolivia, Sur de Brasil y Oeste de Uruguay: Paleobiogeografía y Paleoambientes. In: Aceñolaza FG (ed) Temas de la Biodiversidad del Litoral fluvial argentino, vol 12, INSUGEO, Miscelánea, pp 83–90
Carlini AA, Zurita AE, Miño-Boilini AR (2008) Reseña paleobiogeográfica de los Xenarthra (Mammalia) del Pleistoceno tardío de la región Mesopotámica (Argentina). INSUGEO Miscelánea 17(2):259–270
Coupland RT (1992) Overview of South American grasslands. In: Coupland RT (ed) Natural grasslands: introduction and western hemisphere. Ecosystems of the world 8A, Elsevier, New York, pp 363–366
Cornelis JT, Delvaux B, Titeux H (2010) Contrasting silicon uptakes by coniferous trees: a hydroponic experiment on young seedlings. Plant Soil 336:99–106
del Puerto L, Inda H (2008) Estrategias de subsistencia y dinámica ambiental: análisis de silicofitolitos en sitios arqueológicos de la cuenca de la laguna de Castillos, Rocha, República Oriental del Uruguay. In: Zucol A, Osterrieth M, Brea M (eds) Fitolitos. Estado actual de sus conocimientos en América del Sur Argentina, pp 221–236
Edwards EJ, Smith SA (2010) Phylogenetic analyses reveal the shady history of C4 grasses. Proc Nat Acad Sci USA 107:2532–2537
Ehrenberg CG (1854) Mikrogeologie. Das Erden und Felsen schaffende Wirken des unsichtbar kleinen selbststadigen Lebens auf der Erde. Leipzig, Leopold Voss, 374 p
Epstein E (1994) The anomaly of silicon in plant biology. Proc Natl Acad Sci 91:11–17
Erra G (2011) Criterios metodológicos para el estudio y clasificación de fitolitos cuaternarios. Revista Historia Natural Tercera Serie, Buenos Aires 1:45–62
Erra G, Zucol AF, Kröhling D, Brea M (2006) Análisis fitolíticos en el loess del Pleistoceno Tardío y Holoceno Temprano en la provincia de Entre Ríos: resultados preliminares. In: Actas Congreso Argentino de Cuaternario y Geomorfología, pp 691–699
Erra G, Zucol AF, Kröhling DM (2011) Análisis fitolítico de la Formación TezanosPinto (loess del Pleistoceno Tardío-Holoceno Temprano) en el sector noroeste de su área de distribución en la provincia de Entre Ríos (Argentina). Revista Mexicana de Ciencias Geológicas 28(3):398–412
Erra G, Osterrieth M, Zurita A, Francia A, Carlini AA (2013) Paleoenvironment of the Toropí Formation (Upper Pleistocene), Corrientes province (Mesopotamian region, Argentina): a phytolith approach. Quat Int 287:73–82
Fernández Honaine M, Osterrieth M (2012) Silicification of the adaxial epidermis of leaves of a panicoid grass in relation to leaf position and section and environmental conditions. Plant Biol 14:596–604
Fernández Honaine M, Osterrieth M, Zucol A (2009) Plant communities and soil phytolith assemblages relationship in native grasslands from southeastern Buenos Aires Province, Argentina. Catena 76:89–96
Fernández Honaine M, Zucol A, Osterrieth M (2006) Phytolith assemblage and systematic association in grassland species of the SE Pampean Plains, Argentina. Ann Bot 98:1155–1165
Fernández Honaine M, Zucol A, Osterrieth M (2005) Biomineralizaciones de sílice en *Celtis tala* Planchon (Celtidaceae). Boletín Soc. Argentina Botánica, Buenos Aires 40(3–4):229–239

Fernández Honaine M, Borrelli N, Osterrieth M, del Río L (2013) Amorphous silica biomineralizations in sedges: their relation with senescence and silica availability. Boletín de la Sociedad Argentina de Botánica, Buenos Aires 48:247–259
Florez MMT, Parra SLN (2008) Espectros de fitolitos en tres suelos de la planicie de Puente Largo, Páramo de Frontino, Departamento de Antioquía, Colombia. In: Zucol A, Osterrieth M, Brea M (eds) Fitolitos. Estado actual de sus conocimientos en América del Sur Argentina, pp 111–125
Francia A (2014) Vertebrados cuaternarios de Corrientes, paleoambientes, paleoclimas y diversidad. Un análisis comparativo con faunas continentales equivalentes del Cono Sur de América del Sur. Doctoral Thesis (unpublished), Facultad de Ciencias Exactas y Naturales y Agrimensura, Universidad Nacional del Nordeste, Corrientes, 369 p
Francia A, Carlini AA, Zurita AE, Miño-Boilini AR, Kruck W (2012a) Cronología de las unidades litoestratigráficas aflorantes en el Arroyo Toropí, Provincia de Corrientes, y los registros paleofaunísticos. In: Actas Reunión de comunicaciones Científicas y Tecnológicas [CD-ROM]
Francia A, Carlini AA, Zurita AE, Verzi DH (2012b) Galea (Rodentia, Caviidae) the late Pleistocene of Corrientes Province (Argentina): taxonomic and paleobiogeographic implications. Neues Jahrb Geol Palaontol Abh 266(2):173–184
Francia A, Zurita AE, Carlini AA (2015) How marine isotope stage 3 (MIS3) is reflected in northern Mesopotamia faunal assemblage of Argentina: the Xenarthra Cingulata case. Quat Int 377(2015):126–139. doi:10.1016/j.quaint.2015.03.012
Fredlund G, Tieszen L (1994) Modern phytolith assemblages from the North American Great Plains. Jour Biogeography 21:321–335
Fredlund G, Tieszen L (1997) Calibrating grass phytolith assemblages in climatic terms: application to late Pleistocene assemblages from Kansas and Nebraska. Palaeo 136:199–211
Frenguelli J (1930) Partículas de sílice organizada en el loess y en los limos pampeanos. Células silíceas de Gramíneas. Anales de la Sociedad Científica de Santa Fe, Santa Fe 2:64–109
Galehouse JS (1971) Sedimentation analysis. In: Carver (ed) Procedures in sedimentary petrology. Wiley Interscience, Wisconsin, USA, pp 69–94
Gallego L, Distel RA (2004) Phytolith assemblages in grasses native to Central Argentina. AnnBot 94:1–10
Gomes Coe H, Osterrieth M (2014) Synthesis of some phytolith studies in South America (Brazil and Argentina). Nova Science Publishers, Inc., pp 243–263
Gomes Coe HH, Osterrieth M, Fernández Honaine M (2014) Phytoliths and their applications. In: Coe HH, Osterrieth M (eds) Synthesis of some phytoliths studies in South America (Brazil and Argentina). Botanical Research and Practices. NOVA Publishers, New York, p 262
Grill SC, Morrás HJM (2010) Análisis palinofacial de sedimentos del Cenozoico Tardío en la Pampa Ondulada (Argentina): Primeros resultados. Rev Bras Paleontol 13(3):221–232
Hammer O, Harper DAT, Ryan PD (2001) Past-palaeontological Statistics, version 1.75, 86 p
Herbst R (1971) Esquema estratigráfico de la provincia de Corrientes. Rev Asoc Geol Arg, Buenos Aires XXVI(2):221–243
Herbst R, Alvarez BB (1975) Nota sobre dos Formaciones del Cuaternario en Corrientes. Anales de la Academia Brasilera de Ciencias 47:33–37
Hodson MJ, White PJ, Mead A, Broadley MR (2005) Phylogenetic variation in the silicon composition of plants. Ann Bot 96:1027–1046
Ingram RL (1971) Sieve analysis. In: Carver RE (ed) Procedures in sedimentary petrology Wisconsin. Wiley Interscience, USA, pp 41–68
Iriarte J (2001) Arqueología de las culturas cerámicas del Río Uruguay: retrospectiva y futuras direcciones. In: Arqueología Uruguaya hacia el fin del milenio. Gráficos del Sur, Montevideo, Tomo, pp I355–363
Iriarte J, Power M, Mayle F, Rostain S, Jones H, Watling J, Brownen W, McKey D (2012) Fire-free land use in pre-1492 Amazonian savannas. Proc Nat Acad Sc 19:6473–6479
Iriondo MH (1973) Mineralogía de las arenas de la Formación Yupoi. Rev Asoc Cienc Nat Litoral 4:87–96

Jacobs BF, Kingston JD, Jacobs LL (1999) The origin of grass-dominated ecosystems. Ann Missouri Bot Gard 86:590–643
Jones LHP, Handreck KA (1967) Silica in soils, plants and animals. Adv Agron 19:107–149
Kellogg EA (2001) Evolutionary history of the grasses. Plant Physiol 125:1198–1205
León RJC (1992) Río de la plata grasslands. Regional sub-divisions. In: Coupland RT (ed) Ecosystems of the World 8A: natural grasslands. Elsevier, Amsterdam, pp 376–407
Lowenstam HA (1981) Minerals formed by organisms. Science 211:1126–31
Ma JF, Takahashi E (2002) Soil, fertilizer, and plant silicon research in Japan. Elsevier, Amsterdam, p 281
Ma JF, Yamaji N (2006) Silicon uptake and accumulation in higher plants. Trends Plant Sci 11:392–397
Ma JF, Yamaji N, Mitani-Ueno N (2011) Transport of silicon from roots to panicles in plants. Proc Jpn Acad Ser B 87:377–385
Madella M, Alexandre A, Ball T (2005) International code for phytolith nomenclature 1.0. Ann Bot 96:253–260
Martínez G (2001) Geomorfología y Geología del Cenozoico superior de las cuencas de los arroyos Los Cueros y Seco, vertiente nororiental de las Sierras Septentrionales, provincia de Buenos Aires. Unpublished doctoral thesis. Universidad Nacional del Sur
Massey FP, Ennos AR, Hartley SE (2007) Herbivore specific induction of silica-based plant defences. Oecologia 152:677–683
McNaughton SJ, Tarrants JL, McNaughton MM, Davis RH (1985) Silica as a defense against herbivory and a growth promotor in African grasses. Ecology 66:528–535
Metcalfe CR (1960) Anatomy of the monocotyledons. I. Gramineae. Clarendon Press, Oxford, 731 p
Montti L, Fernandez Honaine M, Osterrieth ML, Graciano Ribeiro D (2009) Phytolith analysis of Chusquea ramosissima Lindm. (Poaceae: Bambusoideae) and associated soils. Quat Int 193:80–89
Moretti LM, Morrás H (2013) New microscopic evidences of the autochthony of the ferrallitic pedological mantle in the Misiones province, Argentina. Lat Am J Sedimentol Basin Anal 20(2): 129–142
Morrás HJ (1997) Origen y mineralogía del material parental de los suelos de la región pampeana. ¿Homogeneidad o heterogeneidad? Res. Primer Taller sobre Sedimentología y Medio ambiente. AAS, Buenos Aires, pp 19–20
Morrás H (2003) Distribución y origen de los sedimentos loéssicos superficiales de la Pampa Norte en base a la mineralogía de arenas. Resultados preliminares. AAS Revista 10(1):53–64
Morrás H (2004) Un nuevo esquema de sedimentación y evolución de los sedimentos loéssicos superficiales en el sur de la Pampa ondulada en base criterios mineralógicos y geoquímicos. In: Actas X Reunión Argentina de Sedimentología (I), pp 108–109
Morrás H, Zech W, Nabel P (1998a) Identificación de distintos materiales parentales de suelos Argiudoles de un sector de la Pampa Ondulada (Castelar, Provincia de Buenos Aires). In: Actas XVI Congreso Argentino de la Ciencia del Suelo, pp 305–306
Morrás H, Zech W, Nabel P (1998b) Composición geoquímica de suelos y sedimentos loéssicos de un sector de la pampa ondulada. Actas Jornadas Geológicas y Geofísicas Bonaerenses 2: 225–232
Morrás H, Moretti L, Píccolo G, Zech W (2009) Genesis of subtropical soils with stony horizons in NE Argentina: autochthony and polygenesis. Quat Int 196:17–159
Motomura H, Mita N, Suzuki M (2002) Silica accumulation in long-lived-leaves of Sasa veitchii (CarrieÁre) Rehder (Poaceae: Bambusoideae). Ann Bot 90:149–152
Motomura H, Fujii T, Susuki M (2004) Silica deposition in relation to ageing of leaf tissues in Sasa veichii (Carriere) Rehder (Poaceae, Bambusoideae). Ann Bot 93:235–248
Nabel PE, Morrás H, Petersen N, Zech W (1999) Correlation of magnetic and lithologic features of soils and quaternary sediments from the ondulating Pampa, Argentina. J S Am Earth Sci 12:311–323
Nabel PE, Morrás HJM, Sapoznik M (2005) Magnetoestratigrafia de sedimentos cenozoicos en el oeste del gran Buenos Aires. RevAsoc Geol Arg 60(2):383–388

Ollendorf AL (1992) Toward a classification scheme of sedge (cyperaceae) phytoliths. In: Rapp G Jr, Mulholland S (eds) Advances in archaeological and museum science. Phytolith Systematics, Emerging Issues. 350 p

Osterrieth ML (1998) Paleosols and their relation to sea level changes during the late quaternary in mar chiquita, buenos aires, Argentina. Quat Int 51 52:43 44

Osterieth M (2000) Silicofitolitos una herramienta para la comprensión de procesos pedológicos del Cuaternario. In: Actas XVII Congreso Arg. De la Ciencia del Suelo. CDR. 4 p

Osterrieth M (2004) Biominerales y biomineralizaciones. In: García Calderón HT (ed) Cristalografía de suelos, Sociedad Mexicana de Cristalografía, México, pp 216–218

Osterrieth M (2006) Silicofitolitos en suelos, paleosuelos y sedimentos. In: Actas III Congreso Argentino de Geomorfología y Cuaternario I:351–366

Osterrieth M (2008a) Silicofitolitos en suelos, paleosuelos y materiales parentales. In: Zucol AF, Osterrieth M, Brea M (eds) Fitolitos. Estado actual de sus conocimientos en América del Sur Argentina, pp 75–85

Osterrieth M (2008b) Silicofitolitos en sedimentos loéssicos de la llanura inter y periserrana de Tandilia, Buenos Aires, Argentina. In: Zucol AF Osterrieth M Brea M (eds) Fitolitos. Estado actual de sus conocimientos en América del Sur, pp 204–215

Osterrieth M, Martínez G (1993) Paleosols on late cainozoic sequences in the northeastern side of tandilia range, Buenos Aires, Argentina. Quat Int 17:57–65

Osterrieth M, Martínez G, Cionchi J (1988) Correlación paleopedológica en secuencias del Cenozoico Superior de Laguna de Los Padres y La Brava. In: Actas II Jorn. Geológicas Bonaerenses. Bahía Blanca, p 4

Osterrieth M, Scampini E, Borrelli N, García Calderón N, Martínez P, Miglioranza K, Bernasconi MV (2002) Geoecología y degradación de Argiudoles típicos afectados por prácticas hortícolas, en áreas sub-urbanas de Buenos Aires, Argentina. In: Actas XVIII Congreso Argentino de la Ciencia del Suelo, p 115

Osterrieth M, Alvarez MF, Madella M (2004) Paleosuelos loéssicos en secuencias sedimentarias del Cuaternario tardío de la planicie fluvioeólica bonaerense. In: Actas X Reunión Argentina de sedimentología (I):123–124

Osterrieth M, Morrás HJM, Alvarez MF (2005) Silicobiolitos en suelos y sedimentos loéssicos de la Pampa Ondulada, Buenos Aires. Ameghiniana 43(4):10R

Osterrieth M, Martínez G, Gutierrez M, Alvarez MF (2008a) Biomorfos de sílice en secuencias pedoarqueológicas del sitio Paso Otero 5, Buenos Aires. Brit Archaelogical Res: 77–90

Osterrieth M, Morrás HJM, Moretti L (2008b) Phytoliths in a pedosedimentary sequence in a small basin in the subtropical province of Misiones, Argentina. In: Abstracts 7th IMPR-4EIF-SPR-CGCyC. Mar del Plata, Argentina, pp 33–34

Osterrieth ML, Madella M, Zurro D, Alvarez MF (2009) Taphonomical aspects of silica phytoliths in the loess sediments of the Argentinean Pampas. Quat Int 193:70–79

Osterrieth M, Fernández Honaine M, Borrelli N, Alvarez MF (2014a) Silicophytoliths in representative soils of the southeast Pampean Plains, Argentina. In: Gomes Coe H, Osterrieth M (eds) Synthesis of some phytolith studies in South America (Brazil and Argentina). Nova Science Publishers, Inc., pp 215–242

Osterrieth M, Borrelli N, Alvarez MF, Fernández Honaine M (2014b) Silicophytoliths and silicon biogeochemical cycle in the pampean plain, Argentina. In: Gomes Coe H, Osterrieth M (eds) Synthesis of some phytolith studies in South America (Brazil and Argentina). Nova Science Publishers, Inc., pp 243–263

Osterrieth M, Borrelli N, Alvarez MF, Fernández Honaine M (2015) Silica biogeochemical cycle in temperate ecosystems of the Pampean Plain, Argentina. J S Am Earth Sci 63:172–179

Paisani J, Pontelli M, Osterieth M, López Paisani M, Sa Fachin, Guerra S, Oliveira L (2014) Paleosols in low-order streams and valley heads in the Araucaria plateau and record of continental environmental conditions in southern Brazil at the end of MIS 3. J S Am Earth Sci 54:57–70

Parry W, Smithson F (1964) Types of opaline silica depositions inthe leaves of British grasses. Ann Bot 28:169–185

Piperno DR (1988) Phytolith analysis: an archaeological and geological perspective. Academic Press, San Diego, p 280
Piperno DR (2006) Phytoliths: a comprehensive guide for archaeologists and paleoecologists. Altamira Press, New York 238 pp
Prychid CJ, Rudall PJ, Gregory M (2004) Systematics and biology of silica bodies in monocotyledons. The Bot Rev 69:377–440
Rolleri EO (1975) Provincias geológicas bonaerenses. In: Relatorio VI Congreso Geológico Argentino, Bahía Blanca, pp 29–53
Rovner I (1971) Potential of opal phytolith for use in paleoecological reconstruction. Quat Res 1:343–359
Schnack EJ, Fasano JL, Isla FI (1982) The evolution of Mar Chiquita Lagoon, Province of Buenos Aires, Argentina. In: Colquhom DJ (ed) Holocene sea-level fluctuations: magnitudes and causes. IGCP61, University of South Carolina, Columbia, pp 143–155
Scillato-Yané GJ, Tonni EP, Carlini AA, Noriega JI (1998) Nuevos hallazgos de mamíferos del cuaternario en el arroyo toropí, corrientes, Argentina. Aspectos Bioestratigráficos, Paleoambientales y Paleozoogeográficos. In: Actas X Congreso Latinoamericano de Geología y VI Congreso Nacional de Geología Económica, vol 1, pp 263–268
Sedov S, Solleiro-Rebolledo E, Morales-Puente P, Arias-Herrería A, Vallejo-Gómez E, Jasso-Castañeda C (2003) Mineral and organic components of the buried paleosols of the Nevado de Toluca, Central Mexico as indicators of paleoenvironments and soil evolution. Quat Int 106–107:169–184
Sedov S, Solleiro-Rebolledo E, Terhors B, Solé J, Flores-Delgadillo ML, Werner G, Poetsch T (2009) The Tlaxcala basin paleosol sequence: a multiscale proxy of middle to late Quaternary environmental change in central Mexico. Rev Mex Cienc Geol 26:448–465
Soil Survey Staff (2010) Claves para la taxonomía de suelos. United States Department of Agriculture, Washington 365 p
Soriano A (1979) Distribution of the grasses and grasslands of South America. In: Numata M (ed) Ecology of grasslands and bamboolands in the world. W. Junk Publisher, The Hague, The Netherlands, pp 84–91
Strömberg CAE (2005) Decoupled taxonomic radiation and ecological expansion of open-habitat grasses in the Cenozoic of North America. Proc Natl Acad Sci USA 102:11980–11984
Strömberg CAE (2011) Evolution of grasses and grasslands. Annu Rev Earth Planet Sci 39:517–544
Strömberg CAE, Werdelin L, Friis EM, Sara G (2007) The spread of grass-dominated habitats in Turkey and surrounding areas during the Cenozoic: phytolith evidence. Palaeogeogr Palaeoclimatol Palaeoecol 250:18–49
Teruggi ME (1955) Algunas observaciones microscópicas sobre vidrio volcánico y ópalo organógeno en sedimentos pampianos. Rev Mus La Plata, La Plata 13(66):17–26
Teruggi ME (1957) The nature and origin of Argentine loess. J Sediment Petrol 27(3):322–332
Tomlinson PB (1961) Anatomy of the monocotyledons. II. Palmae, 1st edn. Oxford University Press, Oxford, 453 p
Tonni EP, Carlini AA, Zurita AE, Frechen M, Gasparini GM, Budziak D (2005) Cronología y Bioestratigrafía de la Unidades del Pleistoceno aflorantes en el Arroyo Toropí, Provincia de Corrientes, Argentina. In: Actas 19° Congreso Brasileiro de Paleontología y 6° Congreso Latino-Americano de Paleontología [CD-ROM]
Twiss C (1992) Predicted world distribution of C3 and C4 grass phytoliths, In: Rapp G Jr, Mulholand SC (eds) Phytolith systematics: emerging issues. Springer, New York, 350 p
Tovar RE, Sedov S, Solís B, Solleiro E (2013) Dark humic alluvial paleosols in Central and Southern Mexico: micromorphological indicators of Late Pleistocene megafauna habitats: Spanish. J Soil Sci 3:217–223
Walkley A, Black CA (1965) Organic carbon. In: Black CA (ed) Method of soils analysis. Madison, American Society of Agronomy 4:1372–1375
Yamaji N, Mitani N, Ma JF (2008) A transporter regulating silicon distribution in rice shoots. Plant Cell 20:1381–1389

Zech M, Zech R, Morrás HJM, Moretti L, Glaser B (2009a) Late Quaternary environmental changes in Misiones, subtropical NE Argentina, deduced from multi-proxy geochemical analyses in a palaeosol-sediment sequence. Quat Int 196:121–136

Zech W, Zech M, Zech R, Peinemann N, Morrás HJM, Moretti L, Ogle N, Kalim RM, Fuchs M, Schad P, Glaser B (2009b) Late Quaternary palaeosol records from subtropical (38°S) to tropical (16°S) South America and palaeoclimatic implications. Quat Int 196:107–120

Zurita A, Lutz A (2002) La Fauna Pleistocena de la Formación Toropí en la Provincia de Corrientes (Argentina). Mastozool Neotrop 9(1):47–56

Zurita A, Ferrero B (2009) Una nueva especie de *Neuryurus* Ameghino (Mammalia, Glyptodontidae) en el Pleistoceno tardío de la Mesopotamia de Argentina. Geobios 42 (5):663–673

Zurita A, Miño-Boilini AR, Francia A, Erra G, Alcaraz MA, Carlini AA (2014) Paleontología y cronología del Cuaternario de las provincias de Corrientes y Formosa, Argentina. Acta Geologica Lilloana, San Miguel de Tucumán, Argentina 26(1):75–86

Index

G.M. Gasparini et al. (eds.), *Marine Isotope Stage 3 in Southern South America, 60 ka B.P.–30 ka B.P.*, Springer Earth System Sciences,
DOI 10.1007/978-3-319-40000-6